T0271173

LASER DYNAMICS

Bridging the gap between laser physics and applied mathematics, this book offers a new perspective on laser dynamics. Combining fresh treatments of classic problems with up-to-date research, asymptotic techniques appropriate for nonlinear dynamical systems are shown to offer a powerful alternative to numerical simulations. The combined analytical and experimental descriptions of dynamical instabilities provide a clear derivation of physical formulae and an evaluation of their significance.

Starting with the observation of different time scales of an operating laser, the book develops approximation techniques to systematically explore their effects. Laser dynamical regimes are introduced at different levels of complexity, from standard turn-on experiments to stiff, chaotic, spontaneous, or driven pulsations. Particular attention is given to quantitative comparisons between experiments and theory. The book broadens the range of analytical tools available to laser physicists and provides applied mathematicians with problems of practical interest, and is invaluable for both graduate students and researchers.

THOMAS ERNEUX is a Professor at the Université Libre de Bruxelles. His current interests concentrate on studying specific laser dynamical phenomena and the applications of delay differential equations in all areas of science and engineering.

PIERRE GLORIEUX is a Professor at the Laboratoire de Physique des Lasers, Atomes et Molécules, Université des Sciences et Technologies de Lille, and a Senior Member of the Institut Universitaire de France. Recently, he has studied spatio-temporal dynamics in liquid crystals and photorefractive oscillators.

LASER DYNAMICS

THOMAS ERNEUX

Université Libre de Bruxelles

PIERRE GLORIEUX

Université des Sciences et Technologies de Lille

CAMBRIDGE
UNIVERSITY PRESS

Shaftesbury Road, Cambridge CB2 8EA, United Kingdom

One Liberty Plaza, 20th Floor, New York, NY 10006, USA

477 Williamstown Road, Port Melbourne, VIC 3207, Australia

314–321, 3rd Floor, Plot 3, Splendor Forum, Jasola District Centre, New Delhi – 110025, India

103 Penang Road, #05–06/07, Visioncrest Commercial, Singapore 238467

Cambridge University Press is part of Cambridge University Press & Assessment,
a department of the University of Cambridge.

We share the University's mission to contribute to society through the pursuit of
education, learning and research at the highest international levels of excellence.

www.cambridge.org
Information on this title: www.cambridge.org/9780521830409

First published 2010

A catalogue record for this publication is available from the British Library

Library of Congress Cataloging-in-Publication data
Erneux, Thomas.
Laser dynamics / Thomas Erneux and Pierre Glorieux.
p. cm.
ISBN 978-0-521-83040-9 (Hardback)
1. Lasers–Mathematical models. 2. Dynamics. I. Glorieux, Pierre. II. Title.
QC688.E76 2010
621.36′6–dc22

2009047384

ISBN 978-0-521-83040-9 Hardback

To Anne and Mijo
for their love and support

Contents

Preface   *page* xi
List of abbreviations   xv

Part I   Basic tools   **1**

1   Rate equations   3
    1.1   Dimensionless equations   4
    1.2   Steady states and linear stability   6
    1.3   Turn-on transients   10
    1.4   Transfer function   17
    1.5   Dynamical system   19
    1.6   Spontaneous emission   26
    1.7   Semiconductor lasers   31
    1.8   Exercises and problems   34

2   Three- and four-level lasers   39
    2.1   Energy level schemes in lasers   40
    2.2   Three-level lasers   41
    2.3   Four-level lasers   49
    2.4   Exercises and problems   57

3   Phase dynamics   59
    3.1   Phase-locking in laser dynamics   59
    3.2   Vectorial laser in presence of Faraday driving   60
    3.3   Adler's equation   64
    3.4   Laser with an injected signal   66
    3.5   Counterpropagating waves in ring class A lasers   70

	3.6	Coupled lasers	75
	3.7	Exercises and problems	79
4		Hopf bifurcation dynamics	84
	4.1	Electrical feedback	87
	4.2	Ikeda system	96
	4.3	From harmonic to pulsating oscillations	103
	4.4	Exercises	104

Part II   Driven laser systems   **109**

5		Weakly modulated lasers	111
	5.1	Driven Adler's equation	112
	5.2	Weakly modulated class B lasers	119
	5.3	Exercises and problems	134
6		Strongly modulated lasers	136
	6.1	Generalized bistability	137
	6.2	Map for the strongly modulated laser	143
	6.3	Dual tone modulation near period-doubling bifurcation	149
7		Slow passage	155
	7.1	Dynamical hysteresis	156
	7.2	Slow passage through a bifurcation point	158
	7.3	Period-doubling bifurcation	168
	7.4	Slow–fast dynamics	170
	7.5	Exercise	170

Part III   Particular laser systems   **173**

8		Laser with a saturable absorber	175
	8.1	LSA parameters	177
	8.2	LSA basic phenomena	178
	8.3	Rate equations	182
	8.4	PQS in CO_2 lasers	197
	8.5	Exercises and problems	208
9		Optically injected semiconductor lasers	213
	9.1	Semiconductor lasers	214
	9.2	Injection-locking	216
	9.3	Adler's equation	217

9.4	Experiments and numerical simulations	220
9.5	Stability of the steady states	222
9.6	Nonlinear studies	229
9.7	A third order Adler's equation	234
9.8	Exercises and problems	237
10	Delayed feedback dynamics	241
10.1	History	241
10.2	Imaging using OFB	255
10.3	Optoelectronic oscillator	261
10.4	Exercises	268
11	Far-infrared lasers	272
11.1	Vibrational bottleneck	273
11.2	Lorenz chaos in the FIR laser	275
11.3	Dual gain line instability	281
11.4	Exercises	292
12	Optical parametric oscillator	294
12.1	Parametric processes	294
12.2	Semiclassical model for the DOPO	298
12.3	Experiments on TROPO-DOPO	301
12.4	TROPO-DOPO and temperature effects	306
12.5	Intracavity singly resonant parametric oscillator	312
12.6	Intracavity SHG	316
12.7	Antiphase dynamics in intracavity SHG	318
12.8	Frequencies	325
12.9	Antiphase dynamics in a fiber laser	328
12.10	Exercises	334
References		336
Index		358

Preface

Many of the physicists studying lasers in laboratories have been confronted by the appearance of *erratic intensity fluctuations* in the laser beam. This type of behavior was already evident *in the early days of the laser (1960s)* when it was found that the intensity of the light generated by the ruby laser displayed irregular spiking. Russian theoreticians showed that equations describing an active medium coupled to an electromagnetic field could display such pulsations. Laser physicists K. Shimoda and C.L. Tang tried to relate these outputs to saturable absorption and mode competition, respectively. But the discrepancy in the values for the instability frequencies, the fact that simple rate equations only predicted damped oscillations, and the development of stable lasers shifted interest towards new topics. About the same time, spontaneous instabilities were found to play key roles in fluid mechanics, chemistry, and the life sciences. Except for some isolated pioneers like L.W. Casperson, laser physicists only understood *in the early 1980s* that the pulsating outputs were not the result of environmental fluctuations but rather originated from the interaction between the radiation field and matter. On June 18–21, 1985, the University of Rochester organized the first International Meeting on "Instabilities and Dynamics of Lasers and Nonlinear Optical Systems" [1]. Two special issues of the *Journal of the Optical Society of America* later appeared [2, 3]. But it took until *the early 1990s* before the idea became widely accepted among physicists that lasers exhibit the same type of bifurcations as oscillating mechanical, chemical, and biological systems [4–7]. The possible laser outputs were then systematically explored by multi-disciplinary groups. Nonlinear laser dynamics became a hot topic of research following similar adventures in the physical and life sciences [8–12].

Early investigations have concentrated on gas and solid state lasers. *Semiconductor lasers* came in the 1990s thanks to an enormous effort in fundamental and applied research. They are the lasers used in most of our current applications. Systematic experimental and theoretical studies of their possible instabilities in a variety of set-ups have been undertaken during the last 20 years and significant progress has been made, to the point where we know how to exploit, avoid, or control them [13, 14].

Laser dynamical instabilities are of interest for a growing number of scientists and engineers, not only laser physicists, but also chemists, biologists, and others in a variety of obviously and not so obviously related fields. Placing pulsating lasers in the framework of *dynamical systems* means that many of the observed instabilities can be investigated using simple classical equations based on material properties rather than design. Twenty years ago, a book largely devoted to laser intensity oscillations using this approach would have been inconceivable without taking account of the quantum mechanical properties of the laser or its cavity design. The visually compelling phenomena observed with laser devices and their potential applications make laser dynamics a subject about which colleagues and graduate students with different experiences seek to become better informed. Special sessions entitled Laser Dynamics now appear at conferences, introductory courses are offered at universities, and research groups have concentrated their main activities on laser stability problems.

The primary objective of this book is to introduce a series of *simple laser dynamical problems* that are the building blocks of our current research in the field. These include a description of the relaxation oscillations of the laser, strongly pulsating outputs following a quick change of a parameter or resulting from a saturable absorber, phase-locking phenomena for a laser subject to an injected signal, resonance phenomena in modulated lasers, and oscillatory instabilities caused by a delayed optical feedback. Topics like the diagnostics of chaotic outputs, ultra-fast optics and mode-locked lasers, or the propagation of spatial solitons in fibers are too broad to be covered in this book.

As is largely the case for engineers and applied scientists, a theoretical model is often considered as a numerical model. The difficulty with this approach is that computation limits insight because of an inability to pose questions properly. We cannot ignore the possibilities offered by our computers but we also need to think about the main objectives of our research. To this end, *asymptotic approaches* [15–17] based on the natural values of the parameters smoothly complement simulations by emphasizing particular properties of our laser. From an applied mathematical point of view the laser rate equations offer challenging (singular) limits requiring the adaptation of known techniques to our laser equations. In this book, we introduce some of these techniques, helping the physicist to highlight the generic character of a specific phenomenon or to compare different lasers through their relevant effective parameters. Key to this approach is the hierarchy of different time scales as they appear in the experimental set-ups and observations. The book explores different laser systems whose descriptions require tools of increasing complexity. In each chapter, both theoretical and experimental points of view are confronted with the goal of finding the underlying physical mechanisms responsible for a specific dynamical output.

The book is *organized into three parts*, namely, I Basic tools, II Driven laser systems, and III Particular laser systems. The first part aims to address how the laser physicist studies simple dynamical outputs by using rate equations and the mathematical tools used for their exploration. There is an extensive discussion of time scales and their relevance in slow-time dynamics. There is confusion in the literature and we hope to clarify some of the questions arising in choosing time scales. Another objective of Part I is to introduce the basic bifurcation transitions that appear in a variety of laser set-ups. To this end, we examine explicit examples and introduce methods in the most friendly way. After many years of teaching the subject of laser dynamics, we have found that this is the best way to introduce bifurcation theory to the physicist. Part II is devoted to specific laser systems that are driven either by a modulated signal or by a slowly varying control parameter. The literature of periodically forced lasers is abundant because modulated lasers are important in telecom applications, but also at a more fundamental level because strongly forced lasers lead to chaotic outputs. This part will not review all that has been realized on driven lasers but rather will emphasize the variety of synchronization mechanisms from weak to strongly modulated. Slow passage problems are a key topic in applied mathematics because they appear in a variety of problems with applications in physics, chemistry, and biology. Surprisingly, many experiments on basic slow passage problems have been realized with lasers or optically bistable devices. Part III is devoted to specific laser set-ups that became important on their own and were motivated by specific applications. Of particular interest is the fact that they each introduce a new dynamical phenomenon, such as the onset of spiking pulses, multimode antiphase dynamics, or instabilities caused by a delayed optical feedback.

The book contains more than enough for *two one-semester courses* and some flexibility is possible in selecting topics. Part I collects simple concepts in both laser physics and nonlinear dynamics such as stability, bifurcation, and multiple time scales that must be understood before exploring Parts II and III. The first two chapters of Part II are linked while the third one on slow passage effects is rather independent. The five chapters of Part III consider specific laser systems and can be read separately. Some of these chapters cover classical areas (such as the laser with a saturable absorber) that are introduced in almost every course. Other chapters concentrate on less known areas (such as the far-infrared laser) which we critically revisit, benefiting from the current capacities of our computers or from new asymptotic investigations. To cover the whole book, the student will need a background in linear algebra and ordinary differential equations. However, the details of the calculations are given each time a new technique is introduced so that the reader less oriented towards theory may follow each step. Similarly, experimental details are introduced in the simplest way and avoid technical descriptions of set-ups. To limit the size of the book, we have combined solved problems and additional material usually relegated to appendices in an associated website, http://www.ulb.ac.be/sciences/ont. This site includes detailed answers to exercises in the book, links to other useful sites, and illustrations of specific mathematical techniques such as the method of matched asymptotic expansions (MAE).

We are very much indebted to many colleagues for help during the years while this book was being written. Over the past 30 years, we greatly profited from collaboration or discussion with our friends at the Laboratoire de Physique des Lasers, Atomes et Molécules and in the department of Optique Nonlinéaire Théorique who shared our enthusiasm comparing experimental and theoretical data. This book pays tribute to the memory of Gilbert Grynberg, Lorenzo Narducci, Yakov I. Khanin, and Fréderic Stoekel who have particularly contributed to many aspects of laser dynamics. Finally, we acknowledge the Belgian National Science Foundation and the Pole Attraction Pole program of the Belgian government for the support we received during the preparation of this book.

Abbreviations

AOM	Acousto-Optic Modulator
BD	Bifurcation Diagram
CCD	Charge Coupled Device (in CCD camera)
cw	continuous-wave
cw and *ccw*	clockwise and counterclockwise
DBR	Distributed Bragg Reflector
DDE	Delayed Differential Equations
DFB	Distributed FeedBack
DROPO	Doubly Resonant Optical Parametric Oscillator
DOPO	Degenerate Optical Parametric Oscillator
ECM	External Cavity Mode
EDT	Exponentially Decaying Terms
EOM	ElectroOptic Modulator
FIR	Far InfraRed (for FIR laser)
FWM	Four-Wave Mixing
KTP	($KTiOPO_4$) Potassium Titanium Oxide Phosphate
LaROFI	Laser Relaxation Oscillation Frequency Imaging
LFF	Low Frequency Fluctuations
LIS	Laser with an Injected Signal
LK	Lang and Kobayashi (for Lang and Kobayashi equations)
LSA	Laser with a Saturable Absorber
MAE	Matched Asymptotic Expansions
MB	Mode Beating
MZ	Mach–Zehnder (for Mach–Zehnder interferometer)
OEO	OptoElectronic Oscillator
OFB	Optical FeedBack

OPA	Optical Parametric Amplification
OPO	Optical Parametric Oscillator
PIN	Positive-Intrinsic-Negative (for PIN photodiode)
PQS	Passive Q-Switching
RE	Regular Emission
RO	Relaxation Oscillations
RF	Radio Frequency
SE	Spontaneous Emission
SHG	Second Harmonic Generation
SL	Semiconductor Laser
SN	Saddle-Node (for the saddle-node bifurcation)
SRE	Standard Rate Equations
SROPO	Singly Resonant Optical Parametric Oscillator
TROPO	Triply Resonant Optical Parametric Oscillator
VCSEL	Vertical Cavity Surface Emitting Laser
YAG	($Y_3Al_{15}O_{12}$)Yttrium Aluminum Garnet

Part I

Basic tools

1

Rate equations

Modeling lasers may be realized with different levels of sophistication. Rigorously it requires a full quantum treatment but many laser dynamical properties may be captured by semiclassical or even purely classical approaches. In this book we deliberately chose the simplest point of view, i.e. purely classical equations, and try to extract analytically as much information as possible. The basic framework of our approach is provided by the *rate equations*.

In their simplest version, they apply to an idealized active system consisting of only two energy levels coupled to a reservoir. They were introduced as soon as the laser was discovered to explain (regular or irregular, damped or undamped) intensity spikes commonly seen with solid state lasers (for a historical review, see the introduction in [18]). These rate equations are discussed and sometimes derived from a semiclassical theory in textbooks on lasers [19–22]. They capture the essential features of the response of a single-mode laser and they may be modified to account for specific effects such as the modulation of a parameter or optical feedback.

The most basic processes involved in laser operation are schematically represented in Figure 1.1. N_1 and N_2 denote the number of atoms in the ground and excited levels, respectively. The process of light–matter interaction is restricted to stimulated emission and absorption. This leads to the following rate equations for the number of laser photons n and the populations N_1 and N_2:

$$\frac{dn}{dT} = G(N_2 - N_1)n - \frac{n}{T_c}, \tag{1.1}$$

$$\frac{dN_2}{dT} = R_p - \frac{N_2}{T_1} - G(N_2 - N_1)n, \tag{1.2}$$

$$\frac{dN_1}{dT} = -\frac{N_1}{T_1} + G(N_2 - N_1)n. \tag{1.3}$$

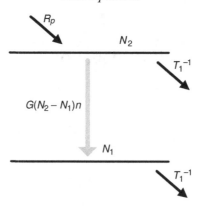

Fig. 1.1 Two-level system. R_p denotes the pumping rate, T_1^{-1} is the decay rate of the populations, and $G(N_2 - N_1)$ is the gain for stimulated emission.

In these equations, G is the gain coefficient for stimulated emission, T_c^{-1} is the decay rate due to the loss of photons by mirror transmission, scattering, etc., T_1^{-1} is the decay rate for each population, and R_p is the pumping rate. Introducing the population difference or population inversion $N \equiv N_2 - N_1$, Eqs. (1.1)–(1.3) reduce to the following two equations for n and N:

$$\frac{dn}{dT} = GNn - \frac{n}{T_c}, \tag{1.4}$$

$$\frac{dN}{dT} = -\frac{1}{T_1}(N - N_0) - 2GNn, \tag{1.5}$$

where $N_0 \equiv R_p T_1$ is the population difference in the absence of laser light. The decay rates T_c^{-1} and T_1^{-1} are identical to the parameters 2κ and $\gamma_\parallel$, respectively, in the "class B" laser equations [23, 6].

In practice, lasing action is realized with three or four energy level systems and the rate equations are more complicated (see Chapter 2). But for many lasers such as Nd^{3+}:YAG, CO_2, and semiconductor lasers, Eqs. (1.4) and (1.5) provide a good description of simple dynamical phenomena such as the laser relaxation oscillations or the build-up of laser radiation following either pump or loss switch. Supplemented by additional terms, these equations are also valid for the description of specific laser instabilities as we shall illustrate in the forthcoming chapters.

1.1 Dimensionless equations

Equations (1.4) and (1.5) depend on four physical parameters, namely, G, T_c, T_1, and N_0. In order to reduce the number of independent parameters, it is worthwhile

Table 1.1 *Characteristic times for common lasers.*

Laser	T_c(s)	T_1(s)	γ
CO_2	10^{-8}	4×10^{-6}	2.5×10^{-3}
solid state (Nd^{3+}:YAG)	10^{-6}	2.5×10^{-4}	4×10^{-3}
semiconductor (AsGa)	10^{-12}	10^{-9}	10^{-3}

to rewrite these equations in dimensionless form (for a dimensionless formulation, see, for example, [24]). Introducing new variables I, D, and t defined as

$$I \equiv 2GT_1 n, \quad D \equiv GT_c N, \quad \text{and} \quad t \equiv T/T_c \qquad (1.6)$$

into Eqs. (1.4) and (1.5), we obtain the following equations for I and D (Exercises 1.8.1 and 1.8.4)

$$\frac{dI}{dt} = I(D-1), \qquad (1.7)$$

$$\frac{dD}{dt} = \gamma(A - D(1+I)) \qquad (1.8)$$

where A and γ are defined by

$$A \equiv GT_c N_0 \quad \text{and} \quad \gamma \equiv T_c/T_1. \qquad (1.9)$$

Compared to the original equations (1.4) and (1.5), Eqs. (1.7) and (1.8) offer two clear advantages. First, we only have two independent parameters instead of the original four parameters. This means that Eqs. (1.7) and (1.8) are simpler to analyze or require fewer numerical simulations. Second, we may estimate these two parameters for different lasers, discover common ranges of values, and possibly propose approximations of the solution based on their respective values.

Table 1.1 gives the order of magnitude of T_c and T_1 for three common lasers. Although their ranges of values are quite different, we note that the ratio $\gamma \equiv T_c/T_1$ is typically a 10^{-3} small quantity. For microchip solid state lasers, γ may even reach 10^{-6} small values. A small γ is a key property of these lasers and, as we shall demonstrate, is responsible for their weak stability properties. On the other hand, A scales the pump in units of the pump at threshold and is typically in the range $1-10$. It barely exceeds 10 in most common lasers although it may reach very high values in specific situations such as the "thresholdless laser" [25]. In addition to solid state lasers, earlier laser studies used He-Ne and Ar gas lasers. For the He-Ne and Ar gas lasers the value of γ is much larger than 1. Consequently, the evolution of the population inversion is very fast until the right hand side of Eq. (1.8) is zero. D then adiabatically follows the intensity as

$$D = \frac{A}{1+I} \tag{1.10}$$

and Eq. (1.7) reduces to

$$\frac{dI}{dt} = \left(\frac{A}{1+I} - 1 \right) I. \tag{1.11}$$

Eq. (1.11) is a first order nonlinear equation. Lasers described by the single equation (1.11) are called "class A" lasers [23, 6]. Moreover, assuming $I < 1$, we may further simplify Eq. (1.11) by expanding $1/(1 + I)$ and obtain

$$\frac{dI}{dt} = (A - 1 - AI)I, \tag{1.12}$$

which exhibits a single quadratic nonlinearity.

There are other ways to non-dimensionalize the rate equations. Here time is measured in units of the photon damping time T_c but T_1 could equally be used to rescale time. It is also possible to introduce $2GT_c n$ and/or $GT_1 N$ as the dimensionless photon and population inversion variables. But the equations resulting from these normalizations are less appropriate for analysis than Eqs. (1.7) and (1.8). As previously emphasized, γ is small and it is mathematically convenient that it appears as a single parameter multiplying the right hand side of one of the two equations. Similar procedures have been applied for classical problems such as the van der Pol equation or the Michaelis–Menten equations in enzyme kinetics [8].

1.2 Steady states and linear stability

The analysis of our model equations starts with the determination of the steady states and their linear stability properties. The results allow us to predict bifurcations, anticipate interesting transient regimes, and possibly propose simplifications of the laser equations. The linear stability analysis is well documented for one- or two-variable systems of ordinary differential equations [26–28]. For higher order systems, we benefit from the Routh–Hurwitz conditions for the stability of the steady states ([26] p. 270, [29] p. 304).

1.2.1 Steady states

The steady state solutions of Eqs. (1.7) and (1.8) satisfy the conditions $dI/dt = dD/dt = 0$ or, equivalently, the following two equations for I and D

$$I(D - 1) = 0, \tag{1.13}$$

$$A - D(1 + I) = 0. \tag{1.14}$$

The possible solutions are (1) the zero intensity solution

$$I = 0 \quad \text{and} \quad D = A, \tag{1.15}$$

and (2) the non-zero intensity solution

$$I = A - 1 \geq 0 \quad \text{and} \quad D = 1. \tag{1.16}$$

The inequality in (1.16) is needed because I is an intensity. We conclude that the desired lasing action is possible only if $A > 1$. The critical point

$$(I, D, A) = (0, 1, 1) \tag{1.17}$$

is called the *laser first threshold* and is a *bifurcation point* because it connects our two steady state solutions. These solutions are represented as a function of the pump parameter A in Figure 1.2. The diagram is called a *bifurcation diagram* because it represents the amplitude of the possible solutions in terms of a control

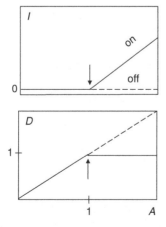

Fig. 1.2 Steady state solutions. Full and broken lines correspond to stable and unstable solutions, respectively. The arrow indicates the bifurcation point at $A = 1$.

or *bifurcation parameter*. In the zero intensity solution (laser OFF), the laser does not emit any radiation and the population difference sets to the value given by the pump ($D = A$). As the pump exceeds its threshold value $A = 1$, a non-zero intensity solution is possible (laser ON) and the laser emits radiation. The amount of emitted energy is proportional to the pump in excess of threshold, i.e. $I = A - 1$. Which of the two solutions will be effectively observed depends on their stability.

1.2.2 Linear stability

In order to analyze the stability of the steady states, we introduce the small deviations u and v defined by

$$u \equiv I - I_s \quad \text{and} \quad v \equiv D - D_s, \tag{1.18}$$

where $(I, D) = (I_s, D_s)$ denotes either OFF (1.15) or ON (1.16) solutions. We insert $I = I_s + u$ and $D = D_s + v$ into Eqs. (1.7) and (1.8), simplify by using the steady state equations (1.13) and (1.14), and neglect the quadratic terms in u and v. We then obtain the following *linearized equations* for u and v

$$\frac{du}{dt} = u(D_s - 1) + I_s v, \tag{1.19}$$

$$\frac{dv}{dt} = \gamma \left(-D_s u - (1 + I_s)v \right). \tag{1.20}$$

It is useful to rewrite these equations in matrix form as

$$\frac{d}{dt} \begin{pmatrix} u \\ v \end{pmatrix} = J \begin{pmatrix} u \\ v \end{pmatrix}, \tag{1.21}$$

where the 2×2 matrix J is called the Jacobian matrix and is defined here as

$$J \equiv \begin{pmatrix} D_s - 1 & I_s \\ -D_s \gamma & -(1 + I_s)\gamma \end{pmatrix}. \tag{1.22}$$

The general solution of Eqs. (1.19) and (1.20) or Eq. (1.21) is a linear combination of two exponential solutions. Introducing $u = c_1 \exp(\sigma t)$ and $v = c_2 \exp(\sigma t)$ into Eqs. (1.19) and (1.20) leads to a homogeneous system of two equations for c_1 and c_2. A nontrivial solution is possible only if the *growth rate* σ satisfies the *characteristic equation* given by

$$\det J - \sigma I = \sigma^2 + \sigma \left[\gamma(1 + I_s) - D_s + 1 \right] + \gamma(1 + I_s - D_s) = 0. \tag{1.23}$$

Stability means that $\text{Re}(\sigma_j) < 0$ ($j = 1, 2$). Then the small deviations u and v will decay to zero. On the other hand, if $\text{Re}(\sigma_j) > 0$ for either $j = 1$ or $j = 2$, u and v will grow exponentially and the steady state is unstable. The stability results are given as follows:

(1) For the zero intensity steady state (1.15), Eq. (1.23) admits the simple solutions

$$\sigma_1 = A - 1 \quad \text{and} \quad \sigma_2 = -\gamma. \tag{1.24}$$

From (1.24), we conclude that the zero intensity steady state is stable if $A < 1$ and unstable if $A > 1$.

(2) For the non-zero intensity steady state (1.16), Eq. (1.23) reduces to the following quadratic equation

$$\sigma^2 + \gamma A \sigma + \gamma(A - 1) = 0. \tag{1.25}$$

To determine the sign of $\text{Re}(\sigma)$, we don't need to solve Eq. (1.25). Indeed, we note that the product of the roots is always positive ($\sigma_1 \sigma_2 = \gamma(A - 1) > 0$) and that the sum of the roots is always negative ($\sigma_1 + \sigma_2 = -\gamma A < 0$). Together, the two inequalities imply that $\text{Re}(\sigma_j) < 0$ ($j = 1, 2$). Thus, the non-zero intensity solution is always stable.

At the bifurcation point (1.17), we note an exchange of stabilities between the zero intensity and non-zero intensity steady state solutions. This is a simple example of a *bifurcation with exchange of stability*. Some dynamical properties linked to the existence of this bifurcation will be examined in Section 1.5.

1.2.3 Damped relaxation oscillations

The linear stability analysis allows us to describe slowly decaying intensity oscillations that are observed in lasers after a sudden excitation such as a loss or gain pulse. Specifically, we solve the quadratic equation (1.25) and obtain

$$\sigma_{1,2} = -\gamma \frac{A}{2} \pm i \sqrt{\gamma(A - 1) - \gamma^2 A^2/4} \tag{1.26}$$

provided $\gamma(A - 1) - \gamma^2 A^2/4 \geq 0$. Expanding the two roots for small γ (A fixed) simplifies (1.26) as

$$\sigma_{1,2} = \pm i \sqrt{\gamma(A - 1)} - \gamma \frac{A}{2} + O(\gamma^{3/2}), \tag{1.27}$$

where the notation $O(\gamma^{3/2})$ means that the correction term is proportional to $\gamma^{3/2}$ (in Section 1.5.2, we examine the limit $A - 1$ small (γ fixed)). The meaning of the two first terms in (1.27) is best understood if we write the general solution for

$u = I - (A-1) = c \exp(\sigma_1 t) + \bar{c} \exp(\sigma_2 t)$, where $\bar{c}$ means the complex conjugate of c. Using (1.27), u can be rewritten as

$$u \simeq C \exp\left(-\gamma \frac{A}{2} t\right) \sin\left(\sqrt{\gamma(A-1)}\, t + \phi\right), \qquad (1.28)$$

where C and ϕ are arbitrary constants determined by the initial conditions. The expression (1.28) implies that the intensity $I = A - 1 + u$ oscillates with a frequency proportional to $\sqrt{\gamma}$ and slowly decays with a rate proportional to γ. The frequency appearing in (1.28), defined by

$$\omega_R \equiv \sqrt{\gamma(A-1)}, \qquad (1.29)$$

is called the laser *relaxation oscillation (RO) frequency* and is a reference frequency for all lasers experiencing intensity oscillations (see Problem 1.8.8 for the RO frequency close to threshold). The quantity

$$\Gamma \equiv \gamma \frac{A}{2} \qquad (1.30)$$

is called the *damping rate* of the laser relaxation oscillations. Note that the expression (1.28) is the product of two functions that exhibit different time scales, namely[1]

$$t_1 = \sqrt{\gamma} t \quad \text{and} \quad t_2 = \gamma t. \qquad (1.31)$$

In summary, the linearized theory reveals that the non-zero intensity steady state is weakly stable for all lasers exhibiting a small γ and that slowly decaying oscillations (RO oscillations) of the laser intensity are possible. Our results are strictly valid for small perturbations of the steady state. But in Section 5.2.1, we show that our conclusions remain valid if we consider arbitrary intensities.

1.3 Turn-on transients

In 1965, Pariser and Marshall [30] investigated the time evolution of the laser intensity using a He-Ne laser pumped by a flash lamp. The laser intensity was assumed to be initially close to zero and the time evolution was fitted using

$$\frac{dE}{dt} = aE - bE^3, \quad E(0) = E_0, \qquad (1.32)$$

---

[1] In physical units, these two time scales are $\sqrt{T_1 T_c}\,T$ and $T_1 T$ respectively. The latter has a simple meaning since it coincides with the lifetime of the population inversion. The former is less obvious since it is the geometrical mean of two lifetimes. It appears because there exists a coupling between the variables I and D.

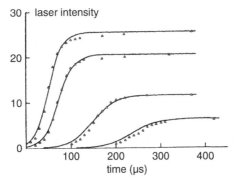

Fig. 1.3 He-Ne gas laser output as a function of time. From the lower to the upper time traces, the pump parameter above threshold is gradually increased. Reprinted Figure 2 with permission from Pariser and Marshall [30]. Copyright 1965 by the American Institute of Physics.

where E represents the electrical field and a and b are positive. This equation is the "class A" laser equation (1.12) with $I = E^2$, $a = (A-1)/2$, and $b = A/2$. Equation (1.32) is a Bernoulli equation that can be solved exactly, leading to the following expression for the intensity I

$$I = \frac{a}{b}\frac{1}{1 - (1 - \frac{a}{bI_0})\exp(-2at)},\tag{1.33}$$

where $I_0 = E_0^2$. The different lines in Figure 1.3 correspond to (1.33) with different values of a and b. Note that a is proportional to the pump parameter above its threshold value. The expression (1.33) tells us that

$$\tau = (2a)^{-1}\tag{1.34}$$

is the time scale of the laser emission. It decreases as a increases (i.e. as the pump increases). Careful statistical studies of the laser build-up using a He-Ne laser [31] and a dye laser [32] complete the earlier investigations [30]. In both cases, Eq. (1.32) was used as the deterministic reference equation.

1.3.1 Typical turn-on experiment

For most common lasers used today in laboratories and in applications (solid state, CO_2, and semiconductor lasers), we switch the pump from a below- to an above-threshold value and observe the time evolution of the intensity. Figure 1.4 shows an example for a Nd^{3+}:YAG laser. We note three distinct regimes:

(1) A long time interval where the laser output power remains very low. In the conditions of Figure 1.4, this extends from the time origin given by the on-switching of pump

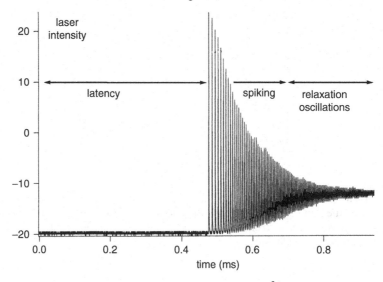

Fig. 1.4 Typical switch-on transient of a Nd^{3+}:YAG laser.

power to about 450 μs. This regime is called the "latency," "lethargy," or "turn-on" regime and is analyzed in Section 1.3.2. The delay of the laser transition is called *"turn-on time"* or *"turn-on delay"* ([33] p. 240, [34] p. 81).

(2) A strongly pulsating intensity regime during which the laser emits a series of sharp spikes separated by periods of very low (almost zero) emission. In Figure 1.4, it extends from 450 μs to about 800 μs.

(3) A regime of damped oscillations as the laser approaches its steady state through exponentially damped sinusoidal oscillations. In Figure 1.4, this goes from 800 μs to 1000 μs. We may compare these oscillations with the decaying oscillations predicted by the linear stability analysis (see Section 1.2) and determine the RO frequency and the RO damping rate.

In terms of the original time T, the relaxation oscillation frequency is defined by $f_R \equiv \omega_R/T_c$ and, using (1.29), we find that

$$f_R \equiv \sqrt{\frac{A-1}{T_1 T_c}}. \qquad (1.35)$$

The expression (1.35) means that the square of the relaxation oscillation frequency (f_R^2) increases linearly with the pump power above the threshold $A - 1$. From the slope of the straight line, we may determine either T_1 or T_c. In [35], the relaxation oscillation frequency is measured for an erbium doped fiber laser. Figure 1.5 represents f_R^2 as a function of the pump power P. The experimental data are then fitted to the expected linear dependence: $f_R^2 = aP + b$,

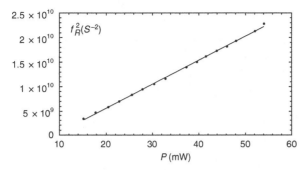

Fig. 1.5 Square of the relaxation oscillation frequency f_R vs. pump power P for an erbium doped fiber laser. Adapted Figure 4 from Sola *et al.* [35] with permission from Elsevier.

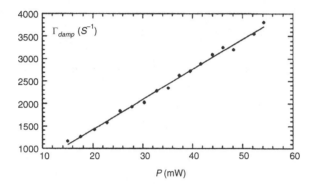

Fig. 1.6 Damping rate as a function of the pump power. Adapted Figure 3 from Sola *et al.* [35] with permission from Elsevier.

where $a = 4.8547 \times 10^{11}\,\mathrm{s^{-2}W^{-1}}$ and $b = -4.3230 \times 10^{9}\,\mathrm{s^{-2}}$. The pump threshold equals $P_{th} = -b/a = 8.9$ mW.

Much less attention has been paid to the damping constant of the relaxation oscillations (1.30). In terms of the original time variable T, the damping rate $\Gamma_{damp} \equiv \Gamma/T_c$ is given by

$$\Gamma_{damp} \equiv \frac{A}{2T_1} \tag{1.36}$$

and its measure provides new information on the laser parameters. For the doped fiber laser studied in [35], the damping rate Γ_{damp} is measured as a function of the pump power P. See Figure 1.6. The experimental data are fitted to the expected linear dependence: $\Gamma_{damp} = a'P + b'$, where $a' = 68\,756\,\mathrm{s^{-1}\,W^{-1}}$ and $b' = -4.2974\,\mathrm{s^{-1}}$. Note that the constant b' is not predicted by (1.36) but comes from the fact that the equation for the inversion of population is slightly different

Fig. 1.7 Relaxation oscillations of a CO_2 laser subject to a square pulse excitation. Upper trace: variations of the laser intensity. Lower trace: loss modulation. Total time scan is 500 μs.

from Eq. (1.5) [35]. For the range of pump power considered, the contribution of b' is less than 1% and could be neglected.

Damped relaxation oscillations are also observed at the output of a CO_2 laser. See Figure 1.7. The laser undergoes step periodic changes of its losses. As can be seen from the long time behavior, these changes are sufficiently small and do not modify the average intensity. On the other hand, the perturbations on the losses are strong enough to initiate relaxation oscillations which disappear after 8–10 cycles. Increasing or decreasing losses induces transients with opposite phases in the laser output. We may understand this behavior by reformulating our rate equations for this particular experiment. Instead of Eqs. (1.7) and (1.8), we now consider

$$\frac{dI}{dt} = I(D - (1 - \varepsilon)), \quad \frac{dD}{dt} = \gamma(A - D(1 + I)), \tag{1.37}$$

$$I(0) = A - 1, \quad \text{and} \quad D(0) = 1, \tag{1.38}$$

where the $-\varepsilon$ accounts for the small decrease of the losses at time $t = 0^+$. Because γ is small, $dD/dt = 0$ in first approximation which means that $D = D(0) = 1$ during the time interval $t = O(1)$. Substituting $D = 1$ into the first equation implies that $dI/dt = I\varepsilon > 0$. This result is in accordance with intuitive thinking since a step decrease of the losses is expected to produce a jump increase of the laser output. Similarly, a small increase ε in the losses will lead to $dI/dt = -I\varepsilon < 0$, i.e. a jump decrease in the laser output.

The simple dependence with respect to the pump displayed by either f_R^2 or Γ_{damp} comes from the two-level rate equations but seems to apply for more complex laser systems exhibiting three, four, or more energy levels (see Chapter 2).

1.3.2 Switching-on or turn-on time

The linearized theory cannot describe the latency regime because the inversion of population does not remain close to the OFF state. However, we may take advantage of the very low values of the intensity and neglect the nonlinear term in

Eq. (1.8) to proceed further in the analytical investigation of the switch-on regime and in particular to obtain the value of the turn-on time. The rate equations (1.7) and (1.8) then reduce to

$$\frac{dI}{dt} = I(D - 1), \tag{1.39}$$

$$\frac{dD}{dt} = \gamma (A_+ - D), \tag{1.40}$$

where $A = A_+ > 1$ denotes the value of the pump above threshold. The initial conditions appropriate for the turn-on experiment are given by

$$I(0) = I_0 << 1 \quad \text{and} \quad D(0) = A_-, \tag{1.41}$$

where $A = A_- < 1$ represents the value of the pump below threshold. The actual value of I_0 is unknown but, as we shall see, the latency regime does not depend on I_0, in first approximation. Eq. (1.40) is linear and its solution is

$$D = (A_- - A_+) \exp(-\gamma t) + A_+. \tag{1.42}$$

Substituting (1.42) into the right hand side of Eq. (1.39), we find that the resulting equation is separable. Its solution is

$$I = I_0 \exp \left(\gamma^{-1} F(\gamma t) \right), \tag{1.43}$$

where $F = F(s)$ is defined by

$$F(s) \equiv (A_+ - 1)s - (A_- - A_+)(\exp(-s) - 1). \tag{1.44}$$

A graphical analysis of (1.44) shows that $F(s)$ is zero at $s = 0$ and $s = s_{on} > 0$, and that $F(s)$ is negative when $0 < s < s_{on}$ and positive when $s > s_{on}$. Recall that γ is small. The expression (1.43) tells us that $I(t)$ is an $O(\exp(-\gamma^{-1}))$ small quantity until $t = t_{on}$ where $t_{on} \equiv \gamma^{-1} s_{on}$. When t is slightly above t_{on}, $I(t)$ suddenly changes from an $O(\exp(-\gamma^{-1}))$ small to an $O(\exp(\gamma^{-1}))$ large quantity whatever the value of $I_0 = O(1)$. The turn-on time satisfies $F(s) = 0$ and its evolution is shown in Figure 1.8 (see Problem 1.8.7 for a turn-on experiment induced by a pump square pulse).

Simplified expressions for t_{on} have been proposed in the literature. If γt_{on} is large, we may neglect the exponential in (1.44) and obtain the expression

$$t_{on} \simeq \gamma^{-1} \left(\frac{A_+ - A_-}{A_+ - 1} \right). \tag{1.45}$$

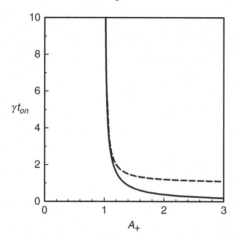

Fig. 1.8 Turn-on time as a function of A_+ ($A_- = 0.8$). The full line is the non-zero root of $F = 0$ where F is defined by (1.44). The dashed line is the approximation (1.45).

The approximation (1.45) is shown in Figure 1.8 by a dashed line. The approximation is good close to threshold but overestimates the actual turn-on time when A_+ increases. If in addition $A_+ - 1 \ll 1$ (near threshold experiments), (1.45) can be further simplified as

$$t_{on} \simeq \gamma^{-1} \left(\frac{1 - A_-}{A_+ - 1} \right). \tag{1.46}$$

This expression is documented in [33] p. 243, for $A_- = 0$. It indicates that t_{on} is an $O(\gamma^{-1})$ large quantity that increases like $(A_+ - 1)^{-1}$ as we approach the laser threshold $A = 1$ from above. In terms of the original time T, the turn-on time $T_{on} = T_c t_{on} = T_c \gamma^{-1} s_{on} = T_1 s_{on}$, where s_{on} is defined as the root of (1.44). Since (1.44) only depends on the initial and final values of the pump (i.e. $A = A_\pm$), T_{on} allows the determination of T_1.

  Mathematically, the turn-on delay is comparable to the delay of a bifurcation transition as the control parameter slowly passes the bifurcation point (see Chapter 7). Indeed, Eq. (1.39) admits a bifurcation point at $D = 1$ where $I = 0$ changes stability. But because $D = D(\gamma t)$ is slowly varying, the actual jump occurs when the integral $\int_0^t (D(\gamma t) - 1) dt$ changes sign and not when $D - 1$ changes sign. The delay of the bifurcation transition is therefore a significant quantity but, physically, it is sensitive to the level of noise and in particular of spontaneous emission always present in a real experiment (see Section 1.6.2).

1.4 Transfer function

The linear stability analysis is intimately connected to the *transfer function* of the system. This quantity, which is widely used by engineers, measures the response of the laser to small-amplitude harmonic modulation of one of its parameters. Theoretically [20, 34], the transfer function is obtained by modulating a parameter in the rate equations and by determining the leading approximation of the solution. It is an interesting function to study because we know that the laser exhibits a weak damping rate and that a resonance is expected when the driving frequency comes close to the laser RO frequency. Specifically, we consider a $2\pi/\omega$-periodic modulation of the pump parameter given by

$$A = A_0 + a \exp(i\omega t) + c.c., \qquad (1.47)$$

where $a \ll 1$ and *c.c.* means complex conjugate. We then seek a $2\pi/\omega$-periodic solution of the rate equations (1.7) and (1.8) of the form

$$I = A_0 - 1 + a \left(p \exp(i\omega t) + c.c. \right) + O(a^2),$$

$$D = 1 + a \left(q \exp(i\omega t) + c.c. \right) + O(a^2). \qquad (1.48)$$

Substituting (1.48) into (1.7) and (1.8), we find that the leading problem is $O(a)$ and provides two equations for p and q given by

$$i\omega \begin{pmatrix} p \\ q \end{pmatrix} - J \begin{pmatrix} p \\ q \end{pmatrix} = \begin{pmatrix} 0 \\ \gamma \end{pmatrix}, \qquad (1.49)$$

where J is the Jacobian matrix (1.22) evaluated at $I_s = A_0 - 1$ and $D_s = 1$. Comparing with the linear stability analysis, we find a similar equation to (1.21) where the contribution of the modulated parameter appears in the right hand side of Eq. (1.49). Eliminating q, we obtain p as

$$p = \frac{1}{1 - \omega^2/\omega_R^2 + 2i\omega\Gamma/\omega_R^2}, \qquad (1.50)$$

where ω_R and Γ are the RO frequency (1.29) and the damping rate (1.30) evaluated at $A = A_0$ (Exercise 1.8.2). p is called the transfer function from current modulation to power output [$p = H(i\omega)$ in [34] with Petermann damping frequency $\omega_d \equiv \omega_R^2/(2\Gamma)$]. With (1.50), we may determine the amplitude and phase of the intensity sinusoidal oscillations. The amplitude of the oscillations is $2a|p|$, where

$$|p| = \sqrt{\frac{1}{\left(1 - \omega^2/\omega_R^2\right)^2 + \left(2\omega\Gamma/\omega_R^2\right)^2}} \qquad (1.51)$$

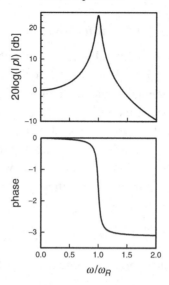

Fig. 1.9 Transfer function. The amplitude and the phase of the transfer fuction p, defined by (1.51) and (1.52), respectively, are represented in terms of ω/ω_R. The values of the fixed parameters are $A = 2$ and $\gamma = 10^{-3}$.

and the phase is given by

$$\phi = \arctan\left[-\frac{2\omega\Gamma}{\omega_R^2 - \omega^2}\right]. \tag{1.52}$$

The amplitude and the phase of p are represented in Figure 1.9 for standard values of the laser parameters. The transfer function (1.51) exhibits a sharp maximum at $\omega/\omega_R \simeq 1$ in a small domain $|\omega/\omega_R - 1|$ proportional to $\Gamma/\omega_R = O(\gamma^{1/2})$. The transfer function is asymmetric showing a smooth increase from $|p| = 1$ to $|p| = (2\Gamma/\omega_R)^{-1}$ at $\omega/\omega_R = 1$ and then a decrease to small values as ω/ω_R further increases. On the other hand, the phase experiences a π jump as ω crosses ω_R. In experiments, the transfer function may be measured in different ways. The standard technique is to monitor the response of the laser to harmonic modulation vs. the excitation frequency. An alternative approach is to excite the laser with a random (noise) signal and to monitor the spectral density of the noise on the laser output. An example of a transfer function obtained by the first technique on a diode pumped Nd^{3+}:YAG laser is shown in Figure 1.10. The transfer function for a semiconductor laser is shown in Figure 1.11 for different bias currents (different A_0). The resonance frequency of the transfer function increases with the bias current (because $\omega_R = \sqrt{\gamma(A_0 - 1)}$ increases with A_0) and the transfer

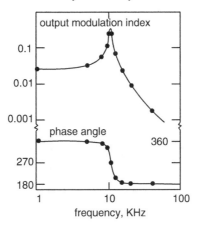

Fig. 1.10 Pump modulation response in a diode-pumped Nd^{3+}:YAG laser (Figure 25.13 of Siegman [20]).

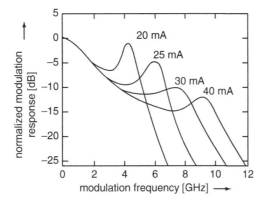

Fig. 1.11 Transfer function for a modulated diode laser. The different curves correspond to different bias currents, or equivalently to different A_0. Reprinted Figure 4.10 of Petermann [34] with kind permission from Springer Science and Business Media.

function becomes smoother near its maximum (because the width of the layer $|\omega - \omega_R| \sim \Gamma = \gamma A_0/2$ increases with A_0).

1.5 Dynamical system

The laser threshold is a good physical example of a bifurcation point, i.e. a point at which the whole dynamics of the system changes. In this section, we investigate the laser bifurcation transition by determining a simplified evolution equation valid near this bifurcation point.

1.5.1 Laser bifurcation

For the rate equations (1.7) and (1.8), the bifurcation occurs at $A = 1$ and is a *transcritical bifurcation*. At such a bifurcation point two steady state solutions, one stable and one unstable, exchange their stability. The OFF state is stable (unstable) below (above) threshold and the ON state is stable (unstable) above (below) threshold. However, the intensity I of the laser output needs to satisfy the physical constraint $I > 0$ and therefore the ON solution does not exist below threshold. We may avoid this constraint by simply reformulating the rate equations in terms of the laser field rather than its intensity. These equations may be derived from the Maxwell–Bloch equations. Under several approximations derived in standard textbooks [6, 21, 36, 37], we obtain equations relating the electric field E, the polarization P induced by the field, and the population difference N. After eliminating adiabatically P, the equation for E is

$$\frac{dE}{dT} = \frac{GNE}{2} - \frac{E}{2T_c}.$$

(1.53)

From $E^* dE/dt + E dE^*/dt$, where E^* is the complex conjugate, we obtain Eq. (1.4) with $n = |E|^2$. In dimensionless form, Eq. (1.53) and Eq. (1.5) with $n = |E|^2$ are given by

$$\frac{d\mathcal{E}}{dt} = \frac{1}{2}\mathcal{E}(D - 1),$$

(1.54)

$$\frac{dD}{dt} = \gamma\left(A - D(1 + \mathcal{E}^2)\right),$$

(1.55)

where $\mathcal{E} \equiv \sqrt{2GT_1} E$. These equations are identical to Eqs. (1.7) and (1.8) if we introduce $I \equiv \mathcal{E}^2$. The OFF state now corresponds to $\mathcal{E} = 0$ and $D = A$ while the ON state is given by $\mathcal{E} = \pm(A - 1)^{1/2}$ and $D = 1$ (see Figure 1.12). The two ON solutions admit the same intensity and have no extra physical meaning compared to our previous analysis of the rate equations. But the nature of the bifurcation transition is different. The bifurcation at $A = 1$ is a *pitchfork bifurcation* which exists only if $A \geq 1$. The linear stability analysis of the steady state solutions (Exercise 1.8.3) indicates that the OFF solution is stable (unstable) below (above) threshold while both ON solutions are stable in their whole domain of existence ($A > 1$). The pitchfork bifurcation is represented in Figure 1.12. Below threshold, only the OFF solution is possible. Beyond threshold, two new solutions corresponding to the ON state are available. The bifurcation is *supercritical* for our laser (the new solutions overlap the unstable basic solution) but it can be *subcritical* for other nonlinear problems (the new solutions overlap the stable basic solution) [8]. For

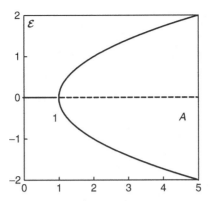

Fig. 1.12 Pitchfork bifurcation for the laser electric field.

elementary bifurcations, the stability of the bifurcating solutions is related to the stability of the basic solution [38]. A supercritical (subcritical) bifurcation then means a bifurcation to stable (unstable) solutions.

1.5.2 Normal form

The laser exhibits typical dynamical features of a steady bifurcation. This can be substantiated mathematically by showing that a simple amplitude equation may capture the essential features of the bifurcation transition. This is already transparent if we review the stability results for the non-zero intensity steady state. The amplitude of the field is given by

$$\mathcal{E} = \pm (A - 1)^{1/2} \tag{1.56}$$

and the characteristic equation admits the following limits for the growth rate σ

$$\sigma_1 \simeq -(A - 1) \quad \text{and} \quad \sigma_2 \simeq -\gamma \tag{1.57}$$

as $(A - 1) \to 0^+$ (γ fixed). The two σ are negative but σ_1 suggests that a small perturbation will slowly decay to zero according to the time scale

$$\tau = (A - 1)t. \tag{1.58}$$

The expressions (1.56) and (1.58) suggest that we seek a long time solution of Eqs. (1.54) and (1.55) in power series of $(A - 1)^{1/2}$ and depending on (1.58) only. This is the first time that we propose to solve a nonlinear problem that exhibits different time scales. The information provided by (1.56) and (1.58) help but we show in the next subsection that the correct limit can be found without preliminary assumptions on the amplitude and time scale of the solution.

Derivation

The key idea is the introduction of a parametric representation of the solution that takes into account its slight change as we increase $A - 1$. Specifically, we seek a solution of the form

$$\mathcal{E} = \varepsilon \mathcal{E}_1 + \varepsilon^2 \mathcal{E}_2 + \ldots$$
$$D = 1 + \varepsilon D_1 + \varepsilon^2 D_2 + \ldots, \tag{1.59}$$

where ε is a small parameter related to $A - 1$ by

$$A - 1 = \varepsilon a_1 + \varepsilon^2 a_2 + \ldots \tag{1.60}$$

In (1.59), the coefficients are assumed to be functions of one or several slow time variables defined by

$$\tau_1 = \varepsilon t, \ \tau_2 = \varepsilon^2 t, \ldots, \ \tau_n = \varepsilon^n t. \tag{1.61}$$

Right now, we don't know if all these slow time variables are really needed. The *method of multiple scales* is a powerful technique [15, 16] but is based on an artifice that is difficult to accept. Even though the solution is a function of t, we shall seek a solution that is a function of all the variables τ_1, τ_2, etc., treated as *independent* variables. Of course, the actual solution is a solution of t only but the solution may be expressed as a product of functions of a single slow time variable.

Because τ_1, τ_2, etc. are treated as independent time variables, we use the chain rule for partial differentiation to compute the derivatives of $\mathcal{E}$ and D, i.e.

$$\frac{dF}{dt} = \varepsilon F_{\tau_1} + \varepsilon^2 F_{\tau_2} + \ldots, \tag{1.62}$$

where F is either $\mathcal{E}$ or D and subscripts mean partial derivatives of F with respect to τ_1, τ_2, etc. Substituting (1.59), (1.60), and (1.62) into Eqs. (1.54) and (1.55) and equating to zero the coefficients of each power of ε leads to a sequence of simple problems for the functions in (1.59). The first three problems are given by

$$O(\varepsilon) : a_1 - D_1 = 0, \tag{1.63}$$

$$O(\varepsilon^2) : \mathcal{E}_{1\tau_1} = \frac{1}{2} D_1 \mathcal{E}_1, \tag{1.64}$$

$$D_{1\tau_1} = \gamma \left(a_2 - D_2 - \mathcal{E}_1^2 \right), \tag{1.65}$$

and

$$O(\varepsilon^3) : \mathcal{E}_{2\tau_1} + \mathcal{E}_{1\tau_2} = \frac{1}{2}(\mathcal{E}_2 D_1 + \mathcal{E}_1 D_2). \tag{1.66}$$

These problems are now solved sequentially.
  Equation (1.63) implies that

$$D_1 = a_1 \tag{1.67}$$

but $\mathcal{E}_1$ is unknown and motivates the study of the next problem.
  With (1.67), Eq. (1.64) is linear and admits an exponential solution of the form $\mathcal{E}_1 = \mathcal{E}_1(0)\exp(\frac{1}{2}a_1\tau_1)$. But if $a_1 > 0$, this solution is unbounded in τ_1 and cannot be accepted. If $a_1 < 0$, the solution approaches zero which contradicts our assumption in (1.59) that $\mathcal{E} = O(\varepsilon)$. We are thus forced to require

$$a_1 = 0. \tag{1.68}$$

Using (1.67), (1.68) then gives

$$D_1 = 0 \tag{1.69}$$

and from (1.64), we find

$$\mathcal{E}_{1\tau_1} = 0. \tag{1.70}$$

Equation (1.70) means that $\mathcal{E}_1$ is now a function of $\tau_2, \tau_3, \ldots$ but no longer of τ_1.
  From Eq. (1.65) with (1.69), we learn that

$$D_2 = a_2 - \mathcal{E}_1^2. \tag{1.71}$$

The function $\mathcal{E}_1$ is still unknown and motivates the analysis of the next problem.
  Equation (1.66) with (1.69) can be rewritten as

$$\mathcal{E}_{2\tau_1} = -\mathcal{E}_{1\tau_2} + \frac{1}{2}\mathcal{E}_1 D_2. \tag{1.72}$$

We note that the right hand side of (1.72) is a constant with respect to time τ_1 and that $\mathcal{E}_2$ only appears in the left hand side. As for the $O(\varepsilon^2)$ problem, we require bounded solutions in τ_1. A bounded solution for $\mathcal{E}_2$ with respect to τ_1 requires that the right hand side is zero, i.e. that $\mathcal{E}_1$ satisfies the equation

$$\mathcal{E}_{1\tau_2} = \frac{1}{2}\mathcal{E}_1 D_2 = \frac{1}{2}\mathcal{E}_1(a_2 - \mathcal{E}_1^2). \tag{1.73}$$

Equation (1.73) is the equation for $\mathcal{E}_1$ that we were looking for. We are now ready to refine our definition of ε. Without loss of generality, we choose $|a_2| = 1$ and $a_j = 0$ $(j \geq 3)$. Then, the expression (1.60) uniquely defines ε as

$$\varepsilon \equiv \sqrt{\frac{A-1}{a_2}}, \tag{1.74}$$

where $a_2 = 1$ $(a_2 = -1)$ if $A - 1 > 0$ (if $A - 1 < 0$). In terms of the original variables, Eq. (1.73) is equivalent to

$$\frac{d\mathcal{E}}{dt} = \frac{1}{2}\mathcal{E}(A - 1 - \mathcal{E}^2), \tag{1.75}$$

which is called the *normal form equation* for the laser bifurcation. This equation can be solved exactly and we may describe the complete time history of the field from its initial condition $\mathcal{E}(0) = \mathcal{E}_i$.

The normal form equation is the simplest equation capturing the main features of the bifurcation. Near the bifurcation point, there exists a small amplitude solution that scales like $(A - 1)^{1/2}$ and depends on the slow time $\tau = |A - 1| t$. This slow time is noted experimentally by longer transients as we approach the laser threshold $(A \rightarrow 1^+)$ and is called *critical slowing down* in the physics community. Such a slowing down is visible on Figure 1.3 where the transients are slower and slower as the laser threshold is approached.

Validity

However, our derivation of Eq. (1.75) suffers from an important deficiency that is not immediately transparent as we review our perturbation analysis. It is not the method that needs to be criticized but our basic assumptions. By deriving Eq. (1.75), we have deliberately ignored the fast time dynamics in t. This is because we found from (1.57) that the second growth rate $\sigma_2 \simeq -\gamma$ implies a contribution of the form of an exponentially decaying function of t which is rapid on the $O((A - 1)^{-1})$ large time interval. But is this observation still correct if γ is small? (See Problem 1.8.10.) We find that Eq. (1.75) is correct but we also find that its validity is limited to a strict vicinity of the bifurcation point. More precisely, we obtain the inequality

$$|A - 1| << \gamma. \tag{1.76}$$

If γ is too small (γ is an $O(10^{-3})$ quantity for many lasers; see Table 1.1), Eq. (1.75) does not provide the simple description of the laser dynamics that we were looking for because it applies only in an extremely restricted region. We need to take into account the small value of γ and realize from the expression of the RO frequency (1.29) that the solution needs an expansion in power series of $\sqrt{\gamma}$. This (singular) perturbation analysis will be described in Section 5.2.1.

1.5.3 Phase space

An alternative method to study the dynamical possibilities of the laser rate equations is to plot different solutions $I = I(t)$ and $D = D(t)$ in the phase-plane (D, I). A *trajectory* is a line $(I(t), D(t))$ which starts at the initial point $(D(0), I(0))$. Trajectories cannot cross each other because each initial point determines a unique trajectory. The phase-plane is a convenient method to analyze all the possible evolutions of the laser. See Figure 1.13. We note that all trajectories spiral around the fixed point $(D, I) = (1, A - 1)$ representing the ON state and that they all escape the OFF state $(D, I) = (A, 0)$. Note that the line $I = 0$ is an exact solution of the laser equations. Trajectories starting close to $I = 0$ remain close to $I = 0$ for a long time before quickly spiraling to the ON state (see Problem 1.8.6 for the trajectories emerging from the saddle-point).

Experimentally, we cannot measure the inversion of population D. But Eq. (1.7) suggests a relation between D and I' given by

$$D = \frac{I'}{I} + 1. \tag{1.77}$$

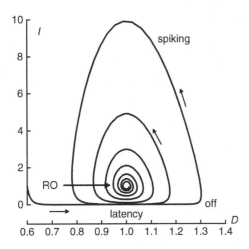

Fig. 1.13 Phase-plane trajectory corresponding to the transient build-up of laser radiation. The trajectory has been determined numerically from Eqs. (1.7) and (1.8) with $\gamma = 5 \times 10^{-3}$, $A = 2$, $I(0) = 1$, $D(0) = 0.6$. It first follows a small intensity regime (latency), then undergoes a series of spikes (spiking), and finally approaches the stable steady state $(D, I) = (1, 1)$ with small amplitude oscillations (RO). Arrows indicate the direction of rotation.

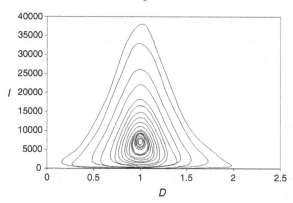

Fig. 1.14 Experimental phase-plane trajectories (I as a function of $D = I'/I + 1$) for the transient switch-on of the Nd^{3+}:YAG laser reported in Figure 1.4.

Since $I' = dI/dt$ is easily determined as we monitor $I(t)$ electronically, an experimental phase-plane is possible by graphing I as a function of $I'/I + 1$. Figure 1.14 displays an example of such trajectories for the turn-on transient of the Nd^{3+}:YAG laser displayed in Figure 1.4.

1.6 Spontaneous emission

In some lasers such as the He-Ne laser and more particularly the semiconductor laser, spontaneous emission (SE) may **not** be neglected even at a simple level of approximation. The rate equations must be modified accordingly to take into account SE as well as the regular emission (RE). In this section, we propose a simple description of SE which reveals important effects such as the laser emission below threshold. More refined models are possible by using stochastic and quantum theories [37, 39] or by introducing Fokker–Planck equations [40].

1.6.1 Imperfect bifurcation and ghosts

An elementary description of SE is possible if we replace the stimulated emission term in the intensity equation, i.e. GnN, by $G(n + 1)N$. We then take into account SE by counting "one extra photon" in the laser mode. The validity of this point is extensively discussed, e.g. in [20]. The modified rate equations now read

$$\frac{dn}{dT} = G(n + 1)N - \frac{n}{T_c},$$

$$\frac{dN}{dT} = -2GnN - \frac{1}{T_1}(N - N_0). \tag{1.78}$$

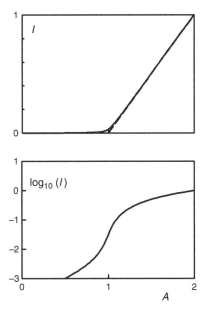

Fig. 1.15 Steady state intensity with spontaneous emission. Top: the branch of steady states given by (1.81) unfolds near the bifurcation point $(A, I) = (1, 0)$ $(b = 10^{-3})$. The broken line is the $b = 0$ non-zero intensity steady state $(I = A - 1)$. Bottom: logarithmic plot emphasizing the change of scale of the intensity near the bifurcation point. The transition layer near $A = 1$ becomes sharper as $b \to 0$.

The modified rate equations are non-dimensionalized in the same way as Eqs. (1.4) and (1.5). The rate equations now are

$$\frac{dI}{dt} = I(D - 1) + bD, \tag{1.79}$$

$$\frac{dD}{dt} = \gamma\,[A - D(1 + I)], \tag{1.80}$$

with $b = 2GT_1$. The modified rate equations (1.79) and (1.80) have only one steady state solution, given by

$$I = \frac{1}{2}\left[(A - 1) + \sqrt{(A - 1)^2 + 4bA}\,\right]. \tag{1.81}$$

The steady laser bifurcation is said to be "imperfect" (see Figure 1.15) meaning that the branch of steady states is smooth near the bifurcation point at $A = 1$.

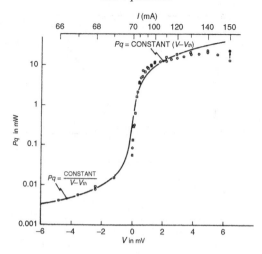

Fig. 1.16 Logarithmic plot of the variations of the intensity vs. pump power for a
GaAs injection laser as the pumping current is increased through the laser bifur-
cation point. Reprinted Figure 5 with permission from H.S. Sommers, Jr. [41].
Copyright 1982 American Institute of Physics.

Analyzing the function (1.81) for small b shows three different scalings for the
intensity. The intensity is $O(b)$ small if $A < 1$ $[I \simeq bA/(1 - A)]$, $O(\sqrt{b})$ small in
the vicinity of the laser threshold $A = 1$ $[|A - 1| = O(\sqrt{b})]$, and $O(1)$ if $A > 1$
$[I \simeq A - 1]$. The change of amplitude from low $O(b)$ values to high $O(1)$ values
is best detected by a logarithmic plot of the intensity. It exhibits more explicitly the
existence of a "threshold" (Figure 1.15 bottom and Figure 1.16) (Exercise 1.8.5).

A consequence of the smooth transition of the steady state branch is the possibility of observing a manifestation of the new state at subcritical values of the pump.
These "ghosts" are particularly spectacular if the bifurcating regime is quite different from the basic steady state solution. This is, for example, the case for other
bifurcation problems where the new state is time-periodic or exhibits a rich spatial
structure.

SE has a strong effect on the laser dynamics both near and below threshold. It is
not restricted to the laser intensity but also modifies the phase of the field and more
generally its spectral properties (linewidth of emission, the number and frequency
of modes). These effects are discussed in [34].

Measurements on the emission below threshold were first performed using
He-Ne lasers and it has been shown to be important in semiconductor lasers. For
semiconductor lasers, the confining effect of the waveguide contributes to the collection of a larger part of the SE, making its contribution more effective than in
open cavity lasers. This effect is readily measured as can be seen from the commercial characteristics of some diode lasers. Figures 1.16 and 1.17 give examples

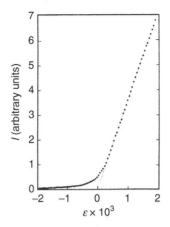

Fig. 1.17 Imperfect bifurcation for a laser in the presence of spontaneous emission, measured for a He-Ne laser. Reprinted Figure 1 with permission from Corti and Degiorgio [42]. Copyright 1976 by the American Physical Society.

of measurements made on a GaAs and a He-Ne laser, respectively. The observed evolutions of the output vs. pump power show remarkable agreement with the predictions of the rate equations.

1.6.2 Dynamical effects

SE not only changes the static characteristics of the laser but it also alters its dynamics. This effect is more subtle than the qualitative change of the steady state near the laser threshold.

The RE tends to produce pulses with a very high contrast ratio and is the dominant dynamical response of the laser as long as it remains of large amplitude. However, if this intensity becomes very small it is expected that processes other than stimulated emission, loss and relaxation, e.g. SE, play a major role. This occurs during the large periods of time separating the intensity spikes as observed at the beginning of the turn-on experiment transient (see Figure 1.18).

We note from Eq. (1.79) that the spontaneous emission bD term feeds continuously the intensity and consequently prevents it from dropping to extremely small values. Mathematically, we may reproduce the analysis described in Section 1.3.2. Instead of (1.43), the intensity during the latency period admits the solution

$$I = I_0 \exp\left(\gamma^{-1} F(\gamma t)\right) + b \exp\left(\gamma^{-1} F(\gamma t)\right) G(\gamma t), \qquad (1.82)$$

where

$$G(s) \equiv \gamma^{-1} \int_0^s \exp\left(-\gamma^{-1} F(s)\right) D(s) ds. \qquad (1.83)$$

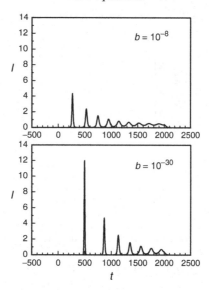

Fig. 1.18 Effect of spontaneous emission as the laser turns on. Numerical solution of Eqs. (1.79) and (1.80). I and t are dimensionless variables. At $t = 0$, A is changed from $A_- = 0.9$ to $A_+ = 1.41$ during the time interval $\Delta t = 2000$. $\gamma = 2.76 \times 10^{-3}$ and $b = 10^{-8}$ (top) or $b = 10^{-30}$ (bottom).

The integral (1.83) can be evaluated for small γ by using Laplace's method [15]. The leading approximation is given by

$$G \simeq \sqrt{\frac{2\pi\gamma^{-1}}{A_+ - 1}} \exp\left(-\gamma^{-1}F(s_c)\right), \tag{1.84}$$

where $s_c \equiv \ln\left[(A_+ - A_-)/(A_+ - 1)\right]$ is the critical time where $D(t)$ passes the bifurcation point. Assuming now that b is an $O(\exp(-\gamma^{-1}))$ small quantity, the second term in (1.82) blows up at $t_j < t_{0n}$ where $s_j \equiv \gamma t_j$ is the root of

$$\ln(b) + \gamma^{-1}\left(F(s) - F(s_c)\right) = 0. \tag{1.85}$$

Provided $s = s_j$ is close to s_c, we find from (1.85) that

$$s_j \simeq s_c + \sqrt{\frac{-2\gamma \ln(b)}{A_+ - 1}}. \tag{1.86}$$

The expression (1.86) indicates that the delay of the jump transition increases only if b decreases exponentially. This exponential sensitivity of the build-up times with respect to spontaneous emission noise led to the development of a sensitive test of the quantum fluctuations in the OFF state ("statistical microscope" [43]).

1.7 Semiconductor lasers

We have seen in the previous sections that quantitative agreement between theory and experiment is possible using rate equations but it often requires a more sophisticated description of the active medium. This is the case for semiconductor lasers (SLs) that we consider in this section.

Traveling through a semiconductor, a single photon can generate an identical photon by stimulating the recombination of an electron–hole pair (see Figure 1.19). Subsequent repetition of this process leads to strong light amplification. However, the competing process is the absorption of photons by generation of new electron–hole pairs. Stimulated emission prevails when more electrons are present at the higher energy level (conduction band) than at the lower energy level (valence band). This situation is called inversion. The photon energy is given by the band gap, which depends on the semiconductor material.

Rate equations appropriate for single mode SLs have been derived in many places [33, 34, 44]. Their formulation is slightly different from the rate equations that were derived for gas or solid state lasers. For historical reasons (the analytical study of the semiconductor laser rate equations came later), the currently used dimensionless rate equations are also different.

In general, the SL rate equations refer to the following equations[2] for the complex amplitude E of the optical field $(E_{opt} = E(\tau) \exp(i\omega_0 \tau))$ and the carrier number N

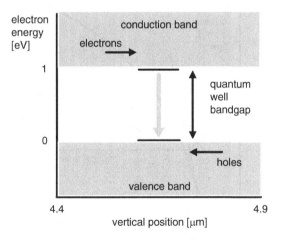

Fig. 1.19 Schematic electron band diagram and carrier transport within an InGaAsP/InP multiquantum well active region (redrawn from Figure 3.1 of Piprek and Bowers [45]).

---

[2] τ (instead of T) is used here to denote the real (physical) time, to keep up with the notations most usually found in the literature.

$$\frac{dE}{d\tau} = \frac{1}{2}(\Gamma G(N) - \tau_p^{-1})E + i(\omega(N) - \omega_0)E, \qquad (1.87)$$

$$\frac{dN}{d\tau} = \frac{J}{e} - \frac{N}{\tau_s} - G(N)|E|^2. \qquad (1.88)$$

In these equations, the optical field is normalized such that the power $|E|^2$ represents the number of photons in the active layer. In Eq. (1.87), the coefficient $\Gamma G(N)$ is defined as the power gain, τ_p is the photon lifetime, and $\omega(N) - \omega_0$ is the detuning between the cavity resonance frequency and the optical frequency ω_0 of the field. The parameter Γ is called the confinement factor and takes into account the fact that only a part of the mode intensity contributes to the gain [33]. In Eq. (1.88), J is the pump current, e is the elementary charge, τ_s is the carrier lifetime, and the term $-G(N)|E|^2$ accounts for the stimulated loss of the carriers.

Numerically computed optical gains show that they vary almost linearly with N if the intensity of the field is not too high. The gain is then commonly approximated as (see Figure 1.20)

$$\Gamma G(N) = \Gamma G_N(N - N_t), \qquad (1.89)$$

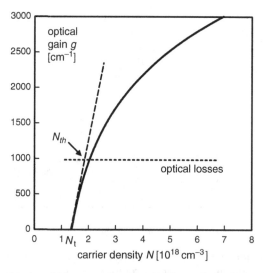

Fig. 1.20 Optical gain $g(N)$ vs. carrier density for an InGaAsP strained quantum well active layer (1.55 μm) at 20°C. The power gain is defined by $\Gamma G(N) = \Gamma v_g g(N)$, where v_g is the photon group velocity ($\sim 10^{10}$ cm s^{-1}) and Γ is the confinement factor (~ 0.1) (redrawn from Figure 3.1 of Piprek and Bowers [45]).

where G_N and N_t are called the gain coefficient and the carrier number at transparency, respectively $(G(N_t)=0)$. In most of the literature, (1.89) is rewritten using the threshold carrier number N_{th} rather than N_t as the reference quantity. N_{th} satisfies the condition $\Gamma G(N_{th}) = \tau_p^{-1}$ and we rewrite Eq. (1.89) as

$$\Gamma G(N) = \tau_p^{-1} + \Gamma G_N(N - N_{th}). \tag{1.90}$$

Similarly, the cavity resonance frequency is linearized around its value at threshold

$$\omega(N) = \omega_{th} + \omega_N(N - N_{th}). \tag{1.91}$$

In (1.91), ω_N is not independent of the gain coefficient G_N. The relation between the two coefficients is given in terms of the linewidth enhancement factor α. The so-called α parameter is defined as the ratio of the real part of the susceptibility χ_p (change in frequency) and its imaginary part (gain). It is given by

$$\alpha = \frac{\mathrm{Re}(\chi_p)}{\mathrm{Im}(\chi_p)} = -\frac{2}{\Gamma}\frac{\omega_N}{G_N}. \tag{1.92}$$

The linewidth enhancement factor typically takes values from 4 to 7 in 1.3 to 1.6 μm InGaAsP lasers and 2.5 to 4 in 0.85 μm GaAs lasers. Assuming $\omega_0 = \omega_{th}$ and introducing (1.90), (1.91), and (1.92) into Eqs. (1.87) and (1.88), we obtain

$$\frac{dE}{d\tau} = \frac{\Gamma G_N}{2}(1 + i\alpha)nE, \tag{1.93}$$

$$\frac{dn}{d\tau} = \frac{J - J_{th}}{e} - \frac{n}{\tau_s} - \left(\frac{1}{\Gamma\tau_p} + G_N n\right)|E|^2, \tag{1.94}$$

where $n \equiv N - N_{th}$ and $J_{th} \equiv N_{th}e\tau_s^{-1}$ are called the inversion and the threshold current, respectively. These equations are now in a form equivalent to our basic rate equations.

We obtain dimensionless equations by introducing the new time t and the new dependent variables Y and Z defined by

$$t \equiv \tau/\tau_p, \quad Y \equiv \sqrt{\frac{\tau_s G_N}{2}}E, \quad \text{and} \quad Z \equiv \frac{\Gamma G_N \tau_p}{2}n. \tag{1.95}$$

In terms of these variables, Eqs. (1.93) and (1.94) become

$$\frac{dY}{dt} = (1 + i\alpha)YZ \tag{1.96}$$

$$T\frac{dZ}{dt} = P - Z - (1 + 2Z)|Y|^2, \tag{1.97}$$

where the new parameters T and P are defined by[3]

$$T \equiv \tau_s / \tau_p \quad \text{and} \quad P \equiv \frac{\tau_s \tau_p G_N \Gamma}{2} \left(\frac{J - J_{th}}{e} \right). \tag{1.98}$$

The fixed parameter T represents the ratio of the carrier and photon lifetimes and is large ($T = 10^2$ to 10^3). P is called the pump parameter above threshold ($|P| = 10^{-2}$ to 1). The equations for $|Y|$ and Z are equivalent to Eqs. (1.7) and (1.8) with $D = 1 + 2Z$, $I = 2|Y|^2$, $A = 1 + 2P$, and $\gamma = T^{-1}$.

If we now determine the steady state solutions of Eqs. (1.96) and (1.97) and analyze their linear stability properties (note that Y is complex), we find that the frequency ω_R and the damping rate ξ of the RO oscillations are simply given by

$$\omega_R \equiv \sqrt{\frac{2P}{T}} \quad \text{and} \quad \xi \equiv \frac{1 + 2P}{2T}. \tag{1.99}$$

This expression is often used to measure SL parameters but it must be handled with care as shown in Problem 1.8.9.

1.8 Exercises and problems

1.8.1 Dimensionless rate equations

Verify the derivation of the dimensionless rate equations (1.7) and (1.8).

1.8.2 Transfer function

Determine the transfer function of a laser described by the rate equations (1.7) and (1.8). Compare the effects of loss and gain modulation given by $T_c = T_c^0(1 + a\exp(i\omega t) + c.c.)$ and $A = A_0(1 + a\exp(i\omega t) + c.c.)$, respectively. After introducing $I = I_s + ap\exp(i\omega t) + c.c.$ and $D = D_s + aq\exp(i\omega t) + c.c.$ into the linearized rate equations, determine p and q. Deduce the transfer function in each case.

1.8.3 Linear stability analysis

Determine the steady states of Eqs. (1.54) and (1.55) and analyze their linear stability properties.

---

[3] These dimensionless parameters are those usually found in the SL literature. Confusion is possible since T was defined as the physical time in our standard rate equations.

1.8.4 Rate equations in terms of power and gain

In the *Encyclopedia of Laser Physics and Technology* [46], the laser rate equations are introduced in a different way. The equations are formulated for the intracavity laser power p and the gain coefficient g,[4] and are given by

$$\frac{dp}{dT} = \frac{g - l}{T_R} p, \tag{1.100}$$

$$\frac{dg}{dT} = -\frac{g - g_{ss}}{\tau_g} - \frac{gp}{E_{sat}}, \tag{1.101}$$

where T_R is the cavity round-trip time, l is the cavity loss, g_{ss} is the small-signal gain (for a given pump intensity), τ_g is the gain relaxation time (often close to the upper state lifetime), and E_{sat} is the saturation energy of the gain medium. Reformulate these equations in the dimensionless form (1.7) and (1.8).

1.8.5 Characteristic equation

Determine the characteristic equation for the steady state solution of Eqs. (1.79) and (1.80). Define the expressions of the RO frequency and analyze its behavior close to the laser threshold ($A - 1 \to 0^+$) for progressively smaller values of b.

1.8.6 Phase-plane and saddle-point

Analyze the separatrices emerging from the saddle-point $(D, I) = (1, 0)$ in the phase-plane. To this end, formulate the equation for the trajectories $I = I(D)$ by dividing Eq. (1.7) and Eq. (1.8). Investigate the limit I small and then the limit γ small.

1.8.7 Turn-on experiment with a pump square pulse

A laser is excited by a pump square pulse from below to above threshold during the short interval of time ($0 < t < t_p$). Under some conditions, the laser may emit a turn-on pulse *after* the pump has been reduced below its threshold [47, 48]. See Figure 1.21. To analyze this phenomenon, assume that the pump parameter A is changed from $A_- < 1$ to $A_+ > 1$ during the time interval $0 < t < t_p$. Assuming then that the intensity remains close to zero, determine the solution for (1) $t \leq t \leq t_+$ when $A = A_+$, and then the solution for (2) $t_+ \leq t < t_{on}$ when $A = A_-$. Combining these two solutions, determine an equation for $t = t_{on}$.

---

[4] The gain measures the strength of optical amplification. It is defined in different ways in the literature. For small gains, it is specified as a percentage, e.g. 3% means an amplification factor of 1.03.

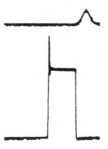

Fig. 1.21 The laser pulse (top) appears after the pump pulse returns to its initial, below-threshold value (bottom). Reprinted Figure 10a with permission from Garreau *et al.* [47]. Copyright 1994 IEEE.

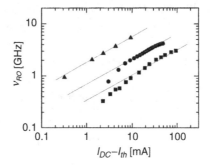

Fig. 1.22 RO frequency as a function of the pump current. ▲: IPAG ($I_{th} =$ 7.4 mA), ●: HLP1400 ($I_{th} = 64.1$ mA), ■: FBH ($I_{th} = 45.2$ mA). The gray lines follow the square-root scaling that is predicted by theory for solitary SLs. Note that the relative deviation $(I_{DC} - I_{th})/I_{th} < 1$ for the first data. Peil's ν_{RO} and $I_{DC} - I_{th}$ are proportional to $\omega_R \tau_p^{-1}$ and P, respectively, where ω_R and P are defined at the end of Section 1.7. From Figure 2.4 of Peil [49].

1.8.8 RO frequency near threshold

In his thesis [49], Michael Peil determined the frequency–current characteristics of the RO frequencies for three different lasers. They are given by a Fabry–Pérot type SL (Hitachi HLP1400) emitting at a center emission wavelength of 840 nm, a telecommunication distributed feedback (DFB) SL (IPAG DFB SL) emitting at a center wavelength of 1551 nm, and another Fabry–Pérot type SL (FBH) emitting at a center wavelength of 786 nm. Figure 1.22 shows the measured RO frequencies from spectra as a function of the pump current. The grey lines correspond to the square-root law (1.29). Note the deviation of the data from this law for small pump currents. M. Peil suggested that this could be the result of spontaneous emission which is dominant close to threshold. But there could be a simpler explanation related to the fact that a log-log plot is used. Consider the exact expression of the RO frequency as provided by the imaginary part of (1.26). Discuss the validity of the approximation (1.29) as $A - 1$ approaches zero. Investigate then the behavior of the exact expression of the RO frequency in a log-log plot.

1.8.9 RO frequency and the design of high-speed SLs

The SL is an important element in fiber optic links since it generates the coherent optical wave that carries the signal. Typical laser wavelengths are 1.3 μm and 1.55 μm, corresponding to the dispersion and absorption minimum, respectively, of silica fibers. The laser frequency is about 200 THz and the RF (10 kHz–300 MHz) or microwave (300 MHz–300 GHz) signal can be modulated onto the laser beam either directly or externally. Direct modulation is simpler to implement than external modulation but the usable bandwidth is limited to a few GHz. Applications of direct analog laser modulation include cable TV, base station links for mobile communication, and antenna remoting. Experiments have shown a resonance peak in the modulation response and these results were well explained by the following laser rate equations

$$\frac{dp}{d\tau} = (G_N(N - N_t) - \tau_p^{-1})p, \tag{1.102}$$

$$\frac{dN}{dt} = \frac{J}{e} - \frac{N}{\tau_s} - G_N(N - N_t)p \tag{1.103}$$

predicting the RO's resonance frequency [50]. Equations (1.102) and (1.103) are equivalent to Eqs. (1.87) and (1.88) (without the confinement factor Γ) rewritten in terms of the output power $p \equiv |E|^2$. The phase of the field depends passively on p and its equation can be ignored. Verify that the steady state power is given by

$$p_0 = \tau_p \left[\frac{J}{e} - \frac{1}{\tau_s} \left(N_t + \frac{1}{\tau_p G_N} \right) \right] = \tau_p \left[\frac{J}{e} - \frac{N_{th}}{\tau_s} \right] \tag{1.104}$$

and that the leading approximation of the RO frequency (in Hz) is

$$f_R = \frac{1}{2\pi} \sqrt{\frac{G_N p_0}{\tau_P}}. \tag{1.105}$$

The modulation bandwidth is widely accepted to be equal to f_R. Equation (1.105) expresses the modulation bandwidth as a simple function of three independent parameters. The differential optical gain constant G_N depends on material properties, the photon lifetime τ_P is related to the device geometry, and the photon density expresses the state of the laser. There are therefore three obvious ways to increase the RO frequency. The gain coefficient can be increased roughly by a factor of five by cooling the laser from room temperature to 77 K. Biasing the laser at higher currents would increase the optical output power density but there is an upper limit where mirror damage occurs. The third way to increase the modulation bandwidth is to reduce the photon lifetime by decreasing the length of the laser cavity.

Compare (1.105) with the expression

$$f_R = \frac{1}{2\pi} \sqrt{\frac{1}{\tau_P \tau_s} \left(\frac{J}{J_{th}} - 1 \right)}, \tag{1.106}$$

which is used extensively. To this end rewrite (1.105) in terms of J/J_{th} where $J_{th} \equiv N_{th} e \tau_s^{-1}$ and show that (1.106) is valid provided that $\tau_P G_N N_t \ll 1$. The expression (1.106) therefore neglects the fact that a large electron density N_t is needed to achieve transparency.

1.8.10 Two-time analysis of the laser rate equations

Determine the solution of Eqs. (1.54) and (1.55) by seeking a solution of the form

$$\mathcal{E} = \varepsilon \mathcal{E}_1(t, \tau) + \varepsilon^2 \mathcal{E}_2(t, \tau) + \ldots$$
$$D = 1 + \varepsilon D_1(t, \tau) + \varepsilon^2 D_2(t, \tau) + \ldots, \tag{1.107}$$

where the small parameter ε is defined by

$$\varepsilon \equiv \sqrt{\frac{A - 1}{a_2}} \tag{1.108}$$

with $a_2 = 1$ ($a_2 = -1$) if $A - 1 > 0$ (if $A - 1 < 0$) and $\tau \equiv \varepsilon^2 t$. Expand the initial conditions $\mathcal{E}(0) = \mathcal{E}_i$ and $D(0) = D_i$ in power series of ε as in (1.107).

2

Three- and four-level lasers

In Chapter 1, we introduced the standard rate equations (SRE) for a laser containing two-level atoms between which stimulated emission is possible. But real lasers exhibit much more complicated energy level schemes. Throughout this chapter, we consider several models in which the light–matter interaction is described by population equations for all the levels involved in the laser operation.[1] Although they are mathematically more complicated, we shall investigate these equations in the same way as in Chapter 1, i.e. by formulating dimensionless equations and by analyzing the stability properties of the steady states.

The basic ingredients of the SRE model used up to here, namely pumping and relaxation processes, play key roles in the efficiency of lasers. But already during the pioneering days of the laser, it appeared important to investigate three- and four-level models in order to obtain more reliable information on quantities such as the power conversion efficiency or the response time. The common extensions of the two-level SRE typically consider an open two-level system, or three- or four-level systems, depending on the nature of the active medium. Restriction to as few as four levels is again a crude simplification of the complex population dynamics occurring in most lasers. But it is rather surprising to see how good these simple kinetic models are. As we demonstrate by studying specific examples, the solution of the three- or four-level rate equations differs only slightly from the solution of the SRE. This reinforces the idea that the dynamical response of many lasers depends on a few dynamical features that are well captured – at least qualitatively – by the two-level SRE.

In practice, the complexity of the level schemes considered depends on the lasing material but also on the desired level of modeling. The description of more complex phenomena such as some specific pulsating instabilities in a CO_2 laser

---

[1] A more rigorous treatment is possible using a semiclassical theory where the Schrödinger and Maxwell equations are considered (see e.g. [36]).

containing a saturable absorber motivated a three-level scheme for the CO_2 active medium. Similarly, the quantitative analysis of long transient behaviors required a model that takes into account the coupling of the lasing levels with their vibrational manifolds in the CO_2 molecule. Both models are analyzed in this chapter.

We begin by a rapid description of several energy level schemes and concentrate on important ones motivated by either historical (ruby laser) or technical (Nd^{3+}:YAG, CO_2 lasers) reasons. We start from the equations appearing in the original papers, formulate dimensionless equations, and discuss the relevance of the parameters. Because some population variables are either small in size or slowly varying in time, they can be eliminated from the rate equations by quasi-steady state approximations. As a result, the number of equations can be reduced. The quasi-steady state approximation (also called adiabatic elimination) is a popular technique in chemistry and biochemistry because it allows major simplifications of the original kinetic equations.

2.1 Energy level schemes in lasers

This section reviews different models commonly used in the literature (for more details, see standard textbooks on lasers, e.g. [20], [51]).

The first laser that exhibited intensity oscillations was a **ruby** (i.e. a $\mathbf{Cr^{3+}}$: $\mathbf{Al_2O_3}$) **laser** and its mode of operation may be understood using a three-level scheme. Such a pumping scheme was suggested by Basov and Prokhorov in 1955 [50] and in a more detailed way in 1956 by Bloembergen [51],[2] for obtaining continuous operation of a maser, the microwave elder brother of the laser (which was only at that time referred to as an "optical maser"). Bloembergen showed that population inversion on a microwave transition could be obtained by pumping with cw radiation on another transition, an idea which was implemented soon afterwards in three-level solid state masers. Practically speaking, the ruby laser was flash pumped and its output was pulsed but the basis of its operation follows Bloembergen's pumping scheme.

The **He-Ne laser**, which was introduced during the same year as the ruby laser, is usually modeled with a two-level scheme. However, in contrast to the model discussed in Chapter 1, the total population of the two levels is not constant since relaxation processes expel the atoms from the level manifold directly involved in the stimulated emission process. This model is called the "open two-level" system (see Exercise 2.4.1).

---

[2] The ideas behind the maser were proposed independently by Nikolai Basov and Alexander Prokhorov in 1955 and by Nicolaas Bloembergen in 1956. Because of this difference in dates, Basov and Prokhorov shared the Nobel Prize with Charles Townes in 1964 while Bloembergen shared it a few years later, in 1981, with Arthur Schawlow.

In solid state lasers such as the **Nd^{3+}:YAG lasers**, many ionic energy levels contribute to pumping and relaxation. Fortunately, in the case of the Nd^{3+} emission at 1.06 μm, the lasing process may be described by a scheme involving four levels only and, as we shall demonstrate, can be reduced to an effective two-level scheme. Other rare-earth doped materials are commonly used to obtain coherent emission in specific wavelength ranges and each one has its own scheme for pumping and relaxation, leaving a series of problems of various complexity (for an overview, see [52]). As far as applications are concerned, the currently important laser is the **Er^{3+} doped fiber laser**, which emits at the wavelength of long-haul fiber telecommunications ($\lambda = 1.55$ μm). It was first described by a three-level model with emission between the intermediate and the ground state, just like the ruby laser. However, because of an accidental resonance of the pump wavelength with another transition, excited state absorption is possible making it often necessary to include more energy levels in the model.

Gas lasers such as the **CO$_2$ and N$_2$O lasers** are qualitatively well described by the two-level SRE introduced in Chapter 1. But, as already mentioned, the SRE model is inadequate to describe instabilities generated by a saturable absorber inserted inside the cavity. The inclusion of additional levels is needed and has been done either by using a three-level model or by considering a "two + two" level model.

Far-infrared lasers are of restricted practical use because of their low emission power and poor efficiency but they often are the only coherent source available in the far infrared (100 μm–1 mm) region of the spectrum. Two simple models are especially relevant. The first one is just a three-level model as discussed in the forthcoming section [55]. The second one is known as the Haken–Lorenz model and will be presented in Chapter 11.

As explained in Chapter 1, emission in **semiconductor lasers** results from electron–hole recombinations between energy bands rather than discrete levels. As a result, unusual dynamical responses are possible but a description in terms of two rate equations is still possible.

2.2 Three-level lasers

Two lasing scenarios are possible for a three-level scheme. See Figure 2.1. In the first case (a), atoms (or molecules or ions) are pumped from the ground state **1** to some excited state **3**. Laser emission occurs between this level and an intermediate state **2**. The atoms relax rapidly from there to the ground state and the cycle repeats. The relaxation from **2** to **1** is fast, implying a low population at level **2**. This then contributes to a large gain between **3** and **2**. In the second case (b), the atoms de-excite from the upper state **3** to an intermediate state **2**. The transition to the ground

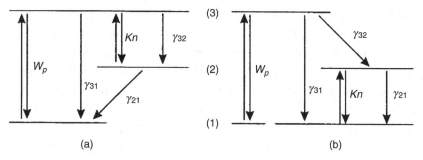

Fig. 2.1 Schematic description of the three-level models. W_p is the pumping rate and the γ_{ij} are the relaxation rates. Kn indicates laser action, where n is the number of photons. (a) CO_2 laser scheme; (b) ruby laser scheme (redrawn from Figure 10 of Dangoisse *et al.* [22]).

state **1** is weakly allowed so that atoms are stored in the intermediate level until sufficient population inversion is accumulated, then laser action occurs between levels **2** and **1**.

We shall consider both cases and show how we may simplify our rate equations by taking advantage of the relative values of the relaxation rates.

2.2.1 Ruby laser

The laser rate equations consist now of equations for the various populations coupled to an equation for the field as Eq. (1.1) in Chapter 1. We first consider the case shown in Figure 2.1(b) which models, for instance, a ruby laser. We assume that pumping is realized from level **1** to level **3**. A double arrow in the figure indicates that the pumping process induces transitions in the two directions from level **1** to level **3** and from level **3** to level **1** (with the same rate W_p in case of coherent pumping). Using Figure 2.1(b), we may formulate the population equations for N_1, N_2, and N_3. They are given by

$$N_1' = \gamma_{21} N_2 - W_p(N_1 - N_3) + Kn(N_2 - N_1) + \gamma_{31} N_3, \qquad (2.1)$$

$$N_2' = \gamma_{32} N_3 - \gamma_{21} N_2 - Kn(N_2 - N_1), \qquad (2.2)$$

$$N_3' = W_p(N_1 - N_3) - \gamma_{32} N_3 - \gamma_{31} N_3, \qquad (2.3)$$

where prime (') means differentiation with respect to time T. Note that $N_1' + N_2' + N_3' = 0$ implying that the total population

$$N_1 + N_2 + N_3 = N_T \qquad (2.4)$$

is a constant. In the case of the ruby laser, the upper laser level lifetime is exceptionally long ($\gamma_{21}^{-1} = 3$ ms [56, 51], $\gamma_{21}^{-1} = 4.3$ ms [20] p. 250). On the other hand, the relaxation rates from level **3** to level **2** or from level **3** to level **1** are fast compared to γ_{21}^{-1} (γ_{32}^{-1} and γ_{31}^{-1} are of the order of 0.1 μs; see [51]: energy level scheme p. 149, constants p. 448). Furthermore, we note the inequalities

$$\gamma_{32}^{-1} << W_p^{-1}, \gamma_{31}^{-1}. \tag{2.5}$$

As soon as one atom is excited from level **1** to level **3**, it will almost instantaneously be de-excited to level **2** and N_3 will remain small. Mathematically, we assume that N_3 is small compared to N_1 and that $\gamma_{31}N_3$ and N_3' are both small compared to $W_p N_1$ (but $\gamma_{32}N_3$ is of the same magnitude as $W_p N_1$). Equations (2.1)–(2.4) then simplify as

$$N_1' \simeq \gamma_{21}N_2 - W_p N_1 + Kn(N_2 - N_1), \tag{2.6}$$

$$N_2' = \gamma_{32}N_3 - \gamma_{21}N_2 - Kn(N_2 - N_1), \tag{2.7}$$

$$0 \simeq W_p N_1 - \gamma_{32}N_3. \tag{2.8}$$

$$N_T \simeq N_1 + N_2. \tag{2.9}$$

Solving Eq. (2.8) for N_3, we obtain

$$N_3 = \frac{W_p N_1}{\gamma_{32}} < 1. \tag{2.10}$$

Using (2.10), Eq. (2.7) is further simplified as

$$N_2' = W_p N_1 - \gamma_{21}N_2 - Kn(N_2 - N_1). \tag{2.11}$$

Introducing the inversion of population $N \equiv N_2 - N_1$ and using (2.6), (2.9), and (2.11), we determine an equation for N as

$$N' = -\gamma_{21}(N + N_T) - W_p(N - N_T) - 2KnN. \tag{2.12}$$

The right hand side of Eq. (2.12) displays the three main processes appearing in laser action. The first term models the relaxation to equilibrium in the absence of pumping: N relaxes towards $-N_T$ since the population accumulates in level **1** under the influence of the single relaxation process. The second term describes the pumping process which creates the inversion of population (if $W_p > \gamma_{21}$): in case of very strong pumping (if $W_p >> \gamma_{21}$), and in the absence of laser emission ($n = 0$), the population accumulates in level **2** ($N = N_T$). The last term indicates the nonlinear coupling between population and intensity as the result of stimulated emission.

Equation (2.12) for N is coupled to an equation for the laser number of photons given by

$$n' = n(-\gamma_c + KN). \qquad (2.13)$$

Equation (2.13) is identical to Eq. (1.4) in Chapter 1 with $K = G$, and $\gamma_c = T_c^{-1}$. Introducing the new variables

$$t \equiv \gamma_c T, \quad I \equiv \frac{2Kn}{\gamma_{21} + W_p}, \quad \text{and} \quad D \equiv \frac{K}{\gamma_c} N \qquad (2.14)$$

into Eqs. (2.12) and (2.13), we obtain

$$\boxed{\begin{aligned} I' &= I(-1 + D), \\ D' &= \gamma\,[A - D(1 + I)] \end{aligned}} \qquad (2.15)$$

where

$$\gamma \equiv \frac{\gamma_{21} + W_p}{\gamma_c} \quad \text{and} \quad A \equiv \frac{(W_p - \gamma_{21})KN_T}{(\gamma_{21} + W_p)\gamma_c}. \qquad (2.16)$$

With these new definitions of γ and A, the system (2.15) is identical to Eqs. (1.7) and (1.8) in Chapter 1.

2.2.2 CO_2 laser

Since the early 1970s, various theoretical models of the CO_2 laser containing a saturable absorber have been proposed with the aim of quantitatively describing the pulsating outputs observed experimentally [57, 58]. Powell and Wolga (1971) [59] introduced a two-level description for the absorber and the active medium which was later studied in detail. However, the two-level model hardly provides the range of parameters in which the pulsating instabilities appear. Arimondo *et al.* (1983) [60] then proposed including the vibrational manifolds to which the lasing and absorbing levels are coupled. The resulting four-level model well reproduced the experimental domain of instabilities but was unable to describe some of the multiperiodic or chaotic regimes observed experimentally. A definite step towards their description came with Tachikawa and coworkers (1986) [61]–[63], who introduced the ground state of the CO_2 molecule as a third level for the active medium. Numerical simulations of the model equations successfully reproduced the variety of periodic and erratic pulsating states [64, 65].

Model

The third-level model of a CO_2 laser is shown in Figure 2.1(a). Assuming independence of the three basic processes (pumping, relaxation, stimulated emission), the population equations for N_1, N_2, and N_3 are now given by

$$N_1' = -W_p(N_1 - N_3) + \gamma_{21}N_2 + \gamma_{31}N_3, \tag{2.17}$$

$$N_2' = \gamma_{32}N_3 - \gamma_{21}N_2 + Kn(N_3 - N_2), \tag{2.18}$$

$$N_3' = W_p(N_1 - N_3) - \gamma_{32}N_3 - \gamma_{31}N_3 - Kn(N_3 - N_2), \tag{2.19}$$

where prime means differentiation with respect to time T. As for the ruby laser, we assume coherent pumping, i.e. the pumping mechanism induces back and forth transitions between levels **1** and **3**. Typical values of the rate constants for a CO_2 laser are listed in Table 2.1.

Note that $N_1' + N_2' + N_3' = 0$ and that Eq. (2.4) is still verified. We would like to take advantage of the relative small values of γ_{32} and W_p compared to either γ_{21} or γ_{31}. The large values of γ_{21} and γ_{31} mean that N_2 and N_3 are small compared to N_1 because they rapidly relax to their equilibrium values. From (2.4), we then conclude that $N_1 \simeq N_T$. With $N_1 = N_T$ and neglecting all $\gamma_{32}N_3$ terms, Eqs. (2.17)–(2.19) considerably simplify as

$$N_1' = -W_p N_T + \gamma_{21}N_2 + \gamma_{31}N_3, \tag{2.20}$$

$$N_2' = -\gamma_{21}N_2 + Kn(N_3 - N_2), \tag{2.21}$$

$$N_3' = W_p N_T - \gamma_{31}N_3 - Kn(N_3 - N_2). \tag{2.22}$$

It is natural to introduce the inversion of population

$$N = N_3 - N_2 \tag{2.23}$$

and express N_2 in terms of N_3 and N as $N_2 = N_3 - N$. From Eqs. (2.20)–(2.22), we obtain the following equations for N_1 and N

$$N_1' = -W_p N_T + \gamma_{21}(N_3 - N) + \gamma_{31}N_3, \tag{2.24}$$

$$N' = W_p N_T - \gamma_{31}N_3 - 2KnN + \gamma_{21}(N_3 - N). \tag{2.25}$$

Since the total population is $N_1 + N_2 + N_3 = N_1 + 2N_3 - N = N_T$, we may express N_3 as

$$N_3 = \frac{N_T - N_1 + N}{2}. \tag{2.26}$$

Substituting (2.26) into Eqs. (2.24) and (2.25), we find

$$N_1' = \gamma_1 N + \gamma_2 (N_T - N_1) - W_p N_T, \tag{2.27}$$

$$N' = -\gamma_1 (N_T - N_1) - 2K n N - \gamma_2 N + W_p N_T, \tag{2.28}$$

where

$$\gamma_1 \equiv \frac{\gamma_{31} - \gamma_{21}}{2} \quad \text{and} \quad \gamma_2 \equiv \frac{\gamma_{21} + \gamma_{31}}{2}. \tag{2.29}$$

Equations (2.27) and (2.28) are coupled to Eq. (2.13) for the number of photons. We now finalize our formulation of the rate equations by introducing the following dimensionless variables

$$t \equiv \gamma_c T, \quad I \equiv \frac{2K n}{\gamma_2}, \quad U \equiv \frac{K}{\gamma_c} N, \tag{2.30}$$

$$W \equiv \frac{K}{\gamma_c \gamma_2} \left[-\gamma_1 (N_T - N_1) + W_p N_T \right]. \tag{2.31}$$

From Eqs. (2.13), (2.27), and (2.28), we obtain

$$\boxed{\begin{aligned} I' &= I(-1 + U), \\ U' &= \varepsilon \left[W - U(1 + I) \right], \\ W' &= \varepsilon \left(A + bU - W \right) \end{aligned}} \tag{2.32}$$

where

$$\varepsilon \equiv \frac{\gamma_2}{\gamma_c}, \quad b \equiv \left(\frac{\gamma_1}{\gamma_2} \right)^2, \quad \text{and} \quad A \equiv \frac{K W_p N_T}{\gamma_c \gamma_2} \left(1 - \frac{\gamma_1}{\gamma_2} \right). \tag{2.33}$$

These are the equations derived by Lefranc *et al.* [65]. In these equations, the population inversion is U which is coupled to a reservoir population W. Both U and W are slow variables because the right hand sides of the equations for U and W are proportional to ε which is a small parameter. A is the control parameter and there are only two fixed parameters, b and ε. Using the values of the parameters given by Lefranc *et al.* [65] (see Table 2.1), we find $b = 0.85$ and $\varepsilon = 0.1375$. In the next two subsections, we determine the steady state solutions and discuss their stability properties.

Table 2.1 *Rate constants for a CO_2 laser (all in s^{-1}).*

	Tachikawa *et al.*[a]	Lefranc *et al.*[b]
γ_{32}	20	10
γ_{21}	380×10^3	289.2×10^3
γ_{31}	10^3	1.2×10^4
W_p	8.5–70	
γ_c	2.5×10^6	1.1×10^6

[a] [63, 64].
[b] [65]. The value of γ_{21} documented in [65] (γ_{10}) is a misprint (M. Lefranc, private communication).

Steady state solutions

From (2.32), we find the following steady state solutions

$$I = 0, \quad W = U = \frac{A}{1 - b}, \tag{2.34}$$

$$I = A + b - 1 \geq 0, \quad U = 1, \quad W = A + b \tag{2.35}$$

corresponding to the OFF and ON states, respectively. The value of b is close to 1 because $\gamma_{31} << \gamma_{21}$. However, we cannot set b equal to 1 because (2.34) is singular at $b = 1$.

From (2.35), we find that the lasing threshold is $A = A_{th} \equiv 1 - b$ suggesting a drastic reduction of the lasing threshold from a two- to a three-level system. This is however not the case because the definition of A is quite different in the two- and three-level problems. Practically, A is not calculated from the physical constants but it is normalized using the threshold pump as a reference since it can be determined experimentally. If this is done, the OFF and ON steady states are $(I, U) = (0, A/A_{th})$ and $(I, U) = (A - A_{th}, 1)$, respectively.

Linear stability analysis

We wish to find how a small perturbation of either (2.34) or (2.35) will grow or decay. To this end, we insert $I = I_s + i$, $U = U_s + u$, and $W = W_s + w$ into Eqs. (2.32) where (I_s, U_s, W_s) denotes either the zero intensity steady state (2.34) or the non-zero intensity steady state (2.35). We simplify the resulting equations using the steady state equations and neglect all nonlinear contributions in i, u, and w. We then obtain the following linear equations for i, u, and w

$$\frac{d}{dt}\begin{pmatrix} i \\ u \\ w \end{pmatrix} = \begin{pmatrix} U_s - 1 & I_s & 0 \\ -\varepsilon U_s & -\varepsilon(1 + I_s) & \varepsilon \\ 0 & \varepsilon b & -\varepsilon \end{pmatrix}\begin{pmatrix} i \\ u \\ w \end{pmatrix}. \tag{2.36}$$

The general solution of Eq. (2.36) is a linear combination of three exponential solutions of the form

$$i = c_1 \exp(\sigma t), \quad u = c_2 \exp(\sigma t), \quad \text{and} \quad w = c_3 \exp(\sigma t), \tag{2.37}$$

where σ is the growth rate and the c_j are constants. Introducing (2.37) into (2.36) leads to a homogeneous system of equations for c_1, c_2, and c_3. A nontrivial solution is possible only if

$$\det \begin{pmatrix} U_s - 1 - \sigma & I_s & 0 \\ -\varepsilon U_s & -\varepsilon(1 + I_s) - \sigma & \varepsilon \\ 0 & \varepsilon b & -\varepsilon - \sigma \end{pmatrix} = 0. \tag{2.38}$$

For the zero intensity steady state (2.34), (2.38) leads to the following characteristic equation for σ

$$\left(\frac{A}{1-b} - 1 - \sigma \right) \left(\sigma^2 + 2\varepsilon\sigma + \varepsilon^2(1 - b) \right) = 0. \tag{2.39}$$

The first root is $\sigma_1 = \frac{A}{1-b} - 1$ and the two other roots satisfy the quadratic equation $\sigma^2 + 2\varepsilon\sigma + \varepsilon^2(1 - b) = 0$. From the sign of the coefficients and since $b < 1$, we conclude that the real part of these roots is always negative. Therefore the stability of the solution is determined by σ_1 only. σ_1 changes sign at $A = A_{th} = 1 - b$ and the solution is stable (unstable) if $A < A_{th}$ (if $A > A_{th}$).

For the non-zero intensity steady state (2.35), (2.38) leads to the following characteristic equation for σ

$$\sigma^3 + C_1\sigma^2 + C_2\sigma + C_3 = 0, \tag{2.40}$$

where the coefficients C_j $(j = 1, 2, 3)$ are defined by

$$C_1 \equiv 2\varepsilon + \varepsilon I_s, \quad C_2 \equiv \varepsilon I_s + \varepsilon^2 A, \quad C_3 \equiv \varepsilon^2 I_s. \tag{2.41}$$

The necessary and sufficient conditions for σ having a negative real part are known as the Routh–Hurwitz conditions [29]. For (2.40), these conditions require the following inequalities on the coefficients C_j

$$C_1 > 0, \quad C_3 > 0, \quad \text{and} \quad C_1 C_2 - C_3 > 0. \tag{2.42}$$

They are easily verified since ε and I_s are both positive. The non-zero intensity solution is always stable.

Relaxation oscillations

As in Chapter 1, we wish to derive a useful analytical approximation of the growth rate taking into account that ε is small. The growth rate for the two-level problem was given by an expansion in powers of $\varepsilon^{1/2}$ and we are going to do the same here. After inserting

$$\sigma = \varepsilon^{1/2}\sigma_0 + \varepsilon\sigma_1 + \dots \tag{2.43}$$

into Eq. (2.40), we equate to zero the coefficients of each power of $\varepsilon^{1/2}$. The first two problems are given by

$$O(\varepsilon^{3/2}) : \sigma_0^3 + I_s\sigma_0 = 0,$$
$$O(\varepsilon^2) : 3\sigma_0^2\sigma_1 + I_s\sigma_1 + (2 + I_s)\sigma_0^2 + I_s = 0 \tag{2.44}$$

and lead to the following solutions

$$\sigma \simeq -\varepsilon, \tag{2.45}$$

$$\sigma \simeq \pm i\sqrt{\varepsilon I_s} - \frac{\varepsilon}{2}(I_s + 1). \tag{2.46}$$

The first term in (2.46) indicates that small perturbations of the steady state exhibit oscillations at a frequency

$$\omega_R = \sqrt{\varepsilon I_s}, \tag{2.47}$$

which is identical to the expression obtained for the two-level system ($\omega_R = \sqrt{\gamma I_s}$). The second term in (2.46) represents the damping rate of the laser relaxation oscillations, i.e.

$$\Gamma = -\frac{\varepsilon}{2}(I_s + 1), \tag{2.48}$$

which is identical to the expression obtained for the two-level system ($\Gamma = -\frac{\gamma}{2}(I_s + 1)$). We conclude that although we cannot reduce the three rate equations (2.32) to the SRE, the relaxation of a small perturbation from the non-zero intensity steady state is the same as for the SRE.

2.3 Four-level lasers

2.3.1 Model

The four-level laser system is a model for a Nd^{3+}:YAG and many similar solid state and dye lasers. An idealized four-level pumping scheme is shown in Figure 2.2.

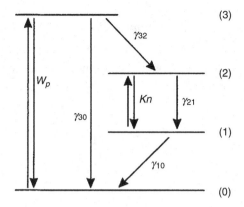

Fig. 2.2 Four-level model. W_p is the pumping rate, the γ_{ij} are relaxation rates, and Kn denotes the lasing action (redrawn from Figure 11 of Dangoisse *et al.* [22]).

From this figure, we derive the following population equations for N_0, N_1, N_2, and N_3

$$N_0' = \gamma_{10}N_1 - W_P(N_0 - N_3) + \gamma_{30}N_3, \tag{2.49}$$

$$N_1' = -\gamma_{10}N_1 + \gamma_{21}N_2 + Kn(N_2 - N_1), \tag{2.50}$$

$$N_2' = \gamma_{32}N_3 - \gamma_{21}N_2 - Kn(N_2 - N_1), \tag{2.51}$$

$$N_3' = W_p(N_0 - N_3) - \gamma_{32}N_3 - \gamma_{30}N_3. \tag{2.52}$$

Again, we verify that $N_0' + N_1' + N_2' + N_3' = 0$ meaning that the total population

$$N_0 + N_1 + N_2 + N_3 = N_T \tag{2.53}$$

is constant. We assume that γ_{32} is much larger than all the other relaxation and pump rates. From (2.52), we then find that N_3 is small and is given by

$$N_3 \simeq W_p N_0 / \gamma_{32}. \tag{2.54}$$

As is the case for most solid state lasers, the upper levels are poorly populated compared to the ground level and from (2.53) we have $N_0 \simeq N_T$. The population equations then reduce to Eqs. (2.50) and (2.51) with $\gamma_{32}N_3 = W_p N_0 = W_p N_T$ as for the three-level system. Coupled to an equation for the number of photons, we have

$$n' = -\gamma_c n + \gamma_{21}Kn(N_2 - N_1) + R_{sp}, \tag{2.55}$$

$$N_2' = S - \gamma_{21}N_2 - \gamma_{21}Kn(N_2 - N_1), \tag{2.56}$$

$$N_1' = -\gamma_{10}N_1 + \gamma_{21}N_2 + \gamma_{21}Kn(N_2 - N_1), \tag{2.57}$$

Table 2.2 *Relaxation rates for a solid state microchip laser [66, 67].*

Parameters	Values
K	7×10^{-6}
γ_c	$7 \times 10^{10} \, \text{s}^{-1}$
γ_{21}	$1.3 \times 10^4 \, \text{s}^{-1}$
γ_{10}	$1.6 \times 10^9 \, \text{s}^{-1}$

where $S \equiv W_p N_T$ is the pump rate, K is defined as the ratio of stimulated to spontaneous emission cross sections, $R_{sp} \equiv N_2 K \gamma_{21}$ is the spontaneous emission rate into the lasing mode (following Siegman, [20] pp. 502–503). Typical values for a $Nd^{3+}:YVO_4$ microchip laser are listed in Table 2.2.

2.3.2 Connection with the two-level model

We note from Table 2.2 that the ratio $\gamma_{21}/\gamma_{10} \sim 10^{-5}$ is small which suggests a simplification of Eqs. (2.55)–(2.57). To this end, we introduce a small parameter ε defined as

$$\varepsilon \equiv \gamma_{21}/\gamma_{10} \tag{2.58}$$

and seek an approximation of N_1 of the form

$$N_1 = \varepsilon N_{11} + \ldots \tag{2.59}$$

Inserting (2.59) into Eq. (2.57) gives N_{11} as

$$N_{11} = N_2(1 + Kn). \tag{2.60}$$

Then introducing $N_1 = \varepsilon N_{11}$ into Eqs. (2.55) and (2.56), we obtain the following two-variable rate equations

$$n' = -\gamma_c n + K \gamma_{21} N_2 [1 - \varepsilon(1 + Kn)] n + R_{sp}, \tag{2.61}$$

$$N_2' = S - \gamma_{21} N_2 - K \gamma_{21} N_2 [1 - \varepsilon(1 + Kn)] n. \tag{2.62}$$

We may reformulate these equations in dimensionless form. Introducing the new variables

$$I \equiv Kn, \quad D \equiv \frac{K \gamma_{21}}{\gamma_c} N_2, \quad \text{and} \quad t \equiv \gamma_c T, \tag{2.63}$$

Eqs. (2.61) and (2.62) become

$$
\boxed{
\begin{aligned}
I' &= I\left[-1 + D\left(1 - \varepsilon(1 + I)\right)\right] + KD \\
D' &= \gamma\left[A - D - D\left(1 - \varepsilon(1 + I)\right) I\right]
\end{aligned}
}
\tag{2.64}
$$

where

$$
\gamma \equiv \frac{\gamma_{21}}{\gamma_c} \quad \text{and} \quad A \equiv \frac{SK}{\gamma_c}.
\tag{2.65}
$$

Equations (2.64) were obtained after eliminating N_1 adiabatically. A better approach is to first formulate dimensionless rate equations and then eliminate one of the population variables.

In the present model, spontaneous emission has been included via the parameter KD.[3] In spite of its small value, it has a important effect on the solution if the intensity is small. Equations (2.64) reduce to the SRE as $\varepsilon \to 0$ and $K \to 0$. The four-level system introduces nonlinear gain saturation and the rate equations exhibit three small parameters, namely $\gamma = 0.2 \times 10^{-6}$, $K = 7 \times 10^{-6}$, and $\varepsilon = 8 \times 10^{-6}$. All three may contribute to the damping of the relaxation oscillations. This can be understood if we analyze the stability properties of the non-zero intensity steady state. From (2.64), we find $(I_s, D_s) = (A - 1, 1) + O(K, \varepsilon)$ and that the characteristic equation leads to the following solution for the growth rate σ

$$
\sigma \simeq \pm i\sqrt{\gamma I_s} - \frac{1}{2}\left[\gamma A + \frac{K}{A - 1} + \varepsilon(A - 1)\right],
\tag{2.66}
$$

where the second term in (2.66) describes the damping rate of the relaxation oscillations. It is given by

$$
\Gamma \equiv -\frac{1}{2}\left[\gamma A + \frac{K}{A - 1} + \varepsilon(A - 1)\right],
\tag{2.67}
$$

where we note the SRE contribution given by $-\gamma A/2$. The two extra terms nevertheless have an influence. $\Gamma(A)$ exhibits a minimum at $A - 1 = \sqrt{\frac{K}{\varepsilon + \gamma}} \simeq 0.92$ which is observed experimentally in [66] from the laser linewidth.

2.3.3 Modified four-level model for CO_2 lasers

An improved four-level model that takes into account rotational bands has been proposed in [58, 60] and further analyzed in [68, 69] because it better describes CO_2 lasers in the presence of slow cavity loss modulations. More specifically, it is

---

[3] The notations of Siegman [20] have been used here. The SE term KD here corresponds to bD in Eq. (1.79).

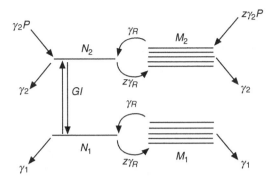

Fig. 2.3 Four-level model for the CO_2 laser. Each lasing level is collisionally coupled to the manifold of rotational energy levels belonging to the same vibrational level.

a two + two model in which the two lasing levels are rotational levels of different vibrational states. The collisional coupling between these lasing levels and the other rotational states of the same vibrational state(s) is explicitly considered. In this section, we present this model starting from the four-level scheme shown in Figure 2.3, derive a dimensionless form for the equations, and deduce from a long time analysis the main effect of the addition of the vibrational manifolds of the rotational levels coupled by relaxation to the lasing levels.

For clarity, we use the same notation as in [70]. As illustrated in Figure 2.3, N_1 and N_2 represent the populations of the two lasing states while M_1 and M_2 are the global populations of the two manifolds of rotational levels. Together with the intensity I of the field, they satisfy the following rate equations

$$I' = -\kappa I + G(N_2 - N_1)I,$$
$$N_1' = -(z\gamma_R + \gamma_1)N_1 + G(N_2 - N_1)I + \gamma_R M_1,$$
$$N_2' = -(z\gamma_R + \gamma_2)N_2 - G(N_2 - N_1)I + \gamma_R M_2 + \gamma_2 P,$$
$$M_1' = -(\gamma_R + \gamma_1)M_1 + z\gamma_R N_1,$$
$$M_2' = -(\gamma_R + \gamma_2)M_2 + z\gamma_R N_2 + z\gamma_2 P. \tag{2.68}$$

Typical values for the parameters are given in Table 2.3.[4]

Introducing the population inversions

$$N = N_2 - N_1 \quad \text{and} \quad M = M_2 - M_1, \tag{2.69}$$

---

[4] The dispersion of the parameter values results from the difficulty of measuring quantities which are defined in the framework of a simplified two + two model knowing that in reality hundreds of levels are involved. Nevertheless, these values will help us to estimate the relative magnitude of the terms appearing in the rate equations. Moreover, laser parameters typically vary by *one order of magnitude* depending on the operating conditions. For example, the pressure inside the active medium, which controls the gain and the relaxation constants, may vary from a few to tens of Torrs for the same laser.

Table 2.3 *Parameters for a two + two level CO_2 laser [70].*

Parameters	Values
$\kappa\,(s^{-1})$	1.35×10^7
$G\,(s^{-1})$	6.7×10^{-8}
$\gamma_1\,(s^{-1})$	8.0×10^4
$\gamma_2\,(s^{-1})$	10^4
$\gamma_R\,(s^{-1})$	7×10^5
P	6.35×10^{14}
z	16

Eqs. (2.68) can be rewritten as

$$I' = -\kappa I + GNI,$$
$$N_1' = -(z\gamma_R + \gamma_1)N_1 + GNI + \gamma_R M_1,$$
$$N' = -(z\gamma_R + \gamma_2)N - 2GNI + \gamma_R M + \gamma_2 P + (\gamma_1 - \gamma_2)N_1,$$
$$M_1' = -(\gamma_R + \gamma_1)M_1 + z\gamma_R N_1,$$
$$M' = -(\gamma_R + \gamma_2)M + z\gamma_R N + z\gamma_2 P + (\gamma_1 - \gamma_2)M_1. \tag{2.70}$$

The special case

$$\gamma = \gamma_2 = \gamma_1 \tag{2.71}$$

is the case studied in [60, 31]. It corresponds to equal relaxation times for the vibrational states associated with the upper and lower energy levels of the lasing transition. Equations (2.70) then separate into two systems of equations. Specifically, I, N, and M satisfy

$$I' = -\kappa I + GNI,$$
$$N' = -(z\gamma_R + \gamma)N - 2GNI + \gamma_R M + \gamma P,$$
$$M' = -(\gamma_R + \gamma)M + z\gamma_R N + z\gamma P, \tag{2.72}$$

while N_1 and M_1 are passively related to N and I and satisfy

$$N_1' = -(z\gamma_R + \gamma)N_1 + GNI + \gamma_R M_1,$$
$$M_1' = -(\gamma_R + \gamma)M_1 + z\gamma_R N_1. \tag{2.73}$$

The difficulty of analyzing the laser rate equations is reduced to three equations making both qualitative and quantitative analyses easier. In the forthcoming section, we formulate dimensionless equations in order to compare parameter values and propose simplifications.

2.3.4 Dimensionless equations

To this end, we introduce the new variables

$$I \equiv \frac{\gamma}{2G}i, \quad N \equiv \frac{\kappa}{G}n, \quad \text{and} \quad M \equiv \frac{\kappa}{G}m. \tag{2.74}$$

From Eqs. (2.72), we obtain[5]

$$i' = \kappa i(-1 + n),$$
$$n' = -(z\gamma_R + \gamma)n - \gamma ni + \gamma_R m + \gamma P_0,$$
$$m' = -(\gamma_R + \gamma)m + z\gamma_R n + z\gamma P_0, \tag{2.75}$$

where

$$P_0 \equiv \frac{G}{\kappa}P. \tag{2.76}$$

Finally, we need to introduce a dimensionless time. If

$$t = \gamma_R T \tag{2.77}$$

Eqs. (2.75) become

$$i' = ki(-1 + n), \tag{2.78}$$
$$n' = -(z + \varepsilon)n - \varepsilon ni + m + \varepsilon P_0, \tag{2.79}$$
$$m' = -(1 + \varepsilon)m + zn + z\varepsilon P_0, \tag{2.80}$$

where prime now means differentiation with respect to t. The dimensionless parameters k and ε are defined by

$$k \equiv \kappa\gamma_R^{-1} \quad \text{and} \quad \varepsilon \equiv \gamma\gamma_R^{-1}. \tag{2.81}$$

Note that time is measured in units of γ_R instead of κ as in our previous formulations, for mathematical convenience only. Using Table 2.3 and $\gamma = 5 \times 10^4$,

---

[5] Equations (2.75) are equivalent to Eqs. (6) in [71] and Eqs. (5) in [69] if we introduce the new variables $R = \frac{m-zn}{z+1}$, $Q = \frac{m+n}{z+1}$, and $\mathfrak{I} = \frac{i}{z+1}$ (Eqs. (6) in [71]) or $y = i$ (Eqs. (5) in [69]).

we find $k = 19.2$ and $\varepsilon = 7.1 \times 10^{-2}$. Mathematically, the SRE equations are obtained by taking the limit $z \to 0$. However, the value of $z = 16$ to 20 is high. In the next subsection, we examine the relaxation properties of a slightly perturbed steady state.

2.3.5 Long time solution

As with Eqs. (2.32), we analyze Eqs. (2.78)–(2.80) by studying the relaxation properties of the non-zero intensity steady state. The analysis is similar to our previous investigation of Eqs. (2.32) and we only summarize the main steps. The non-zero intensity steady state of Eqs. (2.78)–(2.80) is given by

$$i_s = (P_0 - 1)\frac{(z + 1 + \varepsilon)}{1 + \varepsilon}, \quad n_s = 1, \quad m_s = \frac{z(1 + \varepsilon P_0)}{1 + \varepsilon}. \tag{2.82}$$

We determine the characteristic equation for the growth rate σ. It has the form

$$\sigma^3 + (z + 1 + 2\varepsilon + \varepsilon i_s)\sigma^2$$
$$+ (\varepsilon k i_s + \varepsilon + \varepsilon i_s + \varepsilon(z + \varepsilon + \varepsilon i_s))\sigma + k i_s \varepsilon(1 + \varepsilon)$$
$$= 0. \tag{2.83}$$

We solve Eq. (2.83) by taking advantage of the large value of k. Specifically, we seek a solution of the form

$$\sigma = k^{1/2}\sigma_0 + \sigma_1 + \ldots \tag{2.84}$$

We find one real root which is always negative, and

$$\sigma \simeq \pm i\sqrt{k\varepsilon i_s} - \frac{1}{2}(z + \varepsilon + \varepsilon i_s). \tag{2.85}$$

In terms of time measured in units of the cavity lifetime, the growth rate is

$$\sigma \gamma_R \kappa^{-1} = \pm i\sqrt{\frac{\gamma}{\kappa}i_s} - \frac{\gamma}{2\kappa}(1 + i_s) - \frac{\gamma_R}{2\kappa}z. \tag{2.86}$$

The two first terms in the right hand side of (2.86) are identical to the terms appearing for the SRE. The last term is new and contributes significantly to the damping of the relaxation oscillations.

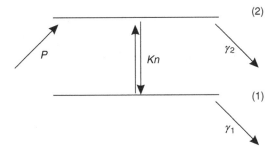

Fig. 2.4 Open two-level laser scheme.

2.4 Exercises and problems

2.4.1 Open two-level system

If the laser emission occurs between two excited states with similar lifetimes, the approximation of a constant total population in the laser levels cannot be used and an open two-level model needs to be considered, as illustrated in Figure 2.4. It serves, for instance, in the case of the He-Ne laser operation at $\lambda = 632$ nm. The evolution equations are

$$\frac{dN_1}{dt} = -\gamma_1 N_1 + (N_2 - N_1)Kn,$$

$$\frac{dN_2}{dt} = P - \gamma_2 N_2 - (N_2 - N_1)Kn,$$

$$\frac{dn}{dt} = Kn(N_2 - N_1) - \gamma_c n.$$

Rewrite these equations in the dimensionless form

$$n_1' = -\varepsilon_1 n_1 + (n_2 - n_1)i,$$
$$n_2' = -\varepsilon_2 n_2 - (n_2 - n_1)i + A,$$
$$i' = -i + (n_2 - n_1)i$$

and determine the stability properties of the steady states.

2.4.2 Two-photon laser

The direct transition from the upper to the lower level of the "laser transition" may be forbidden by electric dipole selection rules, for instance if these states have the same parity. In this situation, there is a possibility of a two-photon transition [72].

The corresponding rate equations for such a process differ in the field–matter interaction: as the emission process involves two photons instead of one, it is easily shown by quantum mechanics that the transition rate evolves as the square of the intensity. Such a laser may be described by the following set of rate equations using the notation of the original paper

$$\frac{dq}{dt} = B^{(2)} q^2 \Delta N - \gamma_c (q - q_{inj}(t))$$

$$\frac{d\Delta N}{dt} = -2B^{(2)} q^2 \Delta N - \gamma (\Delta N - \Delta N_0),$$

where q is the photon density and ΔN the population inversion. γ_c and γ are the relaxation rates for these two quantities. $B^{(2)}$ is the two-photon stimulated emission coefficient and $q_{inj}(t)$ represents the possible injected field at the two-photon frequency that will be assumed to be null here.

Reformulate these equations in dimensionless form and check that the number of effective (useful) constants has been reduced to a single one. Find the steady state solutions and determine their stability properties. Define the laser threshold and discuss the problem this laser may have to start (conclude about the role of q_{inj}).

2.4.3 Asymptotic solution of the characteristic equation

In Section 2.2.2, we proposed seeking a solution of Eq. (2.40) as a power series in $\varepsilon^{1/2}$. Check that the expansion $\sigma = \varepsilon \sigma_1 + \varepsilon^2 \sigma_2 + \dots$ only provides one root which then requires a different scaling for σ.

2.4.4 Dimensionless formulation

Introduce the new variables $I = Kn$, $D = K\gamma_{21} N_2 / \gamma_c$, $N = K\gamma_{10} N_1 / \gamma_c$, and $t = \gamma_c T$ into Eqs. (2.55)–(2.57), and formulate the three equations for I, D, and N. Investigate the conditions for the adiabatic elimination of N.

3

Phase dynamics

Phase dynamical instabilities are described by a single angular variable. The basic phenomenon called *"phase-locking"* occurs as soon as two interacting oscillators have "close enough" frequencies. Adler's equation[1] describes this phenomenon and appears in various areas of laser physics, including polarization dynamics, a laser subject to an injected signal, coupled microlasers, and laser gyros.

3.1 Phase-locking in laser dynamics

In Chapter 1, the response of the laser was described in terms of the light intensity and the population inversion. This description of the laser output in terms of two dependent variables is accurate enough to describe phenomena such as turn-on experiments or the onset of damped ROs. However, there are precise cases where such a description is inadequate.

Specifically, the active medium and the electric field deserve more sophisticated treatments. A better account of the active medium energy levels was considered in Chapter 2. In this chapter, we concentrate on the laser electrical field. We already saw in Chapter 1 that a description based on the intensity only may be misleading and we emphasized the fact that the complex electric field should be used instead of the intensity. Specifically, the laser output depends on the electric field vector $\vec{E} = \mathrm{Re}\{\vec{E}_0 \exp(i(\omega_0 t + \Phi))\}$. Both the phase $\Phi(t)$ and the azimuth, i.e. the direction of the electric field vector $\overrightarrow{E_0(t)}$, may lead to new phenomena as we shall now describe.

We may reasonably wonder why we are interested in laser phase dynamics. Indeed, the phase of the optical field, $\omega_0 t + \Phi$, varies so quickly (10^{-15} s) that, in

---

[1] Robert Adler (1913–2007) is best known as the co-inventor of the television remote control using ultrasonic waves. But in the 1940s, he and others at Zenith Corporation were interested in reducing the number of vacuum tubes in an FM radio. The possibility that a locked oscillator might offer a solution inspired his 1946 paper, "A Study of Locking Phenomena in Oscillators" [79]. Adler's work concerned a single nonlinear phase oscillator. Later the idea was exploited and generalized to describe a number of similar coupled oscillators.

Table 3.1 *Different laser systems where phase-locking is observed.*

Laser	Coupled waves	Coupling
ring[a]	forward/backward waves	back-scattering
vectorial class A[b]	polarization components	Faraday rotation
with injection[c]	injected and laser fields	injection
coupled lasers[d]	waves in each laser	evanescent waves

[a] [6, 74].
[b] [75, 76].
[c] [77].
[d] [78, 79, 80].

most experiments, only the intensity of the laser I may be monitored and the optical phase Φ remains inaccessible. There are however experiments where a phase variable must be considered. This occurs if there exists an optical time/frequency reference such as the phase of another laser beam at a frequency close to that of the laser under study. The relative phase of the two laser fields may then vary slowly. Laser arrays or a laser injected with the field of another laser are typical examples where this relative phase between laser optical fields needs to be taken into account. The nature of the coupled waves and the associated coupling mechanism for different systems are summarized in Table 3.1.

The objective of this chapter is to review several situations in which phase-locking rules the laser dynamics. We begin by studying one of the simplest cases, the vectorial dynamics of a class A laser in the presence of Faraday rotation. We shall introduce Adler's equation and emphasize some specific effects of phase-locking. More complicated examples of phase-locking will then be described.

3.2 Vectorial laser in presence of Faraday driving

3.2.1 Theoretical model

In most lasers, the polarization of the electric field is fixed because polarization-selective elements are inserted in the laser cavity. However, if the laser cavity is made quasi-isotropic, the polarization is free to rotate and subtle polarization dynamics may develop. The choice of a representation for this evolution depends on the properties of the particular system under study. However, the description of the polarization state of light in terms of amplitude, azimuth, and ellipticity (see for example [81]) is quite appropriate for class A monochromatic lasers such as Ar or He-Ne lasers (see Section 1.1). Subject to a longitudinal magnetic field, the atoms of the active medium experience Faraday rotation, i.e. light in the two states

Table 3.2 *Parameters for the anisotropic He-Ne laser.*

Parameters	Values
$|M|$	$2.2 \times 10^5 \, \text{s}^{-1}$
ΓB_0	$\simeq 1.14 \times 10^6 \, \text{rad s}^{-1}$
K_x	$1.101\,4 \times 10^8 \, \text{s}^{-1}$
K_y	$1.106\,1 \times 10^8 \, \text{s}^{-1}$
D_0	$2.4 \times 10^8 \, \text{s}^{-1}$

of opposite circular polarization (*cw* and *ccw*)[2] propagates at slightly different velocities. As a consequence, the azimuth of a linearly polarized light beam rotates as it travels through such an active medium. In a laser cavity, this effect occurs together with other polarization changes caused by other elements (e.g. polarizers, waveplates, or anisotropic crystals inserted inside the laser cavity). Here, we consider the simplest case of a class A laser whose active medium experiences Faraday rotation in the presence of small pure loss (i.e. no refractive index) anisotropies. The laser evolution equations for the azimuth θ and the intensity I are given by [75, 82, 83]

$$\frac{dI}{dt} = -\left[K_x \cos^2(\theta) + K_y \sin^2(\theta) \right] I + \frac{D_0 I}{1 + \zeta I} \tag{3.1}$$

$$\frac{d\theta}{dt} = M \sin(2\theta) + \Gamma(I) B_0, \tag{3.2}$$

where K_x and K_y are the loss (positive) coefficients for the intensities in the x and y polarizations. D_0 is the unsaturated gain coefficient, and $\Gamma(I)$ is the Faraday rotation coefficient, which generally depends on the laser intensity I. B_0 is the magnetic field amplitude and ζ equals 1 or 1/2 for lasers which are homogeneously broadened or not (for the origin of the 1/2 factor see, e.g. [36]). $M < 0$ (see Problem 3.7.1) characterizes the electric field rotation due to the difference in transmissivity for the two orthogonal linear polarizations. Typical values of the parameters for the anisotropic 3.39 μm He-Ne laser [83] are listed in Table 3.2.

We note that the rate constants in Eq. (3.1) are 100 times larger than the terms in Eq. (3.2). This means that I is a fast variable compared to θ and that it quickly reaches a quasi-steady state regime where the right hand side of Eq. (3.1) is close to zero. Setting the right hand side of Eq. (3.1) equal to zero and solving for I, we obtain

---

[2] *cw* and *ccw* stand for clockwise and counterclockwise, respectively. cw is also used for "continuous wave" meaning a constant laser ouput.

$$I = \zeta^{-1} \left[-1 + \frac{D_0}{K_x \cos^2(\theta) + K_y \sin^2(\theta)} \right]. \tag{3.3}$$

For weakly anisotropic lasers, $K_x \simeq K_y$, and (3.3) can be further simplified by using a trigonometric identity. We now have

$$I \simeq \zeta^{-1} \left[-1 + \frac{D_0}{K_x} \right] = \zeta^{-1} 1.18, \tag{3.4}$$

which is a constant. Consequently, $\Gamma(I)$ is constant, and Eq. (3.2) reduces to

$$\frac{d\theta}{dt} = M \sin(2\theta) + \Gamma B_0. \tag{3.5}$$

In the absence of any magnetic field ($B_0 = 0$), Eq. (3.5) admits a stable equilibrium position at $\theta = 0$, and an unstable one at $\theta = \pi/2$ ($M < 0$). On the other hand, a non-zero magnetic field in an isotropic cavity ($K_x = K_y$ and $M = 0$) induces Faraday rotation at an angular velocity $\omega_0 = \Gamma B_0$. We now wish to investigate the case of a non-zero magnetic field ($B_0 \neq 0$) and weak loss anisotropies ($K_x \neq K_y$ or $M \neq 0$). The polarization dynamics of the laser then result from the competition between Faraday rotation and the "restoring force" $M \sin(2\theta)$ which tends to bring the azimuth back to the $\theta = 0$ position. From Eq. (3.5), we note that a steady state solution only exists if $|\Gamma B_0/M| \leq 1$. The laser then approaches a state of constant linear polarization with an azimuth $\theta = \theta_0$ given by

$$\theta_0 = -\frac{1}{2} \arcsin \left(\frac{\Gamma B_0}{M} \right). \tag{3.6}$$

If $|\Gamma B_0/M| > 1$, there are no steady state solutions. In the limit of a strong magnetic field (i.e. $|\Gamma B_0/M| >> 1$), the restoring force is negligible and the system experiences an almost uniform rotation at the rate $\omega_0 = \Gamma B_0$. Between these two regimes (i.e. phase-locking if $|\Gamma B_0/M| \leq 1$ and free rotation if $|\Gamma B_0/M| >> 1$), $d\theta/dt$ exhibits an oscillatory behavior which can be explored analytically (see below). Of physical interest is the frequency of these oscillations which is given by

$$\Omega \equiv \Gamma \sqrt{B_0^2 - B_c^2} \ (B_0 > B_c \equiv M/\Gamma). \tag{3.7}$$

3.2.2 Experiments

The locking phenomenon can be investigated with the He-Ne and He-Xe lasers oscillating at 3.39 μm and 3.51 μm, respectively, if they are subject to a longitudinal magnetic field of the order of a few milliGauss [75]. A tilted plate inserted inside the laser cavity, as shown in Figure 3.1, introduces loss anisotropies. The

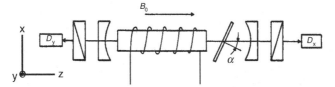

Fig. 3.1 He-Ne laser with an active medium subject to a static magnetic field B_0. A slightly tilted plate introduces weak loss anisotropies in the cavity. α is the tilt angle. Two detectors behind crossed polarizers monitor the evolution of the laser emission. Reprinted Figure 1a from Glorieux and Le Floch [75] with permission from Elsevier.

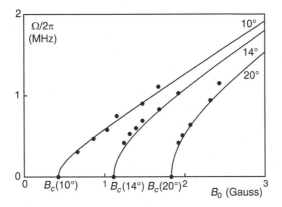

Fig. 3.2 Field azimuth for a He-Ne laser subjected to a static magnetic field and for three different values of the tilt angle α. Dots correspond to the experimental data and the full lines are given by $\Omega = \Gamma(B^2 - B_c^2)^{1/2}$, where $\Gamma = 4.05 \times 10^6 \, s^{-1}$. Redrawn using the data of Figure 1b of Glorieux and Le Floch [75].

tilt angle α controls the amplitude of the anisotropy because the transmission of a tilted plate depends on the polarization of the incident light and on the incidence angle through the Fresnel relations ([81] Chapter 1). A polarizer selects the component of the laser field emitted in one direction.

Monitoring the component of emitted radiation in one polarization direction, i.e. $I_x = I \cos^2(\theta)$, immediately shows that, depending on the strength of the magnetic field, the laser adopts two distinct regimes, as seen in Figure 3.2:

1. For low values of the magnetic field B_0, the laser field is linearly polarized along a fixed direction and its amplitude is constant in time. This corresponds to the phase-locked state.
2. If the magnetic field B_0 exceeds the critical value B_c which depends on the tilt angle α, 100% amplitude modulation is observed on any *polarized* component of the laser output while the *total* intensity exhibits only negligible modulation. This means that the electric field rotates continuously in time, thus producing oscillating linear components.

3.3 Adler's equation

The angular dynamics of the class A laser in the presence of Faraday rotation and weak loss anisotropy is described by Eq. (3.5). This equation is known as Adler's equation and is the simplest equation describing the phase-locking between a nonlinear oscillator and an external periodic drive. It was first derived in connection with the phase-locking of radiofrequency oscillators [84], and has since found application in many other settings, including the depinning of charge density waves, the entrainment of biological oscillators, and the onset of resistance in superconducting Josephson junctions [8]. In its classical form, Adler's equation is given by

$$\frac{d\phi}{dt} = \omega - a\sin(\phi), \tag{3.8}$$

where ϕ is the phase difference between the oscillator and the drive, ω is the frequency detuning, and a is the coupling strength. A system described by Eq. (3.8) can display only two types of long-time behaviors. If $|\omega/a| \leq 1$, all solutions tend to a phase-locked state, where the response oscillator maintains a constant phase difference relative to the driver (phase-locking or synchronization). On the other hand, if $|\omega/a| > 1$, all solutions exhibit phase drift, where the phase difference grows monotonically, with one oscillator periodically overtaking the other (phase drift or rhythm splitting). We briefly analyze these two distinct behaviors.

3.3.1 Phase-locking

If $|\omega/a| < 1$, Equation (3.8) admits two steady-state solutions. They are given by

$$\phi_1 = \arcsin(\omega/a) \quad \text{and} \quad \phi_2 = \pi - \arcsin(\omega/a). \tag{3.9}$$

For a first order differential equation of the form $x' = f(x)$, the growth rate of a small perturbation of the steady state $x = x_0$ is $\sigma = f'(x_0)$. For Eq. (3.8), we find that $f'(\phi_1) = -a\cos(\phi_1) < 0$ and $f'(\phi_2) = -a\cos(\phi_2) > 0$ if $a/\omega > 0$ (similarly, $f'(\phi_1) > 0$ and $f'(\phi_2) < 0$ if $a/\omega < 0$). This implies that there is always a stable and an unstable steady state solution.

3.3.2 Phase drift

If $|\omega/a| > 1$, there are no steady-state solutions and $d\phi/dt$ is a bounded time-periodic function of t. An analytic solution of Adler's equation is possible in this case (see Problem 3.7.2) but is not very instructive. However, the period T of the oscillations has a simpler expression. The period is defined as the time needed for ϕ to vary from 0 to 2π. Using Eq. (3.8), the period is given by the following definite integral

$$T = \int_0^T dt = \int_0^{2\pi} \frac{dt}{d\phi} d\phi = \int_0^{2\pi} \frac{d\phi}{\omega - a\sin(\phi)}. \tag{3.10}$$

The last integral can be solved using a trigonometric substitution (see Problem 3.7.2). We obtain

$$T \equiv \frac{2\pi}{\sqrt{\omega^2 - a^2}}. \tag{3.11}$$

In the absence of the restoring force ($a = 0$), the period equals the angular period $T = 2\pi/\omega$. As a^2 is progressively increased from zero, the period increases and becomes unbounded at $a^2 = \omega^2$. Equivalently, the beating frequency defined as

$$\Omega \equiv 2\pi/T \equiv \sqrt{\omega^2 - a^2} \tag{3.12}$$

is zero at $a = |\omega|$.

From an experimental point of view, it is worthwhile to emphasize three different effects:

1. Well outside the locking range (i.e. $|a/\omega|$ small), we expand (3.12) in Taylor series and obtain $\Omega \simeq \omega(1 - a^2/2\omega^2)$. The *beating frequency is pulled to ω* when the detuning ω becomes large. It is an important effect for the optically injected laser.
2. Close to locking ($|\omega/a| \lesssim 1$), the period T is large meaning that the system becomes very slow. This effect is called "*critical slowing down*" and occurs because we are close to a saddle-node bifurcation point (see Problem 3.7.2). From (3.11), we note that this divergence follows an inverse square-root law.
3. In many experiments, the locking phenomenon is discovered by slowly scanning the detuning back and forth. The typical response is shown in Figure 3.3 where the beating frequency is zero in the locking domain and non-zero as soon as $\omega < -a$ or $\omega > a$.

3.3.3 Long period motion

Although the beating phenomenon is much slower than the optical oscillations, the observation of phase dynamics close to locking ($\omega \lesssim |a|$) is still delicate for lasers. Optothermal systems are much slower than lasers and their phase synchronization has been studied by Herrero *et al.* [85]. They observed regular or chaotic oscillations with a basic recurrence of about 2 Hz in isolated optothermal oscillators. Coupling two such devices leads to phase synchronization which operates on a much slower time scale. Figure 3.4 shows the results of experiments on a set of

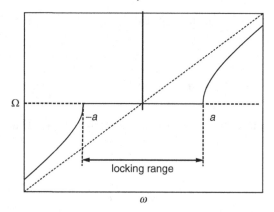

Fig. 3.3 Beating frequency Ω as a function of the detuning ω. The region of zero beat ($-a \leq \omega \leq a$) is the locking range. Outside the locking range, Ω is pulled to the straight line $\Omega = \omega$.

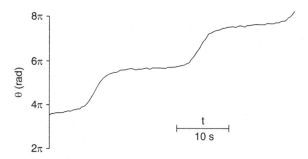

Fig. 3.4 Experimental time evolution of the phase difference between the two optothermal oscillators. Close to locking, the two oscillators are nearly synchronized except when the phase jumps by 2π. Reprinted Fig. 2b from Herrero *et al.* [85]. Copyright 2002 by the American Physical Society.

two bidirectionally coupled devices which oscillate at frequencies of 2.29 Hz and 2.35 Hz, respectively. The relative phase difference varies much more slowly and displays regular 2π phase jumps about every 20 seconds, i.e. close to the inverse of the frequency difference ($=0.06$ Hz) of the two oscillators.

3.4 Laser with an injected signal

In the laser with an injected signal (LIS), the radiation from a "master" oscillator is sent into the cavity of a "slave" oscillator. This is a standard arrangement used to transfer some properties of the master to the slave, and in particular its frequency stability, which is possible if phase-locking can be realized. Of particular interest is the case of high power lasers. The optical spectra of the waves emitted by such lasers are rather broad because the optimization of the output power is usually obtained at the expense of the spectral quality. In order to bypass this difficulty, a low power highly monochromatic laser is injected into the high power one ("optical seeding"). If the two lasers can be phase-locked, the spectrum of the high power laser narrows and makes it suitable for applications demanding spectral purity such as laser cooling of atoms or Doppler velocimetry. The LIS has received renewed interest because it is one of the simplest systems exhibiting dynamical instabilities. However, except for semiconductor lasers, which will be specifically considered in Chapter 9, there have not been many systematic experimental studies [74].

3.4.1 Experiments

Laser injection-locking was first achieved by Stover and Steier using two 6328 Å He-Ne lasers [77]. In their experiment, one laser is injected with a second laser and portions of both laser outputs are combined on a detector. The latter delivers a signal proportional to the beat note (phase drift) or to the interference (phase-locking) of the two lasers. The frequency of either laser is tuned over a few tens of MHz. Figure 3.5 displays the evolution of the detector output vs. laser detuning. In the absence of injection, the beat signal amplitude just follows the response curve of the detection system. The beat is too fast to be resolved in the display conditions and generates a broad oscilloscope trace. In the presence of injection, the beat signal vanishes in the central region where the detuning goes through zero. There the trace displays no beat but a slow continuous variation due to the phase variations of the two fields in the locked regime.

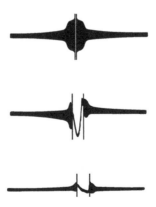

Fig. 3.5 First experimental evidence of phase-locking in a laser with an injected signal. The beat note between the master and slave lasers is displayed vs. their frequency detuning at different injection levels. Phase-locking corresponds to an almost continuous signal in the central part of the trace. From top to bottom, the width of the locked region in the middle increases with the injected power. Adapted Figure 2a, 2b, 2c with permission from Stover and Steier [77]. Copyright 1966 by the American Institute of Physics.

3.4.2 Theory

We concentrate on the He-Ne laser used in the experiments by Stover and Steier. This laser is considered as a class A laser (which also includes the Ar^+ laser). Class A lasers are characterized by a high value of γ and are described by the single equation (1.11) for the intensity of the field $\mathcal{E}$. Introducing $I = \mathcal{E}^2$ and removing the factor 2 by redefining the time variable, (1.11) becomes

$$\frac{d\mathcal{E}}{dt} = \mathcal{E}\left(\frac{A}{1 + |\mathcal{E}|^2} - 1\right). \tag{3.13}$$

If now the laser (called slave) is subject to the injected signal from another laser (called master), Eq. (3.13) exhibits an additional term modeling the injected field. Instead of Eq. (3.13), the equation for the field $\mathcal{E}$ now is

$$\frac{d\mathcal{E}}{dt} = \mathcal{E}\left(-1 + \frac{A}{1 + |\mathcal{E}|^2}\right) + \mathcal{E}_i, \tag{3.14}$$

where $\mathcal{E}_i(t) = X_0 \exp(i\Delta t)$ has amplitude X_0 and frequency detuning $\Delta \equiv \omega_i - \omega_0$.[3] We may eliminate the time dependence of $\mathcal{E}_i$ by introducing the decomposition

$$\mathcal{E} = X \exp(i\Delta t + \phi), \tag{3.15}$$

where X and ϕ are the amplitude and the phase of the slave laser, respectively. Inserting (3.15) into (3.14), we obtain the following two equations for X and ϕ

$$X' = \left(-1 + \frac{A}{1 + X^2}\right)X + X_0\cos(\phi), \tag{3.16}$$

$$\phi' = -\Delta - \frac{X_0}{X}\sin(\phi). \tag{3.17}$$

Eq. (3.16) is an equation for the amplitude of the laser field and includes a source term $X_0 \cos(\phi)$. Equation (3.17) is similar to Adler's equation but admits a coupling term which is inversely proportional to the slave laser field amplitude. If $A > 1$ and if $X_0 << 1$, we find from Eq. (3.16) that X approaches its stable steady state value $X = \sqrt{A-1}$. Substituting this expression into Eq. (3.17), we obtain Adler's equation

$$\phi' = -\Delta - \frac{X_0}{\sqrt{A-1}}\sin(\phi). \tag{3.18}$$

---

[3] In dimensionless form, the electrical field in the laser cavity is $\mathcal{E}(t)\exp(i\omega_0 t)$, where ω_0 is the optical frequency and $\mathcal{E}(t)$ is the slowly varying amplitude. Similarly, the electrical field injected into the laser is $\mathcal{E}_i \exp(i\omega_i t)$, where ω_i is the optical frequency of the master laser.

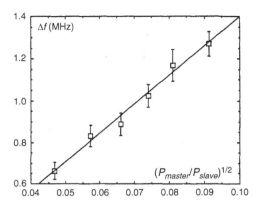

Fig. 3.6 Injection-locking of a Nd^{3+}:YAG laser. The figure represents the locking range Δf obtained by scanning the slave cavity length as a function of the root power ratio $\sqrt{P_{master}/P_{slave}}$. The slave laser power is 4 W (from Figure 2 of Nabors *et al.* [86]).

We assume that $|\Delta|$ scales like $X_0/\sqrt{A-1}$ and Eq. (3.18) predicts that the locking region is

$$|\Delta| \le \frac{X_0}{\sqrt{A-1}}. \tag{3.19}$$

Since X_0 is proportional to the injection field amplitude $|\mathcal{E}_i| = \sqrt{I_{in}}$ and $I_0 = A - 1$ is the intensity of the field in the free running laser, the expression (3.19) means that the locking range is proportional to $\sqrt{I_{in}/I_0}$. If we try to lock a high power slave laser (high I_0) with a low power master laser (low I_{in}), the locking range can be quite narrow. Figure 3.6 shows an experimental verification.

Outside the locking region, the phase is unbounded in time but forces the amplitude of the laser to oscillate. We note that ϕ is a slowly varying function of $S \equiv (X_0/\sqrt{A-1})t$, and determine a solution of Eq. (3.16) in a power series in X_0 that depends on S. We obtain

$$X = \sqrt{A-1} + X_0 \frac{A}{2(A-1)} \cos(\phi(t))dt + O(X_0^2), \tag{3.20}$$

where we note that the first correction term is time-periodic.

The bifurcation diagrams for two different values of Δ are shown in Figure 3.7. The stability analysis of the steady state solutions of Eqs. (3.16) and (3.17) is proposed in Exercise 3.7.3. We find from this analysis that (3.19) is a good approximation of the phase-locking region for small X_0 but that there exist Hopf bifurcation instabilities that are not predicted by Adler's equation. A Hopf bifurcation marks the transition from a steady state to a time-periodic solution

Phase dynamics

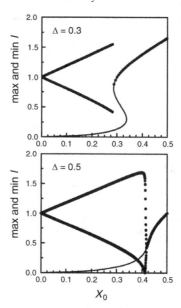

Fig. 3.7 Bifurcation diagrams of the maximum and minimum intensity $I = X^2$. Top: a branch of periodic solutions emerges from $X_0 = 0$ with an unbounded phase. The period progressively increases and becomes infinite at the steady state limit point. Bottom: the branch of periodic solutions emerging at $X_0 = 0$ terminates at a Hopf bifurcation point with a fixed period. The diagrams have been obtained numerically from Eqs. (3.16) and (3.17) with $A = 2$ and $\Delta = 0.3$ (top) or $\Delta = 0.5$ (bottom).

and is characterized by a pair of purely imaginary eigenvalues of the linearized problem. The injection laser is presumably the simplest laser case where such a bifurcation appears.

3.5 Counterpropagating waves in ring class A lasers

Most lasers are built in a standing wave cavity, i.e. a cavity in which the electromagnetic wave bounces back and forth between the mirrors. The superposition of the two opposite traveling waves produces a standing wave and reduces the efficiency of the light–matter interaction at its nodes. Ring lasers have been developed to optimize the energy extraction in the lasing medium. In these lasers, pure traveling waves are possible, i.e. waves circulating in the clockwise or counterclockwise direction. By inserting a Faraday isolator, one of the two rotating waves is eliminated and field nodes no longer exist. The whole lasing medium then uniformly contributes to the laser emission leading to a better efficiency of conversion of the pump into laser power.

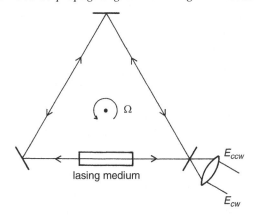

Fig. 3.8 Ring laser gyro. The detector reads the combination of two counter-running waves E_{cw} and E_{ccw}.

Ring cavities are also the basis of most laser gyros [87]. In the ring lasers, both *cw* and *ccw* waves may develop but their frequency degeneracy is lifted through the Sagnac effect induced by the mechanical rotation of the laser frame. The frequency of the beating between these two waves is a measure of the rotation rate of the laser reference frame and, consequently, gives access to the rotation angle of the frame on which the laser is placed. However, the measure of very small rotations (equivalently, very small beat frequencies) is limited by the phase-locking phenomenon of the two counterpropagating traveling waves which occurs when their frequency difference becomes too small. This is the origin of the "blind range" which should be bypassed to allow efficient use of these devices in airplanes or rockets. The blind range due to phase-locking of the *cw* and *ccw* traveling waves has been detected in all the laser gyros and serves as quality evaluator. Ring lasers are not only used for laser gyros. Since they are very sensitive to the coupling of one wave (e.g. *cw*) into the other one (e.g. *ccw*), this effect may be used to measure very small scattering coefficients. For instance, Quintero-Torres *et al.* have demonstrated that intensity back-scattering coefficients as low as 10^{-15} can be detected [88].

A laser gyro is most often based on a triangular (or zig-zag or rectangular)[4] ring cavity in an arrangement as shown in Figure 3.8. The beams emitted in each direction (*cw* and *ccw*) are recombined on the same detector and provide the beat signal from which the information on the rotation angular frequency Ω of the laser frame is deduced.

We consider He-Ne laser gyros where the coupling between the amplitudes of the counter-running waves $\mathcal{E}_{\pm} = E_{\pm}\exp(i\varphi_{\pm})$ is due to back-scattering from the

---

[4] These geometries are chosen for their specific optical properties (astigmatism, stability of optical beams).

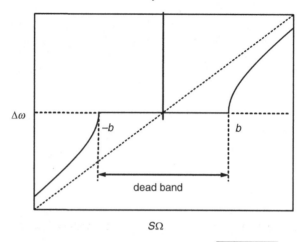

Fig. 3.9 The average frequency difference $\Delta\omega = \sqrt{(S\Omega)^2 - b^2}$ in the unlocked region is shown as a function of $S\Omega$. The straight dashed lines passing through the origin represent the ideal response. The "dead band" or range of input rotation rates for which no frequency difference is observed extends from $\Omega = -b/S$ to $\Omega = b/S$.

mirrors forming the cavity. Back-scattering in the ring laser means that, due to imperfections in the optical path, a small fraction of one of the waves is scattered back in the direction of the opposite traveling wave. The resulting coupling coefficient is usually very small, yet it becomes a dominant effect at low rotation rates, as we shall see. From Lamb's semiclassical theory [36] and provided the coupling between the waves is not too large, it is possible to derive Adler's equation for the phase difference $\psi = \varphi_+ - \varphi_-$ [87]

$$\psi' = S\Omega + b\sin(\psi), \tag{3.21}$$

where Ω is the rotation rate and S is a scale factor. The coefficient b is called the back-scattering coefficient. Locking occurs if

$$|\Omega| < \Omega_{th} \equiv bS^{-1}. \tag{3.22}$$

Ω_{th} is defined as the threshold for locking and hence the minimum detectable rotation rate. Several studies have considered the full laser equations for the amplitudes and the phases of the two fields and have shown that Hopf bifurcation instabilities are possible [89, 6]. But Adler's equation (3.21) for phase-locking remains the reference equation for all laser gyros.

Experimentalists prefer to write Adler's equation as

$$\frac{1}{2\pi}\frac{d\psi}{dt} = f - l\sin(\psi)$$

Table 3.3 *Parameter values for laser systems described by Adler's equation.*

System	Driving f (Hz)	Driver	Lock limit l (Hz)
ring gyro laser[a]	33 240	laser rotation	4 115
supergyro laser[b]	1 513	Earth rotation	~1
ring laser scatterometer[c]	50–1 000	electro-optical	250

[a] [93].
[b] [94].
[c] [88].

so that f and l are expressed in frequency units (Hz) rather than in angular units (rad s^{-1}). Table 3.3 provides typical values for (1) a standard laboratory gyro laser, (2) a "supergyro laser" with top quality components in a super controlled environment, and (3) a ring laser scatterometer.

3.5.1 Dither control of ring laser gyro

Experimentally, several techniques have been developed to overcome the locking problem. One of these techniques is to introduce a mechanical alternating bias where the gyro is rotated alternatively in one direction and the other (this technique is known as "dithering"). This can be done by mounting the gyro on a rotational spring system which is oscillated by means of a piezoelectric transducer. The effect of an alternating bias that changes sinusoidally in time may be described by the modified Adler's equation

$$\psi' = S\Omega + b \sin(\psi) + \alpha \cos(\omega_D t), \tag{3.23}$$

where $\alpha >> b$ and $\omega_D >> b$ are the amplitude and frequency of the oscillating bias, respectively. An approximation of Eq. (3.23) is studied in [87] but in Section 3.5.2 we derive a simpler asymptotic approximation of Eq. (3.23). Assuming $\alpha = O(\omega_D)$ and $\omega_D >> 1$, and averaging the high-frequency oscillations, we find that the average value of ψ is

$$\langle \psi \rangle = \omega_D \phi_0(t) + O(1), \tag{3.24}$$

where ϕ_0 satisfies a new Adler's equation of the form

$$\phi_0' = S\Omega + b J_0(\alpha/\omega_D) \sin(\phi_0). \tag{3.25}$$

Here $J_0(x)$ is the Bessel function of order zero. From this equation, we find that the locking condition is

$$|S\Omega| < |b J_0(\alpha/\omega_D)|. \tag{3.26}$$

Therefore by choosing α/ω_D equal to a root of the zeroth Bessel function, it is possible to make this dead band vanish. This is usually prevented by technical constraints. What can be done, however, is to choose α/ω_D as large as possible since $J_0(\alpha/\omega_D) \sim (\alpha/\omega_D)^{-1/2}$ as $\alpha/\omega_D \to \infty$. And so, the width of the dead zone goes to zero. In mechanically dithered gyros, we may have $\alpha = 190\,\text{kHz}$ and $\omega_D = 200\,\text{Hz}$ giving $\alpha/\omega_D = 950$ and therefore $J_0(\alpha/\omega_D) << 0.02$ [87].

3.5.2 High-frequency asymptotics

We propose to solve Eq. (3.23) for $\alpha = O(\omega_D) >> 1$. To this end, we introduce a small parameter ε defined as

$$\varepsilon \equiv \omega_D^{-1} \tag{3.27}$$

and expand α as

$$\alpha = \varepsilon^{-1}\alpha_0 + \alpha_1 + \dots \tag{3.28}$$

We then seek a solution of the form

$$\psi = \psi_0(T,t) + \varepsilon\psi_1(T,t) + \dots, \tag{3.29}$$

where $T \equiv \varepsilon^{-1}t$ is defined as the fast time of the high-frequency modulations. The assumption of two independent time variables implies the chain rule

$$\psi' = \varepsilon^{-1}\psi_T + \psi_t, \tag{3.30}$$

where subscripts indicate partial derivatives. Inserting Eqs. (3.27)–(3.30) into Eq. (3.23) and equating to zero the coefficients of each power of ε lead to a succession of problems for the unknowns $\psi_0, \psi_1, \dots$ The first two equations are given by

$$O(\varepsilon^{-1}) : \psi_{0T} = \alpha_0 \cos(T), \tag{3.31}$$

$$O(1) : \psi_{1T} = S\Omega + b\sin(\psi_0) + \alpha_1 \cos(T) - \psi_{0t}. \tag{3.32}$$

The solution of Eq. (3.31) is

$$\psi_0 = \alpha_0 \sin(T) + \phi_0(t), \tag{3.33}$$

where ϕ_0 is an unknown function of t. Introducing (3.33) into the right hand side of Eq. (3.32), we apply a solvability condition in order to have a bounded function for ψ_1 with respect to T. This condition is obtained by realizing that the function $\sin(\alpha_0 \sin(T) + \phi_0)$ can be expanded in terms of a Bessel-Fourier series. Specifically,

$$\sin(\alpha_0 \sin(T) + \phi_0) = \cos(\phi_0) \sin(\alpha_0 \sin(T)) + \sin(\phi_0) \cos(\alpha_0 \sin(T))$$
$$= \cos(\phi_0)(2J_1(\alpha_0) \sin(T) + \ldots)$$
$$+ \sin(\phi_0)(J_0(\alpha_0) + \ldots), \tag{3.34}$$

where $J_n(x)$ is the Bessel function of order n. The solvability condition then leads to an ordinary differential equation given by

$$\phi_0' = S\Omega + bJ_0(\alpha_0) \sin(\phi_0). \tag{3.35}$$

3.6 Coupled lasers

Arrays of coupled lasers are of considerable technical importance as high power coherent sources for a number of applications [90]. In order to achieve a single-lobed output profile and at the same time maximize the total system output power, strong phase synchronization and amplitude stability of the individual lasers is desired. Synchronization between lasers is achieved by either injecting a common reference to a series of (laser) amplifiers as in fusion experiments, or by mutual coupling of lasers as in high power laser arrays. So far, CO_2, YAG, and semiconductor laser arrays have been designed and used successfully. However, both experiment and theory have shown that already two single-mode lasers that are stable individually may exhibit pulsating outputs if coupled.

Because the time scale of the intensity fluctuations of solid state lasers is convenient for precise dynamical measurements, laterally coupled YAG microlasers are particularly convenient [78, 79, 91, 92]. Microlasers are tiny lasers – typically 500 μm long – which are implemented in materials such as Nd^{3+}:YAG or YVO_4. In most configurations, they are pumped by radiation delivered by diode lasers connected to optical fibers (see Figure 3.10). This allows for parallel operation of a series of microlasers located on the same chip, opening the way for large-scale optical integration, as with electronic microcircuits. We briefly review some experimental results on laterally coupled microchip lasers and introduce the basic model equations.

3.6.1 Experiments

In a typical experiment [78], the wafer is irradiated by independent laser beams from a Titanium:Sapphire laser. The spacing between these beams can be adjusted

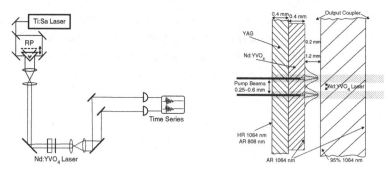

Fig. 3.10 Thermal lensing induced in the Nd:YVO₄ crystal creates two separate laser cavities. The overlap between the fields of these two lasers (i.e. coupling) can be changed by varying the spacing between the incoming beams. Reprinted Figure 1 from Möller *et al.* [78] with permission from Elsevier.

between 0.25 mm and more than 1 mm. Thermal lensing induced in the Nd:YVO₄ crystal creates two stable, separate cavities emitting infrared laser beams. The overlap between these two lasers can be continuously changed by varying the spacing. In the investigated range of distances, there is no appreciable overlap of the pump beams, thus coupling is entirely due to spatial overlap of the infrared laser fields. The individual output intensity time series are recorded with fast photodetectors. In addition to the distance between the pump beams, the frequency detuning between the lasers can be adjusted by tilting the output coupler. Phase synchronization between the lasers is observed by monitoring the fringe pattern of the two beams combined under a small angle with a CCD camera.

At pump powers of about twice the threshold, four regimes can be distinguished (see Figure 3.11):

1. At large spacings and large detunings (region 1 in Figure 3.11), both lasers run independently without any visible phase correlation.
2. At small detuning and small spacings (region 2 in Figure 3.11), the intensities remain steady. Bursts of intensity pulsations appear if we increase the spacing.
3. At very small spacings, there exists a detuning boundary below which intensity pulsations appear. The length of the bursts increases with decreasing spacing, leading to almost continuous, synchronized pulsing (see region 3 in Figure 3.11, and Figure 3.12).
4. At very small detuning (see region 4 in Figure 3.11), the intensity pulsations disappear and the lasers remain steady and phase-locked.

In summary, phase-locking with steady state intensities is observed if the spacing between the laser beams is sufficiently small (coupling sufficiently strong) and if the detuning is sufficiently small. There is a distinct boundary in the detuning vs. spacing diagram where the lasers exhibit synchronous pulsating instabilities.

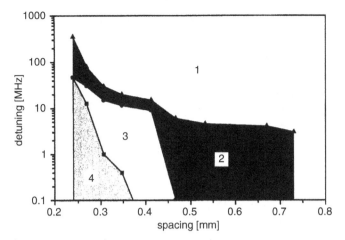

Fig. 3.11 Different regimes of the detuning-spacing parameter space. 1. Steady state intensities and no synchronization. 2. Bursts of intensity oscillations and partial synchronization. 3. Pulsating intensities and strong synchronization (pulsating instability). 4. Steady state intensity and strong synchronization (phase-locking). Reprinted Figure 4 from Möller *et al.* [78] with permission from Elsevier.

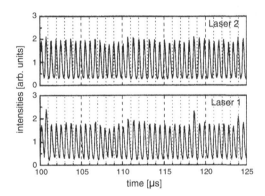

Fig. 3.12 Amplitude instabilities at a laser spacing of 0.23 mm. Reprinted Fig. 3b from Möller *et al.* [78] with permission from Elsevier.

3.6.2 Theory

The coupling between the two lasers arises from the overlap of the two individual electrical fields. In dimensionless form, the two laser rate equations are of the form [78, 79, 91]

$$\frac{d\mathcal{E}_k}{dt} = \mathcal{E}_k \left[D_k - 1 - \kappa \mathcal{E}_j \right] + i\omega_k \mathcal{E}_k, \tag{3.36}$$

$$\frac{dD_k}{dt} = \gamma \left[A - D_k \left(1 + |\mathcal{E}_k|^2 \right) \right], \tag{3.37}$$

where $k = 1$ or 2 and $j = 3 - k$. $\mathcal{E}_k = E_k \exp(-i\phi_k)$ and D_k are the complex field and the inversion of population of laser k, respectively. Time t is measured in units of the cavity constant. ω_1 and ω_2 (angular frequencies) are the detunings of the lasers from a common cavity mode. The lasers are coupled linearly to each other with strength κ, assumed to be small, and the sign of the coupling terms is chosen to account for the observed stable phase-locked state in which the lasers have a phase difference of 180°. Control parameters are the frequency detuning of the lasers ($\Delta = \omega_2 - \omega_1$) and the coupling coefficient κ. We have assumed that both lasers have the same losses and pump. Equations (3.36) and (3.37) are equivalent to five equations for the amplitudes E_k, D_k ($k = 1, 2$) and the phase difference $\Phi \equiv \phi_2 - \phi_1$. However, the connection to Adler's equation can be seen if we consider the particular solution where $E_1 = E_2 = E$ and $D_1 = D_2 = D$. Equations (3.36) and (3.37) then reduce to the following three equations for E, D, and Φ:

$$\frac{dE}{dt} = E\left[D - 1 - \kappa \cos(\Phi)\right], \tag{3.38}$$

$$\frac{dD}{dt} = \gamma\left[A - D\left(1 + E^2\right)\right], \tag{3.39}$$

$$\frac{d\Phi}{dt} = \Delta + 2\kappa \sin(\Phi). \tag{3.40}$$

The last equation is Adler's equation for the phase difference Φ. Since both E and D do not appear in this equation, the variations of Φ are autonomous, and E and D are slaved to Φ, i.e. the phase Φ is driving the laser through the $\cos(\Phi)$ term in Eq. (3.38).

A steady phase (phase-locking) occurs if

$$|\Delta| < 2\kappa \tag{3.41}$$

meaning that the coupling strength needs to be sufficiently large. Since the linearized problem for (3.40) leads to the growth rate $\sigma = 2\kappa \cos(\Phi)$, the stable solution of Eq. (3.40) with $\sigma < 0$ satisfies the two conditions

$$\Delta + 2\kappa \sin(\Phi) = 0 \quad \text{and} \quad \cos(\Phi) < 0. \tag{3.42}$$

Using Eqs. (3.38) and (3.39), we determine the stable steady state as

$$D = 1 + \kappa \cos(\Phi) = 1 - \frac{1}{2}\sqrt{4\kappa^2 - \Delta^2}, \tag{3.43}$$

$$E^2 = \frac{A - 1 + \frac{1}{2}\sqrt{4\kappa^2 - \Delta^2}}{1 - \frac{1}{2}\sqrt{4\kappa^2 - \Delta^2}} \geq 0. \tag{3.44}$$

The situation is however completely different outside the locking region. If condition (3.41) is violated, $\cos(\Phi)$ is a pulsating function of time that is driving the field E. If κ is small, we find from (3.40) that $\Phi = \Delta t + O(\kappa)$ and the remaining equations for E and D are equivalent to the equations of laser subject to periodic loss modulations. See Section 5.2.1. As a consequence, multiple branching of time-periodic intensity regimes is possible and this explains why pulsating intensities are observed if the spacing between lasers increases (coupling decreases).

3.7 Exercises and problems

3.7.1 Rotation induced by loss anisotropy

If loss anisotropy is introduced in the laser cavity, i.e. the fields polarized in the x and y directions are transmitted with different efficiencies t_x and t_y respectively, this results in a rotation of the azimuth of the electric field which tends to align along the axis with lower losses. Show that the azimuth θ of a linearly polarized field is ruled by an evolution equation

$$\frac{d\theta}{dt} = M \sin(2\theta) \text{ with } M = \frac{c}{2L}\left(\frac{t_y}{t_x} - 1\right). \tag{3.45}$$

Solution: The azimuth θ is defined by (see Figure 3.13)

$$\tan(\theta) = \frac{E_y}{E_x}. \tag{3.46}$$

After a single trip into the cavity, the two fields E_x and E_y are reduced by the quantities t_x and t_y, respectively. The new azimuth is now

$$\tan(\theta + d\theta) = \frac{t_y E_y}{t_x E_x}. \tag{3.47}$$

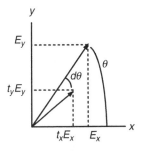

Fig. 3.13 Rotation of the electric field azimuth θ induced by loss anisotropy ($t_x \neq t_y$). x and y refer to the directions transverse to the laser axis as indicated in Figure 3.1.

Taking the difference between (3.47) and (3.46), we find

$$\tan(\theta + d\theta) - \tan(\theta) = \frac{\sin(d\theta)}{\cos(\theta + d\theta)\cos(\theta)} = \left(\frac{t_y}{t_x} - 1\right)\tan(\theta). \qquad (3.48)$$

In the limit $d\theta \to 0$, $\sin(d\theta) \to d\theta$ and $\cos(\theta + d\theta) \to \cos(\theta)$, and the expression (3.48) leads to

$$d\theta = \sin(\theta)\cos(\theta)\left(\frac{t_y}{t_x} - 1\right). \qquad (3.49)$$

The time needed for a single trip into the cavity is $dt = L/c$, where c is the speed of light and L is the length of the cavity. Together with (3.49), we obtain (3.45). If $t_y/t_x < 1$ as in Fig. (3.13), $M < 0$.

3.7.2 Adler's equation

Normal form equation

Derive the normal form equation for Adler's equation (3.8) for ω close to a. To this end, try an expansion of the form

$$\omega = a + \varepsilon\omega_1 + \varepsilon^2\omega_2 \ldots \qquad (3.50)$$

$$\phi = \phi_0(\tau) + \varepsilon\phi_1(\tau) + \ldots, \qquad (3.51)$$

where $\tau \equiv \varepsilon t$ is a slow time variable. ε is a small positive parameter that is related to $\omega - a$. Conditions on the ω_j will be determined by applying solvability conditions. We sequentially find $\phi_0 = \pi/2$, $\omega_1 = 0$, and ϕ_1 satisfying

$$\phi_1' = \omega_2 + \frac{a}{2}\phi_1^2. \qquad (3.52)$$

Oscillation period

Determine the period defined by (3.10) $(\omega > a > 0)$ using the trigonometric substitution $u = \tan(\phi/2)$.

Exact solution

Determine the exact solution of the following Adler's equation

$$\frac{d\phi}{dt} = 1 - a\sin(\phi), \quad \phi(t_0) = -\pi/2 \qquad (3.53)$$

for $|a| < 1$. Introduce first $\theta = (\phi + \pi/2)/2$ and then $y = \tan(\theta)$. The solution in implicit form is given by

$$\sqrt{\frac{1-a}{1+a}}\tan\left(\frac{\phi + \pi/2}{2}\right) = \tan\left(\frac{\sqrt{1-a^2}}{2}(t - t_0)\right). \qquad (3.54)$$

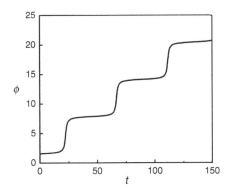

Fig. 3.14 Numerical solution of Adler's equation (3.55) for $\phi(0) = \pi/2$ and $\Lambda = -0.01$.

Solution of Adler's equation close to locking

We wish to solve Adler's equation

$$\phi' = 1 - (1 + \Lambda)\sin(\phi), \quad \phi(0) = \pi/2 \tag{3.55}$$

for small and negative values of the control parameter Λ by using asymptotic methods. The initial condition $\phi(0) = \pi/2$ simplifies the analysis but is not a restriction of the asymptotic theory. The numerical solution shown in Figure 3.14 exhibits successive plateaus separated by relatively fast transition layers. The exact solution of Adler's equation (3.54) is complicated and gives little physical insight into what happens as $|\Lambda|$ approaches zero. If $\Lambda < 0$, the period of the oscillations given by (3.11) is

$$P = \frac{2\pi}{\sqrt{1 - (1 + \Lambda)^2}} \tag{3.56}$$

and it approaches the inverse square-root law

$$P \simeq \frac{2\pi}{\sqrt{-2\Lambda}} \tag{3.57}$$

as $\Lambda \to 0$ (see Figure 3.15). Construct an asymptotic approximation by using the method of matched asymptotic expansions (or MAE) [15]. The basic idea of the method is to construct two distinct solutions, one for the plateaus (the outer solution) and one for the transition layers (the inner solution). Because the period is inversely proportional to $\sqrt{-\Lambda}$, introduce

$$\varepsilon = \sqrt{-\Lambda} \tag{3.58}$$

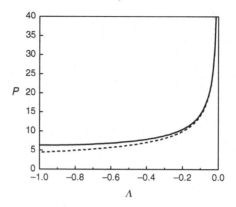

Fig. 3.15 The period of the exact solution of Adler's equation (3.56) (full line) is compared to its approximation, Eq. (3.57) (broken line).

as a small parameter. The outer solution usually refers to the solution that we may obtain by a regular expansion. To this end, seek a solution of the form

$$\phi = \phi_0(s) + \varepsilon\phi_1(s) + \varepsilon^2\phi_2(s) + \dots, \tag{3.59}$$

where s is a slow time variable defined by $s \equiv \varepsilon t$. Note from the expression of $\phi_1(s)$ that it becomes unbounded as $s \to s_c$ where

$$s_c \equiv \sqrt{2}\frac{\pi}{2}. \tag{3.60}$$

This singularity motivates an inner solution of the form

$$\phi = \Phi(S) + \varepsilon\Phi_1(S) + \dots, \tag{3.61}$$

where $S \equiv t - t_c$. Determine this solution and show how it connects with the outer solution.

3.7.3 Class A laser subject to an injected signal

Determine the steady states of Eqs. (3.16)–(3.17) and investigate their stability properties. Derive the locking (saddle-node bifurcation) and Hopf stability boundaries in the (Δ, X_0) diagram. At a Hopf bifurcation, the characteristic equation admits a pair of purely imaginary eigenvalues. Discuss the validity of Adler's locking condition (3.19) for small X_0.

3.7.4 Ring laser with diffraction locking

When apertures smaller than or similar to the beam waist limit the transverse extent of the laser beam, they may also act on standing waves resulting from the

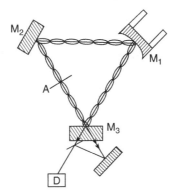

Fig. 3.16 Controlling the position of the standing-wave pattern in a ring laser at rest is possible thanks to diffraction. Modulating the position of the aperture A along the beam axis induces a beat frequency between the two counterpropagating waves in the motionless ring laser (the so-called reverse Sagnac effect). The figure represents the standing-wave structure when the two counterpropagating waves are locked. Mirrors M_2 and M_3 are plane and M_1 is spherical. The output beams are recombined on the detector D by a beam splitter and an extra plane mirror (from [93]).

interference between the counterpropagating beams of a ring laser since these apertures control the nodes of the standing wave (see Figure 3.16). In this situation, phase-locking is due not only to the mirror defects but also to diffraction produced by the diaphragm. Driving the diaphragm allows to "drag" the standing waves. This induces the so-called "reverse Sagnac effect" [93]. In the Sagnac effect, the two waves (*cw* and *ccw*) running in the laser cavity are frequency shifted because they run in a rotating cavity. In the reverse Sagnac effect, the cavity is fixed and the waves inside it are rotating because of the shift of the standing waves' nodes induced by the motion of the internal diaphragm. Adler's equation for ring lasers with such diffracting apertures must be modified to account for this additional effect. It is given by [93]

$$\frac{d\varphi}{dt} = \omega - d\sin(\varphi) - d_a \sin(\varphi - \varphi_0).$$ (3.62)

The competition between the two restoring forces gives rise to new results when d and d_a are of the same magnitude. Show that the new equilibrium positions are shifted by an angle

$$\alpha = \arctan\left(\frac{d_a \sin(\varphi_0)}{d + d_a \sin(\varphi_0)}\right).$$ (3.63)

4

Hopf bifurcation dynamics

A Hopf bifurcation marks the transition from a steady state to a time-periodic solution. We already encountered an example of a Hopf bifurcation in Section 3.4 as we analyzed the laser subject to an injected signal.

The emergence of spontaneous time-dependent regimes in lasers is not a purely academic problem because physicists have been confronted by the appearance of "noise-like" intensity fluctuations in the laser's beam since the beginning of the laser. This type of behavior was evident even during the earlier investigations of the laser in the 1960s [96, 97] where it was found that the intensity of the light generated by the ruby laser displayed irregular spiking, as shown in Figure 4.1. Were these spikes the result of a noisy environment or were they coming from the laser itself? A lot of experiments have been undertaken on the ruby laser under various conditions (see [6]). It eventually appeared that the oscillatory output of the ruby laser resulted from the combined effect of several mechanisms. Research on this topic vanished because of the advent of new lasers whose parameters are much better controlled and therefore capable of delivering cw power or pulses with well-defined and reproducible properties.

For many years, attempts to understand the appearance of such oscillatory instabilities in lasers were limited (for instance, the extensive but isolated effort of Lee W. Casperson to describe the pulsations of the Xe laser [98]), until Hermann Haken showed the equivalence of the laser equations with the Lorenz system [99]. As the latter was known to exhibit deterministic chaos, Haken's work triggered a wave of interest for laser nonlinear dynamics. More importantly, it placed optical systems in the general framework of *dynamical systems* [8]. This means that turbulent flows in fluids, oscillatory chemical reactions, and self-pulsing lasers share similar phenomena that do not depend on the detailed physics or chemical kinetics but rather on simple *bifurcation mechanisms*.

Nowadays, the motivation for studying these bifurcations really depends on the background of the researcher and it extends beyond the community of physicists.

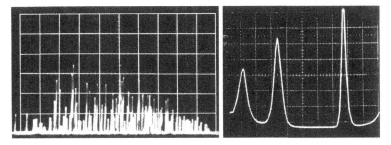

Fig. 4.1 Oscillatory traces of light intensity emitted by a flash-lamp pumped ruby laser. Left: train of irregular pulses, horizontal scale 50 μs/division, vertical scale 0.5 V/division. Right: enlarged view at higher temporal resolution (2 μs/division), vertical scale 50 mV/division. Reprinted Figure 10.10 of Lauterborn and Kurz [95] with permission. Copyright Springer-Verlag 1995, 2003.

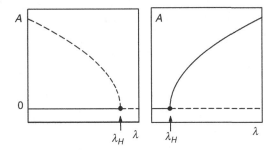

Fig. 4.2 Amplitude of the oscillations as a function of the control parameter λ. Left and right figures represent a subcritical and a supercritical Hopf bifurcation, respectively.

For some engineers, oscillatory instabilities are viewed as a limitation on the performance of the optical device that must be avoided or controlled. For example, we wish to control chaotic fluctuations in the intensity of diode lasers because they limit their ability to detect information stored on compact discs [100]. In contrast, other researchers have put the unstable behavior to good use making practical devices such as low-jitter high-frequency generators for communication or even chaotic oscillators for an optical cryptosystem [101].

A Hopf bifurcation denotes the appearance of a periodic solution in the neighborhood of a steady state whose stability changes due to the crossing of a conjugate pair of eigenvalues over the imaginary axis.[1] The Hopf theorem states that if this cycle coexists with the steady state solution, it is unstable and vice versa, as illustrated in Figure 4.2. A *supercritical Hopf bifurcation* leads to a branch of *stable periodic solutions* overlapping a branch of unstable steady states (Figure 4.2 right).

---

[1] I.e. the characteristic equation has a pair of roots with zero real parts $\sigma = \pm i\omega$.

A *subcritical Hopf bifurcation* leads to a branch of *unstable periodic solutions* overlapping a branch of stable steady states (Figure 4.2 left). In the latter case, the branch of periodic solutions may fold back at a larger amplitude and give rise to a branch of stable periodic solutions. However, this evolution at large amplitudes is not predicted by Hopf theory which is purely local (i.e. valid near the bifurcation point $\lambda = \lambda_H$). In the vicinity of the bifurcation, the oscillations are nearly harmonic in time and, in general, the amplitude A of the oscillations grows like $(\lambda - \lambda_H)^{1/2}$. Other bifurcation behaviors, such as a vertical bifurcation or a different scaling law for A, are not ruled out by the Hopf bifurcation theorem.

The Hopf bifurcation was rediscovered in the 1970s when new oscillatory phenomena were found in fluid, chemical, and biological systems.[2] Mathematicians interested in proving the existence of specific solutions became interested by their stability. Hopf's bifurcation paper appeared in 1942 in German [104] and was translated in 1970 [105]. In his paper, Hopf says "I scarcely think that there is something new in the above theorem. The methods were developed by Poincaré perhaps 50 years ago ... " Thus, as Louis N. Howard [106] commented, "Hopf himself might not entirely agree with the current usual designation of the result as the *Hopf bifurcation theorem* or the description of the kind of oscillatory bifurcation to which it refers as *Hopf bifurcation*. Still, Hopf's clear formulation and presentation of the result was a significant contribution, and he was perhaps one of the first to understand clearly some features of it, particularly with regard to the stability properties of the periodic solution."

How good the Hopf asymptotic solution is as $\lambda - \lambda_H$ increases depends on the nonlinear system. In the last section, we showed that a Hopf solution which is purely local may be limited in the strict vicinity of $\lambda = \lambda_H$ for systems exhibiting different time scales, as with many of our lasers.

A Hopf bifurcation is the simplest mechanism leading to nonlinear oscillations in many dynamical systems, but **not** for a single mode class B laser. As we already know from Chapter 1, these lasers exhibit slowly decaying oscillations called *relaxation oscillations* that do not result from a change of stability of a reference steady state. However, we may sustain and even amplify these "relaxation oscillations" by weakly modulating a parameter such as the pump or loss parameter. We concentrate on this important topic in Section 5.2.1.

Another way to compensate for the damping of the relaxation oscillations is to apply a positive feedback to the laser. The two laser rate equations for the field

---

[2] Although physicists and mathematicians gathered several times at conferences from 1973 to 1977, the New York Academy of Sciences conference on Bifurcation Theory and Applications in Scientific Disciplines held from October 31 to November 4, 1977 [102], was the first meeting attended by a larger number of scientists with different backgrounds [103].

in the cavity and the inversion of population are now supplemented by a third equation for the voltage of the feedback loop. *The three coupled first order differential equations* exhibit multiple steady states and Hopf bifurcations that we analyze in Section 4.1. In Section 4.2, we consider the case of a passive resonant cavity subject to delayed optical feedback, which provided the first clear identification of a Hopf bifurcation in nonlinear optics. This problem was analyzed in Japan by Kensuke Ikeda [108] and had a considerable historical impact. Today, similar devices are built as sources of periodic or chaotic outputs for uses such as transmitting digital information. This device is accurately modeled by *a scalar delay differential equation*. We show how this may be reduced to the equation for a *map* which appears to be very efficient in obtaining analytical approximations of the periodic solutions.

4.1 Electrical feedback

Negative electrical feedback is frequently used in laser design, e.g. to establish a regulatory loop in order to achieve stable output or to protect the laser from burnout. For example, semiconductor lasers (SLs) have to be operated at a high current density and have a very low forward resistance when lasing action occurs, so they are at risk of destroying themselves due to thermal runaway. Their operating light density can also rise to a level where the end mirrors can begin melting. As a result their electrical operation must be carefully controlled. This means that not only must a laser's current be regulated by a "constant current" circuit but optical negative feedback must generally be used as well – to ensure that the optical output is held to a constant safe level. In the most usual feedback scheme, laser diodes have a silicon PIN photodiode built right into the package, arranged so that it automatically receives a fixed proportion of the laser's output. The output of this monitor diode can then be used to control the current fed through the laser by the constant current circuit, for stable and reliable operation [107]. Other types of feedback (feedback on the cavity length, or optical feedback) have also been successfully used for stabilizing the laser frequency. A schematic diagram of a semiconductor laser controlled by optoelectronic feedback is shown in Figure 4.3. In this section we concentrate on the case of electrical feedback controlling the cavity losses, and the delay of the feedback does not play a major role. Section 4.2 and Chapter 10 are specifically devoted to the important effects of a delayed optoelectronic or optical feedback.

Feedback may lead to a variety of dynamical instabilities. This question was raised for the first time in 1986 for a CO_2 laser with an intracavity electro-optic modulator (EOM) by Arecchi *et al.* [108, 109]. The same problem was recently revisited for a Nd:YAG laser with the more convenient acousto-optic modulator

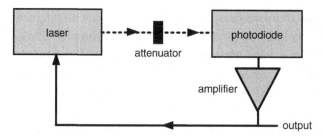

Fig. 4.3 Schematic diagram of optoelectronic feedback. The optical power emitted by the laser is detected by a photodiode with a fixed bandwidth. The electrical output is fed back to the laser through an amplifier.

(AOM) [110]. Both lasers are class B lasers and are described by the same rate equations for the intensity of the laser field I, the inversion of population D, and an additional equation for the voltage V of the feedback loop. The latter accounts for the limited bandwidth of the feedback loop. In dimensionless form, the three evolution equations are of the form [110]

$$I' = I\left[D - 1 - a\sin^2(V)\right], \tag{4.1}$$

$$D' = \gamma[A - D(1 + I)], \tag{4.2}$$

$$V' = \beta(B + fI - V), \tag{4.3}$$

where γ and A have the same meaning as in the SRE. a scales the maximum loss introduced by the modulator, the damping rate β of the feedback loop is normalized by the cavity decay rate, B is the bias voltage applied to the modulator amplifier, and f is the scaling of the feedback gain, i.e. it measures the relation between the intensity incident on the photodiode and the voltage delivered by the differential amplifier. In general, B is the control parameter and the bifurcation diagram of the possible long-time regimes is studied for different values of f which may be adjusted through the detector preamplifier.

In this system, the bias B sets a reference value for the voltage applied to the modulator, and consequently allows the operation point of the laser to vary. The feedback enters through the term fI in (4.3). Positive or negative feedback depends on the relative sign of f and B. Assuming $B > 0$, Eq. (4.3) indicates that $V > 0$ is favored in absence of feedback. As $I > 0$, $f > 0$ implies that the feedback increases V, i.e. increases the losses through $-a\sin^2(V)$ since modulators are generally operated at $V < \pi/2$. Therefore $f > 0$ implies negative feedback in the classical sense that an increase in laser output is transduced as less efficiency for laser action.

Table 4.1 *Values of the parameters for a CO_2 laser and a Nd:YAG laser.*

Parameters	Symbol	CO_2 laser [a]	Nd:YAG laser [b]
cavity loss	κ	9.6×10^6 s^{-1}	6.6×10^7 s^{-1}
population decay	$\gamma_{\parallel}$	3×10^4 s^{-1}	4.166×10^3 s^{-1}
	$\gamma \equiv \gamma_{\parallel}/\kappa$	3.125×10^{-3}	6.31×10^{-4}
pump parameter	A	1.66	1.85
feedback damping rate	β_0	5×10^5 s^{-1}	6.28×10^5 s^{-1}
	$\beta \equiv \beta_0/\kappa$	5.21×10^{-2}	9.51×10^{-3}
feedback amplitude	a	5.8	0.052
feedback gain[c]	f	-0.8 to 0	0.75

[a] [111].
[b] [110].
[c] f is denoted by $-r$ in [108].

Typical values of the fixed parameters are listed in Table 4.1. In spite of their different nature, these two lasers have very similar relaxation frequencies (compare $\sqrt{\kappa\gamma_{\parallel}}$). The difference by a factor of 100 in a is compensated by the change by a factor of 5 in the damping factor β.[3]

The dynamical effects of the electrical feedback on both lasers have been extensively studied by R. Meucci and different coworkers since the end of the 1980s. In [109], periodic and chaotic intensity oscillations were observed and interpreted as resulting from the presence of several stable and unstable steady states, and, in particular, a saddle-node which is responsible for the appearance of Shilnikov chaos. The shape of the different signals has been carefully investigated in the vicinity of each bifurcation with special attention on those related to Shilnikov dynamics. In [111], a more global approach in the parameter space is proposed and a subcritical Hopf bifurcation has been observed. Quasi-sinusoidal oscillations are obtained near the bifurcation. They evolve into spikes in regions further from the bifurcation (see Figure 4.4). Two distinct bifurcation diagrams have been examined for low or high feedback gain $|f|$, respectively (see Figure 4.5). In [110], the dynamics associated with two different Hopf bifurcations are studied in detail with the idea of producing a chaotic function generator.

Note that the discussion on the laser energy level schemes that we developed in Chapter 2 also applies to the present problem. As far as qualitative agreement is looked for, the two-level model applies quite well to both CO_2 and YAG lasers. But it must be kept in mind that a quantitative agreement may be obtained with the CO_2 laser only if more refined models are used. In addition, we need to

[3] By dividing Eq. (4.1) by the parameter a and introducing the new time $s = at$, we obtain a term β/a multiplying the right hand side of Eq. (4.3). It is this ratio that controls the decay of V.

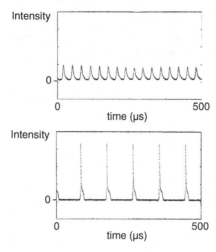

Fig. 4.4 Laser intensity vs. time for $f = -0.25$. The oscillations are harmonic near the Hopf bifurcation point (top) but become pulsating as we increase B (bottom). Reprinted Figure 2 from Wang *et al.* [111] with permission from Elsevier.

take into account the nonlinearity of the detector response [112]. In the following sections, we concentrate on the steady state solutions and discuss two instability mechanisms (saddle-node bifurcation and Hopf bifurcation) and how they rule the bifurcation diagrams as observed in the experiments.

4.1.1 Steady-state solutions

Here, we limit our review to the steady states and their bifurcation points for $B > 0$.[4] Equations (4.1)–(4.3) admit a zero intensity solution given by

$$I = 0, \quad D = A, \quad \text{and} \quad V = B, \tag{4.4}$$

and a non-zero intensity solution given in parametric form as

$$D = 1 + a \sin^2(V), \tag{4.5}$$

$$I = -1 + \frac{A}{1 + a \sin^2(V)}, \tag{4.6}$$

$$B = f - \frac{fA}{1 + a \sin^2(V)} + V, \tag{4.7}$$

where V is the parameter. Their stability properties can be obtained from the linearized equations, which we do not formulate.

---

[4] The case $B < 0$ can be analyzed by noting the change $(B, f, V) \to (-B, -f, -V)$.

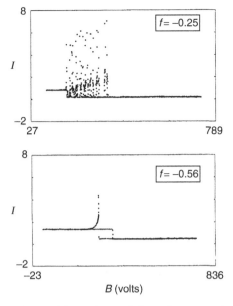

Fig. 4.5 Laser intensity vs. bias voltage B for a low negative ($f = -0.25$, top) and a high negative value of the feedback gain f ($f = -0.56$, bottom). For a low negative f, strongly pulsating oscillations are observed as we pass a Hopf bifurcation point. For a high negative f, oscillations disappear but the non-zero and the zero intensity steady states may coexist for a short range of value of B. The large spike at the left boundary of the bistable domain corresponds to the jump transition between the lower and upper branches. The jump transition between the upper and lower branches at the right boundary of the bistable domain produces negligible transient. Reprinted Figure 1c, 1d from Wang et al. [111] with permission from Elsevier.

The zero intensity solution (4.4) is stable if $A - 1 - a \sin^2(B) < 0$ or, equivalently, if

$$a > A - 1 \quad \text{and} \quad B > B_c \equiv \arcsin\left(\sqrt{\frac{A-1}{a}}\right). \tag{4.8}$$

If $a < A - 1$, the zero intensity solution is always unstable. The critical point $B = B_c$ denotes a bifurcation point from the zero intensity steady state to the non-zero intensity steady state.

For the non-zero intensity steady state (4.5)–(4.7), the characteristic equation for the growth rate σ is

$$\sigma^3 + C_1 \sigma^2 + C_2 \sigma + C_3 = 0, \tag{4.9}$$

where the coefficients are defined by

$$C_1 = \gamma(1+I) + \beta, \tag{4.10}$$

$$C_2 = \gamma I D + aI \sin(2V)\beta f + \gamma\beta(1+I), \tag{4.11}$$

$$C_3 = \beta\gamma I \left[D + af \sin(2V)(1+I)\right]. \tag{4.12}$$

The real part of σ is negative provided the Routh–Hurwitz conditions are satisfied [29]. Violation of one of these conditions leads to two stability boundaries corresponding to saddle-node and Hopf bifurcations, respectively, which we detail in the next two sections.

4.1.2 Steady or saddle-node bifurcation

The condition for a steady bifurcation or a saddle-node bifurcation point is $C_3 = 0$ (one zero eigenvalue, $\sigma = 0$). Using (4.12), we find that this condition is realized either if $I = 0$ or if

$$D + af \sin(2V)(1+I) = 0. \tag{4.13}$$

The first case corresponds to the steady bifurcation point located at $B = B_c$ and documented in (4.8). The second case corresponds to a saddle-node bifurcation or limit point (see Exercise 4.4.1). Using (4.5) and (4.6), we eliminate D and $1 + I$ in (4.13) and obtain

$$f = -\frac{\left(1 + a \sin^2(V)\right)^2}{aA \sin(2V)}. \tag{4.14}$$

In Figure 4.6, we represent (4.14) in terms of the steady state intensity I as a function of f (for curve SN, note: I is related to V by (4.6)). The diagram shows that there are three different domains of f where zero, two, and one saddle-node bifurcation points are possible.

4.1.3 The Hopf bifurcation

The Hopf bifurcation conditions can be determined by introducing $\sigma = i\omega$ into Eq. (4.9). From the real and imaginary parts, we obtain

$$C_1 C_2 - C_3 = 0 \quad \text{and} \quad C_2 > 0. \tag{4.15}$$

The first condition simplifies as

$$\gamma^2(1+I)\left[DI + \beta(1+I)\right] + \beta^2\left[af I \sin(2V) + \gamma(1+I)\right] = 0. \tag{4.16}$$

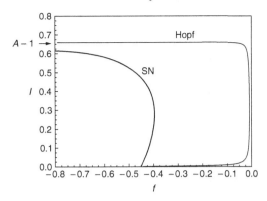

Fig. 4.6 Saddle-node (SN) and Hopf bifurcation (Hopf) stability boundaries in terms of the steady state intensity I and f. They are given by (4.14) and (4.17), respectively. The values of the fixed parameters are: $\gamma = 3.125 \times 10^{-3}$, $A = 1.66$, $\beta = 5.21 \times 10^{-2}$, and $a = 5.8$.

We may again eliminate I and D by using (4.5) and (4.6) and find the following expression relating f and V

$$f = -\frac{\gamma^2 A \left[A - 1 - a \sin^2(V) + \beta A(1 + a \sin^2(V))^{-1} \right] + \beta^2 \gamma A}{\beta^2 a(A - 1 - a \sin^2(V)) \sin(2V)}. \tag{4.17}$$

The Hopf stability boundary (4.17) is shown in Figure 4.6 in terms of the steady state intensity I and feedback factor f (for curve Hopf, note: I is related to V by (4.6)). The Hopf line which also satisfies $C_2 > 0$ emerges from the SN line at a very low intensity ($I = O(\gamma)$) from a critical point satisfying the two conditions $C_3 = C_2 = 0$ (a double zero eigenvalue, $\sigma_1 = \sigma_2 = 0$). From left to right in Figure 4.6, the Hopf line remains nearly constant at low intensities, suddenly turns near $f = 0$, and then saturates to an almost constant intensity from right to left. This behavior can be anticipated by noting that β and γ are small parameters. Assuming $\gamma = O(\beta^2)$,[5] the leading term in Eq. (4.16) is $O(\beta^2)$ and given by

$$aI \sin(2V) f = 0. \tag{4.18}$$

Equation (4.18) implies that either (1) $I = 0$, (2) $f = 0$, or (3) $\sin(2V) = 0$. Case (1) anticipates the low-intensity part of the Hopf bifurcation line; case (2) predicts the vertical line at $f = 0$; case (3) is verified if $V = 0$ which implies $I = A - 1$, i.e. the horizontal Hopf bifurcation line in Figure 4.6 (case (3) is also verified if $V = \pi/2$, which implies $I = (A - 1 - a)/(1 + a)$, but the condition $I > 0$ cannot be realized with the values of the parameters used in Figure 4.6).

[5] This is required to balance the two terms in (4.16).

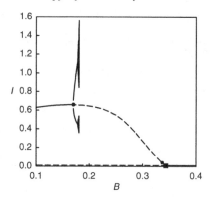

Fig. 4.7 Low-gain bifurcation diagram. The figure represents the maxima and minima of the long time stable solutions. A square marks the steady state bifurcation point at $(I, B) = (0, B_c)$. Two dots at $B = 0.167$ and 0.34 indicate Hopf bifurcation points. The diagram has been determined numerically from Eqs. (4.1)–(4.3) with $\gamma = 3.125 \times 10^{-3}$, $A = 1.66$, $\beta = 5.21 \times 10^{-2}$, $a = 5.8$, and $f = -0.25$.

4.1.4 Bifurcation diagrams

We illustrate our stability results by studying bifurcation diagrams in two cases representing relatively low ($f = -0.25$) and high ($f = -0.6$) gain and using B as the control parameter. The values of the other parameters are the same as in Figure 4.6.

For $f = -0.25$, a single branch of steady states emerges from zero at $B = B_c \simeq 0.344$ as shown in Figure 4.7. Two Hopf bifurcation points bound a domain of unstable steady states. The Hopf bifurcation point with the higher steady state intensity ($B \simeq 0.167$ in Figure 4.7) is located at $I_H \simeq A - 1$ according to our previous analysis. Using then (4.7) with $V = 0$, we find

$$B_H \simeq f(1 - A), \tag{4.19}$$

which gives $B_H \simeq 0.165$. The Hopf bifurcation is supercritical and, as can be shown from an analysis of $\omega^2 = C_2 = C_3/C_1$, exhibits oscillations close to the laser relaxation frequency $\omega_R = \sqrt{\gamma(A - 1)}$ (see Exercise 4.4.2). The role of the feedback is therefore to sustain the RO oscillations and β does not play a major role, in first approximation. The high intensity Hopf bifurcation is followed by a Torus bifurcation to quasi-periodic oscillations.[6] More complex bifurcations appear as we further increase B but they are not analyzed here. The oscillations are then strongly pulsating in time and are of similar nature to the passive Q-switch

---

[6] A Torus bifurcation is a bifurcation from periodic to quasi-periodic oscillations characterized by two noncommensurable frequencies.

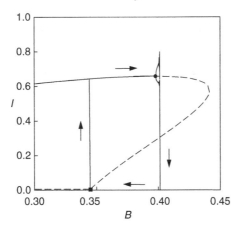

Fig. 4.8 High-gain bifurcation diagram. The figure represents the maxima and minima of the long time stable solutions. A square and a dot mark the steady state and the Hopf bifurcation points, respectively. A hysteresis cycle indicated by arrows is possible by increasing and then decreasing B. The diagram has been determined numerically from Eqs. (4.1)–(4.3) with $\gamma = 3.125 \times 10^{-3}$, $A = 1.66$, $\beta = 5.21 \times 10^{-2}$, $a = 5.8$, and $f = -0.6$.

oscillations for a laser with a saturable absorber (see Chapter 8). The pulsating character of the oscillations results from the small value of γ, forcing the laser to operate on two distinct time scales. Note that the Hopf bifurcation point does not depend on β, in first approximation. This motivates an adiabatic elimination of the variable V and a simplification of Eqs. (4.1)–(4.3) (see Section 4.3).

The second Hopf bifurcation point appears at a very low intensity and is located close to the steady bifurcation at $B = B_c \simeq 0.344$. The physical mechanism responsible for this Hopf bifurcation is quite different and depends on both γ and β (see Exercise 4.4.3). In Figure 4.8, the values of the parameters are the same as in Figure 4.6 but $f = -0.6$. The system exhibits bistability (coexistence of two stable steady states) and a hysteresis cycle is observed as we increase or decrease B beyond the interval 0.35–0.40. Note that the Hopf bifurcation point at $B_H = 0.396$ is well approximated by (4.19). As we progressively increase B, the transition to the zero intensity steady state does not occur at the steady state limit point but from the Hopf bifurcation branch.

The two numerical diagrams shown in Figure 4.7 and Figure 4.8 reproduce the main features of the experimental ones (Figure 4.5), namely a Hopf bifurcation leading quickly to a large domain of irregular regimes for low gain, and hysteresis between steady states at large gain. As we see in these two examples, the transition from the steady state to harmonic oscillations may be quite abrupt. Hopf bifurcation theory is a local theory which is only valid in the vicinity of the bifurcation

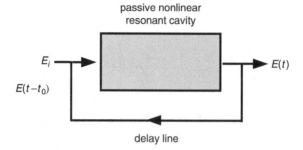

Fig. 4.9 Schematic diagram of Ikeda system. A passive cavity is subject to an injected field E_i. Part of the output $E(t)$ is reinjected into the cavity as $E(t - t_0)$ after it has undergone a long delay t_0.

point. It doesn't tell us if pulsating or square-wave regimes may appear as we deviate from it.

4.2 Ikeda system

In 1979, Ikeda considered a nonlinear absorbing medium containing two-level atoms placed in a ring cavity and subject to a constant input of light. If the total length of the cavity is sufficiently large, the optical system undergoes a time-delayed feedback which destabilizes its steady state output. See Figure 4.9. From the Maxwell–Bloch equations, Ikeda derived a set of coupled differential-difference equations [113] (this derivation is simpler if we start from the Maxwell–Debye equations for highly dispersive media [114]; see [5] p. 122, [7] p. 39). Then introducing more assumptions, Ikeda formulated the following scalar delay differential equation (DDE) [114, 21]

$$\tau \phi' = -\phi + A^2 \left[1 + 2B \cos \left(\phi(t - t_D) - \phi_0\right)\right], \qquad (4.20)$$

where the growth of ϕ depends both on its value at time t and on its value at time $t - t_D$. Here the delay represents the round-trip time along the optical path. Using (4.20), Ikeda then showed numerically that periodic, multiperiodic, and chaotic outputs are possible. In 1983, the experimental system was realized by his colleagues with a train of light pulses injected in a long single-mode optical fiber [115], but this physical system is poorly described by Eq. (4.20). Efforts to develop an experimental device that is accurately modeled by a simple DDE like Eq. (4.20) immediately followed the early work of Ikeda and today quantitative comparisons between experiments and theory are possible.

In Besançon (France), work has been done on a delayed optical system where the dynamical variable is the wavelength [116]. An improved device using a tunable DBR laser was then realized [117, 118]. This experience led to the

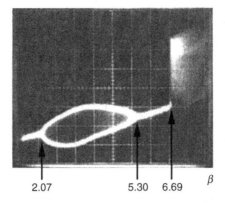

Fig. 4.10 Experimental bifurcation diagram for $\Phi = 0$ (from Figure 4a of Larger *et al.* [119]).

development of a system based on coherence modulation. The dynamical variable is the optical-path difference in a coherent modulator driven electrically by a nonlinear delayed feedback loop [118]. The system is realized from a Mach–Zehnder coherence modulator powered by a short coherence source and driven by a nonlinear feedback loop that contains a second Mach–Zehnder interferometer and a delay line. In dimensionless variables, the response of the system is well described by [118]

$$\frac{\tau}{t_d}x' = -x + \beta\left[1 + \frac{1}{2}\cos(x(t-1) + \Phi)\right], \qquad (4.21)$$

where x is proportional to the optical-path difference and the dimensionless time t is the original time t' normalized by the delay t_d. The parameter τ measures the relaxation of the system in the absence of delay. The bifurcation parameter β is proportional to the gain in the feedback loop, which can be varied. The phase Φ also can easily be changed electrically by means of a bias voltage.

The experimental bifurcation diagram for $\Phi = 0$ is shown in Figure 4.10. It has been obtained by progressively changing β from small to large values. No attempt has been made to find if other attractors are possible in the same range of values of β (for example, by decreasing β from a high to a small value). This experimental bifurcation diagram was obtained by recording the extrema of the oscillations. Steady operation provides a single-valued output as, for example, for $\beta < 2.07$. The emergence of a cycle at a Hopf bifurcation is revealed by the appearance of a double value oscilloscope trace. Quasiperiodic or chaotic regimes are associated with continuous bands of values for the extrema. Two successive Hopf bifurcation points are visible at $\beta = 2.07$ and 5.30, marking the beginning and the end of sustained oscillations with one maximum and one minimum. A third point at $\beta = 6.69$

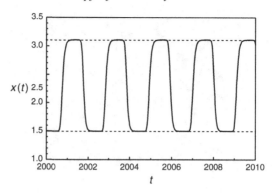

Fig. 4.11 Long time periodic solution of the delay differential equation. The oscillations are nearly square wave with a period close to 2. The values of the parameters are $\Phi = 3$ and $\tau/t_d = 0.05$. The dotted lines at $x = 1.5$ and $x = 3.1$ are the values predicted by the equation of the map valid as $\tau/t_d \to 0$.

marks the sudden transition to chaotic oscillations exhibiting irregular maxima and minima.

Before analyzing the bifurcation diagram, we numerically investigate the solutions of Eq. (4.21) and find that periodic solutions are typically square-wave with a period close to 2 (see Figure 4.11). The square-wave form of the oscillations results from the fact that the ratio $\tau / t_D = 0.05$ is small. We may then neglect the left hand side of (4.21) and obtain an equation for a map relating the extrema of the square wave $x_n = x(t - 1)$ and $x_{n+1} = x(t)$ and given by

$$x_{n+1} = \beta \left[1 + \frac{1}{2} \cos(x_n + \Phi) \right]. \tag{4.22}$$

A steady state solution of the DDE (4.21) corresponds to a Period 1 fixed point of the map $(x_{n+1} = x_n)$. A periodic solution of the DDE (4.21) corresponds to a Period 2 fixed point of the map $(x_{n+2} = x_n)$.

The bifurcation diagram of the fixed points of Eq. (4.22) has been studied numerically and is shown in Figure 4.12 for $\Phi = 0$. Different initial conditions have been used in order to find all possible stable attractors. The numerical bifurcation diagram indicates that, in addition to the branches found experimentally, there is another domain ($\beta \simeq 4.64$–5.5) of periodic and chaotic oscillations emerging from a third Hopf bifurcation located on the upper branch of steady states.

The experimental values of β for three observed bifurcation transitions are compared to the numerical estimates obtained from Eq. (4.22). See Table 4.2. The excellent quantitative agreement means that (1) the optical system closely mimics the Ikeda differential equation and (2) the reduction of the DDE to the equation for a map is fully justified. This motivates some additional work on Eq. (4.22).

Table 4.2 *Experimental and numerical estimates of the first three bifurcations agree quantitatively. The first two correspond to Hopf bifurcations, and the third to a limit point.*

Results	$\beta(\Phi = 0)$		
numerical	2.06	5.06	6.59
experimental	2.07	5.30	6.69

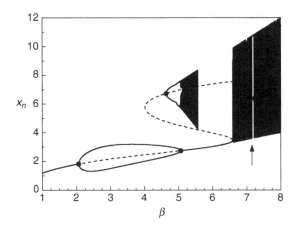

Fig. 4.12 Stable fixed points of the map. The broken line is the branch of steady states. Dots mark three Hopf bifurcation points. Transition to chaos occurs as β surpasses the first limit point of the steady states. The upper steady state branch admits a Hopf bifurcation leading to a cascade of bifurcations. It ends as the minimum reaches the unstable branch of steady states. The arrow indicates a small window of periodic solutions.

4.2.1 Limit and Hopf bifurcation points

We consider $\Phi = 0$ and analyze the stability of the Period 1 fixed points of Eq. (4.22). Exercise 4.4.6 considers the case of arbitrary values of Φ. Introducing $x = x_{n+1} = x_n = x$, we find from (4.22) the implicit solution

$$\beta = \frac{x}{1 + \frac{1}{2}\cos(x)}. \tag{4.23}$$

The linearized problem for the small perturbation $u_n = x_n - x$ is then given by

$$u_{n+1} = -\beta \frac{1}{2}\sin(x)u_n, \tag{4.24}$$

which is a linear difference equation. Substituting $u_n = r^n$ into Eq. (4.24), we find r as

$$r = -\beta \frac{1}{2}\sin(x). \tag{4.25}$$

Table 4.3 *Hopf bifurcation points for increasing positive values of β. The first two points mark the beginning and the end of the isolated branch of periodic solutions (isola). The third one is located on the upper branch of steady states.*

	β_H	x_H
1	2.06	1.81
2	5.06	2.74
3	4.64	6.73
4	18.53	9.32

Stability requires $|r| < 1$ because $u_n \to 0$ as $n = 1, 2, \ldots \to \infty$. There are two possible stability changes.

(1) The first possibility is $r = 1$, which requires

$$1 = -\beta \frac{1}{2} \sin(x). \tag{4.26}$$

Eliminating β using (4.23), (4.26) reduces to the following equation for x only

$$1 + \frac{1}{2} \cos(x) + x \frac{1}{2} \sin(x) = 0. \tag{4.27}$$

From Eq.(4.23), we determine $d\beta/dx$ and find that the condition $d\beta/dx = 0$ for a steady state limit point is identical to (4.27). Thus, the condition $r = 1$ marks the change of stability of the steady state at a limit point.

(2) The second possibility is $r = -1$, which requires

$$1 = \beta \frac{1}{2} \sin(x). \tag{4.28}$$

The small perturbation $u_n = r^n$ with $n = 1, 2, \ldots$ is alternatively equal to -1 or $+1$ and exhibits an oscillatory behavior. This condition marks the transition from a Period 1 to a Period 2 fixed point and is equivalent to the Hopf bifurcation point of the DDE. From (4.28) and using (4.23), we obtain

$$1 + \frac{1}{2} \cos(x) - x \frac{1}{2} \sin(x) = 0, \tag{4.29}$$

which needs to be solved numerically. The first four roots of Eq. (4.29) with positive values of β are listed in Table 4.3.

The position of the Hopf bifurcations is extremely well predicted analytically. We may proceed further and obtain information on the amplitude of the oscillations.

4.2.2 Hopf bifurcation approximation

The Hopf bifurcation of Eq. (4.22) leads to square-wave oscillations that we may further analyze. Specifically, a Period 2 fixed point satisfies the condition $x_2 = x_0$ and from two iterations of Eq. (4.22) we determine the following conditions for x_0 and x_1

$$x_1 = \beta \left[1 + \frac{1}{2} \cos(x_0) \right], \tag{4.30}$$

$$x_0 = \beta \left[1 + \frac{1}{2} \cos(x_1) \right]. \tag{4.31}$$

We wish to find the solution of these equations near the Hopf bifurcation point (x_H, β_H) where x_H is a root of Eq. (4.29) and β_H is obtained from (4.23) with $x = x_H$ or, equivalently, using (4.29)

$$\beta_H = \frac{2}{\sin(x)}. \tag{4.32}$$

Specifically, we seek a perturbation solution of the form

$$x_j = x_H + \varepsilon u_{j1} + \varepsilon^2 u_{j2} + \dots, \tag{4.33}$$

where ε is proportional to the small deviation $\beta - \beta_H$ and is defined by[7]

$$\beta - \beta_H = \varepsilon^2 c \ (c = \pm 1). \tag{4.34}$$

Introducing (4.33) and (4.34) into Eqs. (4.30) and (4.31), and equating to zero the coefficients of each power of ε, leads to a sequence of linear problems to solve. The first three problems are given by

$$O(\varepsilon) : u_{11} = -u_{01}; \tag{4.35}$$

$$O(\varepsilon^2) : u_{12} = -u_{02} - \cot(x_H)\frac{u_{01}^2}{2} + cx, \tag{4.36}$$

$$u_{02} = -u_{12} - \cot(x_H)\frac{u_{11}^2}{2} + cx; \tag{4.37}$$

$$O(\varepsilon^3) : u_{13} = -u_{03} - \cot(x_H)u_{01}u_{02} + \frac{u_{01}^3}{6} - cu_{01}, \tag{4.38}$$

$$u_{03} = -u_{13} - \cot(x_H)u_{11}u_{12} + \frac{u_{11}^3}{6} - cu_{11}. \tag{4.39}$$

---

[7] We have anticipated that $\beta - \beta_H$ is proportional to ε^2 for mathematical clarity.

The solution of Eq. (4.35) is

$$u_{01} = A \quad \text{and} \quad u_{11} = -A, \tag{4.40}$$

where A is an unknown amplitude. The solution of Eqs. (4.36) and (4.37) is

$$u_{02} = B \quad \text{and} \quad u_{12} = -B - \cot(x_H)\frac{A^2}{2} + cx_H, \tag{4.41}$$

where B is a new unknown amplitude. A is still undetermined, so we consider the next problem. Subtracting the two $O(\varepsilon^3)$ equations, we eliminate u_{13} and u_{03} and obtain a condition for A given by

$$\left(\frac{(\cot(x_H))^2}{2} + \frac{1}{3} \right) A^3 - c(\cot(x_H)x_H + 2)A = 0. \tag{4.42}$$

In terms of the original variables the nontrivial solution of Eq. (4.42) is

$$A^2 = \frac{6(\cot(x_H)x_H + 2)}{(3\cot(x_H)^2 + 2)} \left(\frac{\beta - \beta_H}{\beta_H} \right) \geq 0. \tag{4.43}$$

In Figure 4.13, we compare $x_n = x_H \pm \varepsilon A$ with the numerical bifurcation diagram. The amplitude of the analytical solutions emerging at the bifurcation point changes like the square root of $\beta - \beta_H$.

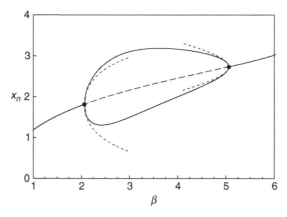

Fig. 4.13 The Hopf bifurcation approximation (dashed lines) is compared to the numerical bifurcation diagram (full lines).

The analysis of the Hopf bifurcation presented here takes advantage of the equation for a map which is valid for sufficiently large delay. But Hopf perturbation theory can be applied to all types of equations provided a change of stability through a pair of purely imaginary eigenvalues is observed. Because the two bifurcating branches are overlapping the unstable steady state they are supercritical and the Hopf theorem guarantees their stability in the vicinity of the bifurcation points.

4.3 From harmonic to pulsating oscillations

In its original formulation, the Hopf bifurcation describes the transition from a steady state to nearly harmonic oscillations. But the bifurcation that we analyzed from the map (4.22) actually corresponds to the emergence of square-wave oscillations. Is this a contradiction of our understanding of a Hopf bifurcation? It isn't. We have to remember that the equation for the map is derived for large delay (i.e. t_d/τ large) and for arbitrary but finite $O(1)$ amplitude. As we numerically solve the DDE (4.21), we observe nearly harmonic oscillations very close to the Hopf bifurcation ($\beta - \beta_H << \tau t_d^{-1}$). But as soon as $\beta - \beta_H = O(\tau t_d^{-1})$, the small amplitude harmonic oscillations continuously change to square-wave oscillations [120]. These small amplitude square-wave oscillations then match the square-wave oscillations of the map as $\beta - \beta_H >> \tau t_d^{-1}$. This dramatic change of the waveform near the Hopf bifurcation point is typical of dynamical systems that exhibit several time scales.

This is also the case for the laser subject to an electrical feedback problem because γ is small. In the limit $\gamma \to 0$, V is faster than D which suggests eliminating V by a quasi-steady state approximation. Setting the right hand side of Eq. (4.3) to zero leads to $V = B + fI$. Substituting this expression into Eq. (4.1) leads to the following two-variable equations for I and D:

$$I' = I\left[D - 1 - a\sin^2(B + fI)\right], \tag{4.44}$$

$$D' = \gamma\left[A - D(1 + I)\right]. \tag{4.45}$$

The bifurcation diagram of the solutions of Eqs. (4.44) and (4.45) is shown in Figure 4.14. It exhibits a sudden increase in amplitude near $B = 0.17$ which marks the transition from oscillating to pulsating oscillations. Figure 4.15 shows the oscillations in intensity I for three different values of B. As for the delay differential equation, Eqs. (4.44) and (4.45) exhibit a transition layer for $B - B_H = O(\gamma)$. Using this scaling, we may further analyze the solution by a different asymptotic analysis as in [121, 122, 123, 124].

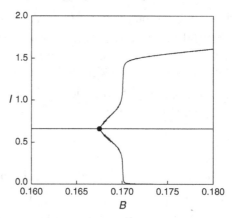

Fig. 4.14 The bifurcation diagram for the laser subject to an electrical feed-back. The Hopf bifurcation at $B = 0.1675$ leads to harmonic oscillations in a small vicinity of the Hopf bifurcation. Near $B = 0.17$, the oscillations increase in amplitude and become strongly pulsating. The bifurcation diagram has been obtained numerically from Eqs. (4.44) and (4.45) with $a = 5.8$, $f = -0.25$, $\gamma = 3.125 \times 10^{-3}$, and $A = 1.66$.

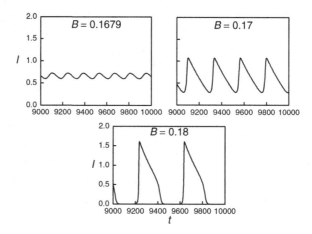

Fig. 4.15 The harmonic oscillations near the Hopf bifurcation gradually change into saw-tooth and then triangular oscillations. Same values of the fixed parameters as in the previous figure.

4.4 Exercises

4.4.1 Saddle-node bifurcation

One effect of the electrical feedback is to generate multiple steady states. Use (4.7) and verify that Eq. (4.14) locates a saddle-node bifurcation.

Solution: at a saddle-node bifurcation or limit point $dB/dV = 0$. Using the expression of $B = B(V)$ for the non-zero intensity steady state, we find

$$\frac{dB}{dV} = \frac{af A \sin(2V)}{(1 + a \sin^2(V))^2} + 1.$$

Therefore $dB/dV = 0$ leads to

$$f = -\frac{(1 + a \sin^2(V))^2}{aA \sin(2V)},$$

which is the expression obtained from the condition of a zero eigenvalue.

4.4.2 Frequency of the limit-cycle oscillations

Investigate the frequency of the high intensity Hopf bifurcation point assuming $\gamma = O(\beta^2)$. If we consider $\omega^2 = C_2$, the leading order problem gives $\omega^2 = aI \sin(2V) f \beta$, which is zero with $V = 0$. To avoid a higher order analysis of the expression $\omega^2 = C_2$, consider the equivalent expression $\omega^2 = C_3/C_1$.

Solution: the high intensity Hopf bifurcation admits the approximation $V \simeq 0$ and $I \simeq A - 1$. From $\omega^2 = C_3/C_1$ with $\gamma = O(\beta^2)$, we obtain

$$\omega \simeq \sqrt{\gamma(A - 1)},$$

which we recognize as the RO frequency.

4.4.3 Low intensity Hopf bifurcation

For low values of f, a single branch of steady states admits two Hopf bifurcation points. Analyze the low intensity Hopf bifurcation point assuming $\gamma = O(\beta^2)$.

Solution: the branch of steady state close to $B = B_c$ is given by

$$I = -\frac{a \sin(2B_c)(B - B_c)}{A + af \sin(2B_c)}$$

and the Hopf bifurcation condition gives

$$I_H = -\frac{\gamma^2 A + \beta^2 \gamma}{\beta^2 af \sin(2B_c)}.$$

4.4.4 Multiple steady states

In [125], the number of possible steady states is investigated. From (4.7) and using the stability diagram in Fig. 4.6, show that a case of tristability is possible for a small range of values of f. Figure 4.16 illustrates the case of tristability.

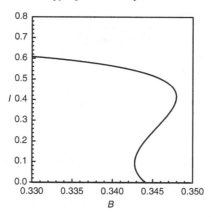

Fig. 4.16 Tristability. The steady states are obtained using Eqs. (4.5)–(4.7) with $A = 1.66$, $a = 5.8$, and $f = -0.42$.

4.4.5 Laser with feedback on the cavity length

Many CO_2 lasers use a feedback loop to stabilize the laser output frequency. A typical set-up maximizes the laser output to lock the laser at the wavelength corresponding to maximum emission. Chen *et al.* [126, 127] investigated the case of a laser subject to feedback on the cavity length. The model equations are similar to Eqs. (4.1)–(4.3) except that the two rate equations for I and D are now supplemented by a third equation which describes the relaxation of cavity length as a function of the feedback. In the original paper, these equations are given by

$$\frac{dI}{dt} = -2kI + \frac{2kAID}{1 + \delta^2},$$

$$\frac{dD}{dt} = \gamma_\| \left[1 - D - \frac{ID}{1 + \delta^2} \right],$$

$$\tau \frac{d\delta}{dt} = -(\delta - \delta_0) - BI.$$

Formulate these equations in the dimensionless form

$$\frac{dI}{ds} = -I + \frac{AID}{1 + \delta^2},$$

$$\frac{dD}{ds} = \gamma \left[1 - D - \frac{ID}{1 + \delta^2} \right],$$

$$(2k\tau)\frac{d\delta}{ds} = -(\delta - \delta_0) - BI,$$

where $\gamma = \gamma_\|/(2k)$. Determine the steady states and their stability properties.

4.4.6 Double Hopf bifurcation and the eye bifurcation diagram

Consider Φ as a parameter and analyze the bifurcation diagram of the Period 2 fixed point solutions near a double Hopf bifurcation point of Eq. (4.22).

Solution: the basic steady state satisfies the condition $x_{n+1} = x_n = x$. From Eq. (4.22), we obtain $x = x(\beta)$ in the implicit form

$$\beta = \frac{x}{1 + \frac{1}{2}\cos(x + \Phi)}. \tag{4.46}$$

The linearized problem for the small perturbation $u_n = x_n - x$ is then given by

$$u_{n+1} = -\beta \frac{1}{2}\sin(x + \Phi)u_n. \tag{4.47}$$

Substituting $u_n = r^{-1}$ into Eq. (4.47), we find the condition

$$1 = \beta \frac{1}{2}\sin(x + \Phi). \tag{4.48}$$

Using (4.46), we may eliminate β in Eq. (4.48) and obtain a single equation for x only

$$1 + \frac{1}{2}\cos(x + \Phi) - x\frac{1}{2}\sin(x + \Phi) = 0, \tag{4.49}$$

which needs to be solved numerically. Eq. (4.49) admits a double root for a particular value of Φ. The condition for a double root is determined by taking the derivative of (4.49) with respect to x. We find

$$\sin(x + \Phi) + x\frac{1}{2}\cos(x + \Phi) = 0. \tag{4.50}$$

Using (4.50), we eliminate $\sin(x + \Phi)$ in (4.49) and obtain $\cos(x + \Phi)$ as

$$\cos(x + \Phi) = -\frac{4}{2 + x^2}. \tag{4.51}$$

Substituting (4.51) into (4.50), we find $\sin(x + \Phi)$ as

$$\sin(x + \Phi) = \frac{2x}{2 + x^2}. \tag{4.52}$$

Using the trigonometric identity $\sin^2(x + \Phi) + \cos^2(x + \Phi) = 1$, we determine an equation for $x = x^*$ only. It admits the simple solution

$$x^* = 12^{1/4} \simeq 1.86. \tag{4.53}$$

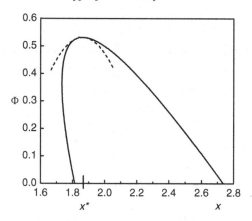

Fig. 4.17 Hopf bifurcation points in terms of the steady state value x. As $\Phi \to \Phi^* \simeq 0.53$, two Hopf bifurcation points come closer together and disappear as $\Phi > \Phi^*$. The parabola is the local approximation and is given by $\Phi = \Phi^* - \frac{2}{x^*}\left(\frac{x^{*2}}{4} + 2\right)(x - x^*)^2$.

From (4.46), (4.51), and (4.52), $\beta = \beta^*$ and $\Phi = \Phi^*$ are given by

$$\beta^* = \frac{2 + x^{*2}}{x^*} \simeq 2.94, \tag{4.54}$$

$$\Phi^* = -x^* + \pi - \arctan\left(\frac{x^*}{2}\right) \simeq 0.53. \tag{4.55}$$

Figure 4.17 shows the Hopf bifurcation line in the Φ vs. x diagram. For each Φ, there exist two Hopf bifurcation points, provided $\Phi < \Phi^*$. The critical point $\Phi = \Phi^*$ corresponds to a double Hopf bifurcation point. From Eqs. (4.30) and (4.31), supplemented with Φ, we may now analyze the Period 2 fixed point solutions in the vicinity of $\beta = \beta^*$ and $\Phi = \Phi^*$.

Part II

Driven laser systems

5

Weakly modulated lasers

Class B lasers naturally exhibit damped relaxation oscillations and, as for any nonlinear oscillator, their responses to a time-periodic modulation of a parameter are rich and varied. The study of forced oscillators itself has a long history. Systematic studies started with Edward Appleton (1922) and Balthasar van der Pol (1927) who showed that the frequency of a triode generator can be entrained by a weak external signal with a slightly different frequency. These studies were of high practical importance because such generators became basic elements of radio communication systems. The next impact on the development of the theory of forced oscillators came from the Russian school when control engineering became an emerging discipline. Alexandr Aleksandrovich Andronov (1901–1952) was a key figure in the development of mathematical techniques for driven oscillators, yet his name, and his contributions to control theory and nonlinear dynamics, are much less well known in the West than they deserve to be [128]. As we shall demonstrate later in this chapter, these analytical techniques are totally appropriate for our laser problems.

Today, lasers and fiber optic cables have replaced the electronic amplifying tubes and cables. Light signals are modulated with the information to be sent into fiber optic cables by lasers. Telephone fiber drivers may be solid state lasers the size of a grain of sand and consume a power of only half a milliwatt. Yet they can send 50 million pulses per second into an attached telephone fiber and encode over 600 simultaneous telephone conversations.

Studies on driven lasers started soon after their discovery when it was found that the laser output dramatically peaks as the modulation frequency comes close to the RO frequency. In the early 1980s, laser physicists became fascinated by new dynamical instabilities appearing if the modulation amplitude is sufficiently high. These instabilities were investigated in laboratories for different lasers by changing either the modulation frequency or the modulation amplitude and acting on different laser parameters (cavity loss, length etc.). Bifurcation diagrams were generated showing different routes to chaos (subharmonic sequence, incommensurate frequencies, chaotic bursts), various forms of pulsations were analyzed (harmonic to spiking), and non-autonomous rate equations were investigated numerically in order to simulate the experimental observations.

Before we examine the case of a modulated class B laser described by two coupled non-autonomous first order differential equations, we concentrate on lasers subject to a modulated magnetic field. As shown in Section 3.2, the response of this laser can be elegantly described by Adler's first order differential equation.

5.1 Driven Adler's equation

5.1.1 Weakly nonlinear and arbitrary modulation

In Section 3.2, we examined the case of a laser with pure loss anisotropies subject to a DC longitudinal magnetic field B_{dc}. We showed that the polarization angle θ with respect to the x axis (see Figure 5.1) satisfies Adler's equation. Cotteverte *et al.* [83] further studied the laser polarization dynamics by considering the additional effect of an AC longitudinal magnetic field $B_{ac} \cos(\omega_{ac} t)$. Using the same approximations as detailed in Section 3.2, the response of the laser is well described by the following periodically driven Adler's equation

$$\frac{d\theta}{dt} = M \sin(2\theta) + \gamma \left[B_{dc} + B_{ac} \cos(\omega_{ac} t) \right], \tag{5.1}$$

where $M < 0$ measures the electrical field rotation due to the difference in transmittivity of the plane plate for the two polarizations. As in [83], we shall consider the case of weak anisotropies (i.e. $|M|/\omega_{dc} \ll 1$, where $\omega_{dc} = 2\gamma B_{dc}$). γ is defined as the saturated Faraday rotation coefficient. Introducing the dimensionless time $s \equiv \omega_{ac} t$ and dividing Eq. (5.1) by $\omega_{dc} \equiv 2\gamma B_{dc}$, Eq. (5.1) takes the form

$$\sigma \frac{d\theta}{ds} = -\varepsilon \sin(2\theta) + \frac{1}{2} + \delta \cos(s), \tag{5.2}$$

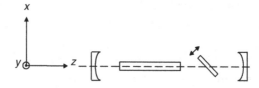

Fig. 5.1 A laser containing only loss anisotropies (controlled by a tilted plate) exhibits only one stable eigenstate polarized along the x axis.

where the dimensionless parameters σ, ε, and δ are all positive and are defined by

$$\sigma = \frac{\omega_{ac}}{\omega_{dc}}, \quad \varepsilon = -\frac{M}{\omega_{dc}}, \quad \text{and} \quad \delta = \frac{\gamma B_{ac}}{\omega_{dc}}. \tag{5.3}$$

δ is our control parameter which is progressively increased from zero by varying B_{ac}. The small parameter ε motivates a perturbation analysis.

The regular perturbation solution (see Exercise 5.3.1) shows the possibility of resonances at $\sigma = n^{-1}$ ($n = 1, 2, 3, \ldots$) as well as oscillations of frequencies 1 and $|\sigma - n^{-1}|$, respectively. We concentrate on these interesting cases by assuming

$$\sigma = n^{-1}(1 + \varepsilon\alpha), \tag{5.4}$$

where $\alpha = O(1)$ is a new parameter which measures the offset from pure resonance. We now proceed as in our previous perturbation problems: we seek a solution in power series of ε, formulate a sequence of linear problems for the unknown functions, and apply solvability conditions. Specifically, we seek a two-time solution of the form $\theta = \theta_0(s, \tau) + \varepsilon\theta_1(s, \tau) + \ldots$ where $\tau = \varepsilon s$. Using the chain rule $d\theta/ds = \theta_s + \varepsilon\theta_\tau$, the first two problems for θ_0 and θ_1 are given by

$$\theta_{0s} = n\left[\frac{1}{2} + \delta\cos(s)\right], \tag{5.5}$$

$$\theta_{1s} = -\alpha\theta_{0s} - \theta_{0\tau} - n\sin(2\theta_0). \tag{5.6}$$

The solution of Eq. (5.5) is

$$\theta_0 = \frac{ns}{2} + \delta n\sin(s) + \Theta(\tau), \tag{5.7}$$

where Θ is an unknown function of the slow time τ. Substituting (5.7) into the right hand side of Eq. (5.6), we obtain

$$\theta_{1s} = RHS \equiv -\alpha n\left[\frac{1}{2} + \delta\cos(s)\right] - \Theta_\tau - n\sin\left[ns + 2\delta n\sin(s) + 2\Theta\right]. \tag{5.8}$$

The homogeneous problem $\theta_{1s} = 0$ admits a constant solution. Therefore, the right hand side needs to satisfy the solvability condition

$$\frac{1}{2\pi}\int_{-\pi}^{\pi} RHS(s, \tau)ds = 0. \tag{5.9}$$

This condition leads to a new Adler's equation

$$\frac{d\Theta}{d\tau} = -\alpha\frac{n}{2} + n\sin(2\Theta)J_n(2\delta n), \tag{5.10}$$

where the coefficient of $\sin(2\Theta)$ is a periodic function of the control parameter δ. We have used the integral expressions for the Bessel functions, namely,

$$\int_{-\pi}^{\pi} \sin(ns - z\sin(s))ds = 0,$$

$$\int_{-\pi}^{\pi} \cos(ns - z\sin(s))ds = 2\pi J_n(z).$$

The critical values of α that limit the locking domain satisfy the condition $\sin(2\Theta) = \pm 1$, and using Eq. (5.10) with $\Theta' = 0$, we obtain

$$\alpha = \pm 2J_n(2\delta n). \tag{5.11}$$

The lines defined by (5.11) delimit the region where a periodic solution of frequency ω_{ac} is possible (inside the "Christmas tree" in Figure 5.2). Cotteverte *et al.* [83] obtained excellent quantitative agreements when they compared experimental and theoretical diagrams in terms B_{ac} vs. ω_{ac}. They also observed higher order resonances ($n = 2$ and 3) but these domains of subharmonic periodic solutions become progressively smaller as n increases, as we may expect from (5.11) (i.e. $J_n(2\delta n) \sim n^{-1/2}$ as $n \to \infty$).

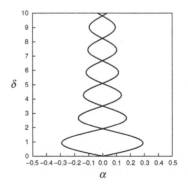

Fig. 5.2 Analytical locking region near the first resonance ($n = 1$). δ is represented as a function of $\alpha \equiv \varepsilon^{-1}(\sigma - 1)$ using (5.11). Inside the "Christmas tree", a locked periodic regime is possible. Outside, the oscillations are quasi-periodic.

5.1.2 *Strongly nonlinear and weak modulation*

Cotteverte *et al.* [83] also investigated other limiting cases but always kept a weak nonlinearity ($|M|/\omega_{dc} \ll 1$). What happens if we consider the case of a strongly nonlinear Adler's equation? Our previous analysis of (5.1) took advantage of the weak nonlinearity in order to obtain an analytical solution of the nonlinear problem. Here we consider M to be of the same order of magnitude as γB_{dc} and we need to find a different route to an analytical solution. Specifically, we shall consider the case of a weak modulation, i.e. $B_{ac} \ll B_{dc} = O(M\gamma^{-1})$. This motivates a different dimensionless formulation of the original problem. Introducing the new variables

$$s = 2\gamma B_{dc}t \quad \text{and} \quad \phi = 2\theta, \tag{5.12}$$

Eq. (5.1) can be rewritten as

$$\frac{d\phi}{ds} = 1 - a\sin(\phi) + \varepsilon\sin(\omega s), \tag{5.13}$$

where the new parameters are defined by

$$a = \frac{M}{\gamma B_{dc}}, \quad \varepsilon = \frac{B_{ac}}{B_{dc}}, \quad \text{and} \quad \omega = \frac{\omega_{ac}}{2\gamma B_{dc}}. \tag{5.14}$$

The numerical bifurcation diagram is shown in Figure 5.3 and exhibits time-periodic locking near $a = 0.5$ where Adler's frequency

$$\omega_0 \equiv \sqrt{1 - a^2} \tag{5.15}$$

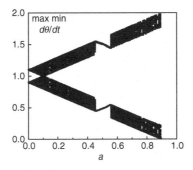

Fig. 5.3 Numerical bifurcation diagram of the modulated Adler's equation (5.13). The extrema of $d\theta/dt$ are shown as a function of a ($0 \le a < 1$). The values of the parameters are $\omega = 0.866$ and $\varepsilon = 0.1$. Locking occurs near $a = 0.5$ where Adler's frequency $\omega_0 = \sqrt{1 - a^2}$ is close to ω. Outside the locking region, the oscillations are quasi-periodic.

is close to ω. Regular perturbation methods fail when resonance occurs and we propose a singular perturbation technique to obtain expressions which are accurate even when resonance appears [129].

A change of variable

We wish to capture this resonance phenomenon by a perturbation analysis valid for ε small. To this end, we shall benefit from the fact that Adler's equation admits an exact solution (see 3.7.2). We therefore introduce the function $\Phi(s)$ defined as the exact solution of Adler's equation

$$\frac{d\Phi}{ds} = 1 - a\sin(\Phi), \quad \Phi(0) = -\pi/2 \tag{5.16}$$

and $a > 1$. We next change ϕ in (5.13) into ψ by means of the relation

$$\phi = \Phi(\psi), \tag{5.17}$$

where $\psi(s, \varepsilon)$ is an unknown function of s to be found. Introducing (5.17) into Eq. (5.13), we have

$$\frac{d\phi}{ds} = \frac{d\Phi}{d\psi}\frac{d\psi}{ds} = 1 - a\sin(\Phi(\psi)) + \varepsilon\sin(\omega s). \tag{5.18}$$

Using (5.16), we obtain an equation for $\psi(s)$ of the form

$$\frac{d\psi}{ds} = 1 + \frac{\varepsilon\sin(\omega s)}{1 - a\sin(\Phi(\psi))}. \tag{5.19}$$

This equation can be rewritten as (see exercise 5.3.2)

$$\frac{d\psi}{ds} = 1 + \frac{\varepsilon}{\omega_0^2}\left[\sin(\omega s) - \frac{a}{2}\sin(\omega s - \omega_0\psi) - \frac{a}{2}\sin(\omega s + \omega_0\psi)\right]. \tag{5.20}$$

Eq. (5.20) is exact and no approximations have been used in reducing (5.13) to (5.20). Although (5.20) looks more complicated than (5.13), the linearity of Eq. (5.20) at $\varepsilon = 0$ makes it easy to work with. Moreover, possible cases of resonance clearly appear in the terms multiplying ε: because $\psi = s$, in first approximation, the three sine functions in the right hand side of Eq. (5.20) exhibit the modulation frequencies ω and $\omega \pm \omega_0$.

Two-time solution

We now propose to solve Eq. (5.20) by a two-time perturbation method. The two times are suggested by the right hand side of Eq. (5.20), motivating the necessity of a slow time $\tau \equiv \varepsilon s$ to take into account the long time behavior of the solution. Specifically, we seek a solution of the form

$$\psi = \psi_0(s, \tau) + \varepsilon \psi_1(s, \tau) + \ldots \tag{5.21}$$

and require that the corrections ψ_1, ψ_2 are bounded functions for s large. Using the chain rule $d\psi/ds = \psi_s + \varepsilon \psi_\tau$ and substituting the expansion for ψ yields the following sequence of problems

$$\psi_{0s} = 1, \tag{5.22}$$

$$\psi_{1s} = -\psi_{0\tau} + \frac{1}{\omega_0^2} \left[\sin(\omega s) - \frac{a}{2} \sin(\omega s - \omega_0 \psi_0) - \frac{a}{2} \sin(\omega s + \omega_0 \psi_0) \right]. \tag{5.23}$$

The solution of (5.22) is $\psi_0 = s + \alpha(\tau)$, where the function α is undetermined at this stage of the perturbation analysis. Substituting the expression for ψ_0 into the right hand side of Eq. (5.23), we obtain the solution

$$\psi_1 = -s\frac{d\alpha}{d\tau} - \frac{1}{\omega_0^2} \left[\frac{\cos(\omega s)}{\omega} - \frac{a}{2} \frac{\cos((\omega - \omega_0)s - \omega_0 \alpha)}{\omega - \omega_0} \right.$$
$$\left. - \frac{a}{2} \frac{\cos((\omega + \omega_0)s + \omega_0 \alpha)}{\omega + \omega_0} \right] + C, \tag{5.24}$$

where C is an integration constant. If ω is not equal to ω_0, $-\omega_0$, or 0 (i.e. if resonance does not occur at first order), then the requirement that ψ_1 must never become unreasonably large implies that $d\alpha/d\tau = 0$. α is thus equal to a constant α_0. Equivalently, the solvability condition for Eq. (5.23) requires that the average of the right hand side is zero, meaning $d\alpha/d\tau = 0$. On the other hand, cases of resonance need to be treated carefully.

Resonance

We wish to investigate the case where ω is close to ω_0. Using (5.24), we note that the expansion (5.21) becomes nonuniform if $|\omega - \omega_0| = O(\varepsilon)$. This motivates the expansion of the modulation frequency ω as

$$\omega = \omega_0 + \varepsilon\sigma + \ldots \tag{5.25}$$

Substituting (5.21) and (5.25) into (5.20) leads to Eq. (5.22) for ψ_0 and the following equation for ψ_1

$$\psi_{1s} = -\psi_{0\tau} + \frac{1}{\omega_0^2}\left[\sin(\omega_0 s) - \frac{a}{2}\sin(\sigma\tau - \omega_0\alpha) - \frac{a}{2}\sin(2\omega_0 s + \omega_0\alpha)\right].$$

(5.26)

After substituting $\psi_0 = s + \alpha(\tau)$ into Eq. (5.26), we apply the solvability condition requiring that the average of the right hand side is zero. This then leads to an equation for α given by

$$\frac{d\alpha}{d\tau} = -\frac{a}{2\omega_0^2}\sin(\sigma\tau - \omega_0\alpha).$$

(5.27)

Eq. (5.27) is equivalent to a new Adler's equation. Introducing $\beta = \sigma\tau - \omega_0\alpha$ into Eq. (5.27) gives

$$\frac{d\beta}{d\tau} = -\sigma + \frac{a}{2\omega_0}\sin(\beta),$$

(5.28)

which implies locking if

$$\left|\frac{2\sigma\omega_0}{a}\right| \leq 1.$$

(5.29)

We now recall from (5.25) that $\sigma = \varepsilon^{-1}(\omega - \omega_0)$ and that ω_0 is Adler's frequency (5.15). Eq. (5.29) then is of the form

$$\left|\frac{2(\omega - \omega_0)\omega_0}{\varepsilon a}\right| \leq 1$$

(5.30)

requiring that ω remains sufficiently close to ω_0 for locking. If now ω is fixed and a is the control parameter as in Figure 5.3, we expand a as

$$a = a_0 + \varepsilon a_1 + \ldots,$$

(5.31)

where a_0 satisfies the resonance condition $\omega = \omega_0(a)$, i.e.

$$a_0 = \sqrt{1 - \omega^2}.$$

(5.32)

Inserting (5.31) into (5.30) then leads to the simple condition

$$|a_1| \leq \frac{1}{2}.$$

(5.33)

For the bifurcation diagram in Figure 5.3, $\omega = 0.866$ and $\varepsilon = 0.1$. From (5.32), we determine $a_0 = 0.5$ which then implies, using (5.33), that the boundaries of the locking domain are $a = 0.5 \pm 0.05$. These values are in excellent quantitative agreement with the locking domain shown in Figure 5.3. As a first order solution has been found, we obtain that the locking range ($\sim a_1 \varepsilon$) increases linearly with the forcing amplitude ($\sim \varepsilon$).

5.2 Weakly modulated class B lasers

5.2.1 Nearly conservative oscillations

The limit γ small of the class B laser linearized rate equations (1.19) leads to a solution of the form of weakly damped sinusoidal oscillations. The analysis is valid only for small deviations from the equilibrium state. Here we explore the same limit but allow for large deviations with respect to the equilibrium state. By contrast to Chapter 1 where we analyzed the linearized equations, we now consider the full nonlinear laser equations (1.7) and (1.8).

To this end, we introduce the initial conditions

$$I(0) = I_i \quad \text{and} \quad D(0) = D_i \tag{5.34}$$

and propose to determine an asymptotic solution of Eqs. (1.7) and (1.8) valid in the limit γ small. Setting $\gamma = 0$ into Eqs. (1.7) and (1.8) leads to the following problem for I and D

$$\frac{dI}{dt} = I(D_i - 1) \quad \text{and} \quad \frac{dD}{dt} = 0. \tag{5.35}$$

The second equation implies $D = D_i$. Solving then the first equation gives $I = I_i \exp((D_i - 1)t)$: the intensity either decays to zero or grows exponentially depending on the value of D_i. Both cases contradict our earlier observation of a stable non-zero intensity steady state. The limit γ small is a singular limit because the $\gamma = 0$ problem does not lead to a physical solution.

Singular perturbation problems are difficult problems because there exist no systematic techniques to resolve them. The simplest way to resolve our γ small difficulty is to reformulate the laser equations in a form where the small parameter γ does not multiply any right hand side. This can be realized by introducing the new time

$$s \equiv \omega_R t, \tag{5.36}$$

where ω_R is the RO frequency (Eq. (1.29)) and the new dependent variables x and y are defined by (see Exercise 5.3.3)

$$x \equiv \frac{D-1}{\omega_R} \quad \text{and} \quad y \equiv \frac{I-(A-1)}{A-1}. \tag{5.37}$$

The variables x and y are deviations of D and I from their non-zero intensity steady state values. The coefficient ω_R dividing $D-1$ in (5.37) is required when we insert (5.36) into Eq. (1.7) and try to balance left and right hand sides. With (5.36) and (5.37), Eqs. (1.7) and (1.8) become

$$\frac{dx}{ds} = -y - \varepsilon^2 x[1 + (A-1)(1+y)], \tag{5.38}$$

$$\frac{dy}{ds} = (1+y)x, \tag{5.39}$$

where ε^2 is a small parameter defined by

$$\varepsilon^2 \equiv \sqrt{\frac{\gamma}{A-1}} << 1. \tag{5.40}$$

Note that γ now appears in ε^2 and that the limit γ small is no longer singular for (5.38) and (5.39). Indeed we can find an implicit form of a bounded solution as we shall see now.

Setting $\varepsilon = 0$ in (5.38) leads to

$$x' = -y \quad \text{and} \quad y' = (1+y)x, \tag{5.41}$$

which admit bounded periodic solutions. Indeed, (5.41) is conservative and admits a one parameter family of periodic solutions. This can be demonstrated by determining the first integral

$$C = \frac{x^2}{2} + y - \ln(1+y), \tag{5.42}$$

where C is the constant of integration. It is called the energy of the conservative system (5.41) and will be used as we examine the possibility of subharmonic solutions (see Section 6.1.2). The expression (5.42) is obtained by dividing the equations for x and y in (5.41) and by noting that the resulting problem for $y = y(x)$ is separable. From (5.42), we then find

$$x = \pm\sqrt{2(C - y + \ln(1+y))}, \tag{5.43}$$

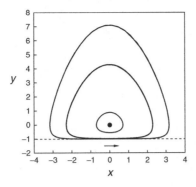

Fig. 5.4 Laser nonlinear oscillations. Each orbit is a periodic solution of the laser conservative equations (5.41). Near the center, the frequency of the oscillations is close to the relaxation oscillation ω_R. As the amplitude increases, the frequency becomes smaller. The dotted line is the invariant line $y = -1$.

which allows us to draw closed orbits in the phase plane (x, y) for all values of C $(0 < C < \infty)$. See Figure 5.4. Each orbit corresponds to emission of a laser spike with a given amplitude. The horizontal part of the trajectory exploring the vicinity of $y = -1$ is associated with the long period of very small intensity between successive spikes.

Key properties of the laser spiking such as the interpulse period and the pulse intensity may be related to the energy. Introducing $y = -1$ into the first equation in (5.41) and integrating from $x = -\sqrt{2C}$ to $x = \sqrt{2C}$ (the two extrema of x according to (5.43)), the interpulse period is $P_{int} = 2\sqrt{2C}$. On the other hand the maximum value of the intensity y occurs at $x = 0$ and from (5.42), we find $y_M \simeq C$ for C large. The interpulse period and pulse intensity are therefore correlated and this observation was already noted for the irregular pulses emitted by a ruby laser: the larger the temporal separation between one pulse and its predecessor, the larger its peak intensity (see Figure 4.1 right). As we simply demonstrated, this relation follows the scaling law $y_M = P_{int}^2/8$, in first approximation.

Because ε is small, the laser equations (5.38) and (5.39) form a nearly conservative system of equations. This fundamental property of the laser rate equations was discovered in early studies of the laser equations, which are interesting to review. In 1966, Basov *et al.* [130] (see also Morozov [131]) had already found that a large number of laser problems can be formulated as

$$u'' - u^{-1}u'^2 + u(u - 1) + \varepsilon^2 f(u, u') = 0. \tag{5.44}$$

Eq. (5.44) can be obtained from Eqs. (5.38) and (5.39) by introducing $u \equiv 1 + y$ and by formulating a second order differential equation for u only. In addition, Basov *et al.* [130] showed that the reduced problem ($\varepsilon = 0$) is conservative, and later, Morozov [131] used an elementary averaging theory to determine the long time effect of the function $f(u, u')$. In 1985, the same idea was explored, independently, by Oppo and Politi [132]. Specifically, they reformulated Eqs. (1.7) and (1.8) as a second order differential equation describing the motion of a particle in a Toda potential. This equation is given by

$$u'' + V'(u) + \varepsilon^2 h(u, u') = 0, \tag{5.45}$$

where the single-well potential $V(u)$ and the nonlinear dissipation term $h(u, u')$ are defined by

$$V(u) \equiv \exp(u) - u, \tag{5.46}$$

$$h(u, u') \equiv u' \left[1 + (A - 1) \exp(u) \right]. \tag{5.47}$$

Eq. (5.45) can be obtained from Eqs. (5.38) and (5.39) by introducing the new variable $u \equiv \ln(1 + y)$ and by formulating a second order differential equation for u only. Again, an averaging method is proposed in [132] in order to describe the long time behavior of the laser oscillations. Eqs. (5.38) and (5.39) offer the advantage that x and y are defined as deviations of D and I from their steady state values, which allows an immediate physical interpretation. On the other hand, Eq. (5.45) allows the application of modern averaging techniques developed for a large class of second order and nearly Hamiltonian systems (see [16, 133, 134]).

5.2.2 Pump and loss modulations

A periodic modulation of a laser can be realized by modulating either the pump or the cavity losses. We note from the laser equations that modulating the cavity losses changes the evolution equation of the intensity while modulating the pump modifies the evolution equation of the population inversion. Since the population inversion for all class B lasers is much slower than the intensity of the field, we expect different responses depending on the modulated parameter. As we shall demonstrate below, this difference can be highlighted by introducing the relaxation oscillation frequency into the rate equations.

The case of pump modulation is straightforward. Controlling the pump implies that

$$A_m = A(1 + mF(t)), \tag{5.48}$$

where A and m represent the pump mean value and the modulation amplitude, respectively. The function $F(t)$ is $2\pi/\omega$-periodic and has zero mean but can be sinusoidal, pulsating, or square wave. Introducing (5.48), (5.36), and (5.37) into Eqs. (1.7) and (1.8) leads to the following modification of Eq. (5.38)

$$\frac{dx}{ds} = \delta f(s) - y - \varepsilon^2 x[1 + (A - 1)(1 + y)] \tag{5.49}$$

while Eq. (5.39) remains unchanged. The function $f(s) = F(s)$ is $2\pi/\sigma$-periodic and δ and σ are defined by

$$\boxed{\delta \equiv \frac{A_m}{A - 1}} \quad \text{and} \quad \sigma \equiv \frac{\omega}{\omega_R}. \tag{5.50}$$

In the case of loss modulation, we need to go back to the original rate equations (1.4) and (1.5) with a time-dependent cavity decay rate: $\overline{T}_c^{-1}(T) = T_c^{-1}(1 + mH(\Omega T))$. After introducing the dimensionless time $t \equiv T/T_c$, Eq. (1.7) for the intensity becomes

$$\frac{dI}{dt} = I(D - 1 - mG(t)), \tag{5.51}$$

where $G(t) = H(t)$ is $2\pi/\omega$-periodic and $\omega \equiv \Omega T_c$. We next insert (5.36) and (5.37) into Eqs. (5.51) and (1.8) and find

$$\frac{dy}{ds} = (1 + y)(x - \delta g(s)), \tag{5.52}$$

where $g(s) = G(s)$ is $2\pi/\sigma$-periodic and δ and σ are defined by

$$\boxed{\delta \equiv \frac{m}{\omega_R}} \quad \text{and} \quad \sigma \equiv \frac{\omega}{\omega_R}. \tag{5.53}$$

Eq. (5.38) remains unchanged. Note the presence of ω_R in the normalized modulation amplitude δ for the case of loss modulation.

We may now compare the effects of these two modulations. Since $\omega_R \sim \sqrt{\gamma} \sim 10^{-2}$ is small for most class B lasers, assuming $\delta = O(1)$ requires that m is $O(1)$ in the case of pump modulation (see (5.50)) while m is $O(\omega_R)$ small in the case of loss modulation (see (5.53)). In other words, a significant control of the modulation amplitude ($\delta = O(1)$) is more easily achieved via loss modulation. But we also need to take into account technical factors. Pump modulation is easily achieved in diode pumped Nd^{3+}:YAG/YVO$_4$ lasers by changing the pump diode current, while this technique is difficult for CO_2 lasers which are pumped by a DC or RF discharge. On the other hand, loss modulation is easily achieved by inserting an acousto- or electro-optic modulator inside the cavity of a CO_2 laser.

Since the availability and performance of optical components are crucial for experiments, practical factors will generally determine the most appropriate modulation technique.

5.2.3 Weak pump modulation near the fundamental resonance and hysteresis

In this section, we concentrate on the primary resonance ($\sigma \sim 1$) and show that the typical response of the laser is bistable even if the modulation amplitude is small. This is the first case where we need to develop a nonlinear theory to describe the phenomenon. Specifically, we consider (5.48) with m small but $F(t)$ arbitrary and apply a singular perturbation method. A singular perturbation problem arises when the regular perturbation method is no longer uniformly valid. The technique might break down, for example, either for large t values (as for our laser problem; see Exercise 5.3.4) or in the presence of boundary or interior layers. Another difficulty comes from the fact that we wish to consider values of σ close to 1 and δ small which imply that we need to scale these two parameters with respect to our small parameter ε. The first difficulty will be resolved by introducing the natural strained coordinate $S = \sigma(\varepsilon)t$. The second difficulty will be resolved by introducing general expansions of $\sigma - 1$ and δ in power series of ε and by systematically applying solvability conditions.

Although we could directly analyze Eqs. (5.49) and (5.39), the algebra is simpler if we have a second order differential equation rather than a system of two first order equations. To this end, we use the same procedure as for the formulation of Eq. (5.45), i.e. we eliminate one of the dependent variables by introducing $u = \ln(1 + y)$. We then obtain $du/ds = x$ from Eq. (5.39) and, using Eq. (5.49), we determine an equation for u only, given by

$$u'' = \delta f(s) + 1 - \exp(u) - \varepsilon^2 u' \left[1 + (A - 1)\exp(u)\right], \qquad (5.54)$$

which is Eq. (5.45) with an extra modulation term, and prime means differentiation with respect to time s.

Analysis of the successive orders

If $\delta = \varepsilon = 0$, the linearized problem for $u = 0$ is $u'' + u = 0$, which admits a 2π-periodic solution. This motivates looking for a small amplitude, $2\pi/\sigma$-periodic solution of Eq. (5.54) for δ small and σ close to 1. Specifically, we seek a 2π-periodic solution of Eq. (5.54) of the form

$$u(S, \varepsilon) = \varepsilon u_1(S) + \varepsilon^2 u_2(S) + \dots, \qquad (5.55)$$

where $S \equiv \sigma s$. With this definition of S, the frequency σ will appear as a parameter in the equation for u and no longer in the modulation function $f(s)$. The two parameters δ and σ need to be expanded in power series of ε. At this stage of our analysis, we don't know their leading scaling with respect to ε and expand them in the most general way as

$$\delta(\varepsilon) = \varepsilon\delta_1 + \varepsilon^2\delta_2 + \dots \quad \text{and} \quad \sigma(\varepsilon) = 1 + \varepsilon\sigma_1 + \varepsilon^2\sigma_2 + \dots, \quad (5.56)$$

where the coefficients $\delta_1, \delta_2, \dots, \sigma_1, \sigma_2, \dots$ need to be determined by applying solvability conditions. Introducing $S \equiv \sigma s$ and (5.56) into Eq. (5.54) and equating the coefficients of each power of ε to zero leads to a sequence of linear problems for $u_1, u_2, \dots$ that we investigate now.

The leading order problem is $O(\varepsilon)$ and is given by

$$u_1'' + u_1 = RHS \equiv \delta_1 f(S), \quad (5.57)$$

where prime means differentiation with respect to time S and *RHS* means the right hand side. We note that the homogeneous problem admits the two solutions $\exp(\pm iS)$ (or equivalently, $\sin(S)$ and $\cos(S)$). In order to have a bounded periodic solution, the right hand side *RHS* needs to satisfy a solvability condition which is given by

$$\int_0^{2\pi} RHS \exp(\pm iS)dS = 0. \quad (5.58)$$

Assuming that

$$I \equiv \frac{1}{2\pi} \int_0^{2\pi} f(S)\exp(iS)dS \neq 0, \quad (5.59)$$

i.e. the main Fourier component of the modulation is different from zero as can reasonably be expected for periodic modulation, (5.58) requires that

$$\boxed{\delta_1 = 0.} \quad (5.60)$$

Equation (5.57) then admits the periodic solution

$$u_1 = \alpha \exp(iS) + c.c., \quad (5.61)$$

where *c.c.* means complex conjugate and α is an unknown complex amplitude. In order to determine α, we need to investigate the higher order problems.

The next problem is $O(\varepsilon^2)$ and, with (5.60), it is given by

$$u_2'' + u_2 = RHS \equiv -\frac{1}{2}u_1^2 + \delta_2 f(S) - 2\sigma_1 u_1''. \tag{5.62}$$

We note that $u_1^2 = \alpha^2 \exp(2iS) + c.c. + 2\alpha\bar{\alpha}$ only exhibits subharmonic functions. This means that the condition (5.58) reduces to the following linear equation for α

$$\delta_2 I + 2\sigma_1\alpha = 0. \tag{5.63}$$

Solving for α, we find that α is unbounded if $\sigma_1 = 0$. Thus the only physical solution is

$$\boxed{\delta_2 = \sigma_1 = 0} \tag{5.64}$$

and α arbitrary. Since α is still unknown, we examine the next problem for u_3. To this end, we first determine the solution of Eq. (5.62) with (5.64). It has the form

$$u_2 = (\alpha_2 \exp(iS) + c.c.) + \left(\frac{\alpha^2}{6}\exp(2iS) + c.c.\right) - \alpha\bar{\alpha}, \tag{5.65}$$

where α_2 is a new unknown complex amplitude so that we must solve for the next order problem.

The problem for u_3 is $O(\varepsilon^3)$ and is given by

$$u_3'' + u_3 = RHS \equiv \delta_3 f(s) - u_1 u_2 - \frac{1}{6}u_1^3 - Au_1' - 2\sigma_2 u_1''. \tag{5.66}$$

Applying (5.58) now leads to the following nonlinear equation for α

$$\delta_3 I + \frac{1}{3}\alpha^2\bar{\alpha} - i\alpha A + 2\sigma_2\alpha = 0. \tag{5.67}$$

We analyze its solution by introducing the decomposition $\alpha = R\exp(i\theta)$ and assuming $I = J\exp(i\phi)$. We obtain the following equations for R and θ.

$$\delta_3 J \cos(\theta - \phi) + \frac{1}{3}R^3 + 2\sigma_2 R = 0,$$

$$-\delta_3 J \sin(\theta - \phi) - AR = 0. \tag{5.68}$$

Eliminating the trigonometric functions, we obtain

$$(\delta_3 J)^2 = R^2 \left[A^2 + \left(\frac{1}{3}R^2 + 2\sigma_2\right)^2 \right] \tag{5.69}$$

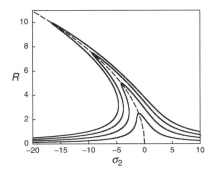

Fig. 5.5 Hysteresis of periodic solutions. $A = 2$, $f(S) = \cos(S)$, implying $J = 1/2$. The branches of solutions are represented for $\delta_3 = 10, 20, 30$, and 40. As δ_3 increases, the middle and upper branches increase in amplitude and come closer to the parabola $\sigma_2 = -R^2/6$ (broken line).

or, equivalently, the implicit solution $\sigma_2 = \sigma_\pm(R)$ where

$$\sigma_\pm(R) \equiv -\frac{1}{6}R^2 \pm \frac{1}{2R}\sqrt{(\delta_3 J)^2 - R^2 A^2} \qquad (5.70)$$

$(R \leq \delta_3 J/A)$. Knowing R, we may find θ using one of the two equations in (5.68).

The problem is now solved. In summary, the solution is nearly harmonic in time and is given by $u \simeq \varepsilon R \exp(i(S + \theta)) + c.c.$, where R satisfies Eq. (5.70). R is implicitly given vs. δ_3 and σ_2, which measure the amplitude and the detuning of the modulation through the scalings $\delta = \delta_3 \varepsilon^3$ and $\sigma = 1 + \sigma_2 \varepsilon^2 \left(\varepsilon = \left(\frac{\gamma}{A-1}\right)^{1/4}\right)$. Since in many lasers γ is typically $10^{-2} - 10^{-3}$, the resonance is very sharp ($u \sim \gamma^{1/4}$).

Bifurcations

We next examine the bifurcation properties of R in order to obtain quantitative information on the main properties of the resonance curves. The solution $R = R(\sigma_2)$ may be multiple if the modulation amplitude is sufficiently high. See Figure 5.5. We determine the location of the limit points by analyzing the condition $d\sigma_2/dR = 0$. This condition is never satisfied if $\sigma_2 = \sigma_+$. On the other hand, if $\sigma_2 = \sigma_-$, we obtain a quadratic equation for $(\delta_3 J)^2$, which is given by

$$(\delta_3 J)^4 - (\delta_3 J)^2 \frac{4R^6}{9} + A^2\frac{4R^8}{9} = 0. \qquad (5.71)$$

As we progressively increase δ_3, two real solutions for R appear at a critical point $\delta_3 = \delta_{3c}$. At this point there is a vertical tangency in the curve $R(\delta)$, and then the critical point satisfies the condition

$$d\delta_3/dR^2 = 0. \qquad (5.72)$$

Differentiating (5.71) with respect to R^2 and using (5.72), we obtain a second condition relating R^2 and δ_3. Then eliminating R^2 using (5.71), we find $\delta_3 = \delta_{3c}$ as

$$\delta_{3c} \equiv \frac{2}{J}\sqrt{\frac{2A^3}{\sqrt{3}}} \simeq \frac{2.15}{J}A^{3/2}. \tag{5.73}$$

The critical modulation amplitude δ_{3c} is proportional to $A^{3/2}$, which means that the lowest value of δ_{3c} is at the laser threshold $A = 1$. However, the laser should not operate too close to its threshold since noise may eliminate the hysteresis phenomenon. We also note that δ_{3c} is inversely proportional to J which depends on the modulation function. The lowest value of δ_{3c} means a high J. In the case of a sinusoidal modulation of the form $f(S) = \cos(S)$, we find $J = 1/2$. In the case of a bang-bang modulation $f(S) = \pm 1$, we obtain $J = 4/2\pi > 1/2$. The bang-bang modulation thus leads to a smaller δ_{3c}.

If $A = 2$ and $J = 1/2$, we find from (5.73) that $\delta_{3c} \simeq 12.16$, which is numerically relatively large. This suggests determining the large δ_3 (or equivalently, the large R) approximations of the two roots of Eq. (5.71). We obtain $R_1 \simeq (3\delta_3 J/2)^{1/3}$ and $R_2 \simeq \delta_3 J/A$. Near $R = R_1$, the laser experiences a jump-up transition while near $R = R_2$, we expect a jump-down transition. Inserting these expressions into $\sigma_-(R)$, given by (5.70), we obtain

$$\sigma_{2up} \simeq -\frac{1}{2}\left(\frac{3\delta_3 J}{2}\right)^{2/3} \quad \text{and} \quad \sigma_{2down} \simeq -\frac{1}{6}\left(\frac{\delta_3 J}{A}\right)^2. \tag{5.74}$$

In terms of the original variables $\sigma - 1 = \varepsilon^2 \sigma_2$ and $\delta = \varepsilon^3 \delta_3$, and for $J = 1/2$, the two limit points (5.74) are

$$\sigma_{up} - 1 \simeq 0.41\delta^{2/3} \quad \text{and} \quad \sigma_{down} - 1 \simeq -0.04\left(\frac{\delta}{\varepsilon A}\right)^2. \tag{5.75}$$

Equation (5.75) tells us that the σ_{down} limit point changes much faster than the σ_{up} limit point as δ increases. This is clearly seen experimentally in Figure 5.6: σ_{down} decreases significantly as the modulation amplitude increases while the decrease of σ_{up} is moderate. In Figure 5.7, the experimental estimations of the two limit-point frequencies are compared to the theoretical predictions. The experimental values for σ_{up} compare well with the theoretical prediction but the experimental values of σ_{down} deviate from the theoretical line as δ increases and saturate at a fixed value. This saturation is the result of noise that is always present in the experiment. As δ increases, the bifurcation diagram near the σ_{down} limit point becomes very sharp (see Figure 5.5) and small perturbations of the upper stable branch may lead to jump-down transitions before the limit point is reached. The saturation level is

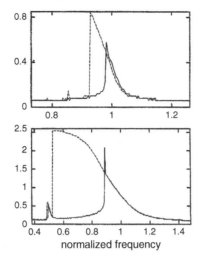

Fig. 5.6 Experimental hysteresis cycle obtained by slowly changing the modulation frequency forward (full line) and then backward (broken line). The laser is a single mode Nd^{3+}:YAG laser subject to a periodically modulated pump. The figure represents the envelope of the maxima of the rapid oscillations as a function of the normalized frequency $\sigma = \omega/\omega_R$ near the onset of hysteresis (top), and far away from the onset of hysteresis (bottom). The additional smaller jump near $\sigma = 1/2$, seen when increasing σ, signals another resonance. Reprinted Figure 1 with permission from Celet et al. [135]. Copyright 1998 by the American Physical Society.

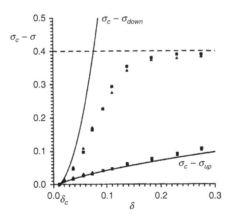

Fig. 5.7 Size of the bistable domain. The figure represents the limit point frequencies as a function of the modulation amplitude. The full lines are the exact numerical solutions of the limit-point equations with $A = 2$, $J = 1/2$, $\varepsilon = 0.11$. The squares and triangles are the experimental points. The critical point $(\delta, \sigma) = (\delta_c, \sigma_c)$ is defined as the point where hysteresis appears. The dashed line indicates the saturation level set by noise. Reprinted Figure 2 with permission from Celet et al. [135]. Copyright 1998 by the American Physical Society.

then a function of the level of noise. In [135], this was demonstrated quantitatively by adding noise to the pump and observing a lower saturation.

5.2.4 Subharmonic modulation and period doubling bifurcation

In the previous section, we considered the case of a near resonant modulation (frequency $\sigma \simeq 1$) and analyzed the conditions for hysteresis with an arbitrary modulation function $f(s)$. In this section, we analyze the period doubling bifurcation generated by a subharmonic modulation of the form $f(s) = \cos(\sigma s)$ where $\sigma \sim 2$. To this end, we use the same perturbation technique as in the previous section but anticipate a bifurcation from a $2\pi\sigma^{-1}$ periodic solution to a $4\pi\sigma^{-1}$ periodic solution.

As for the case $\sigma \simeq 1$, we first expand the parameters δ and σ as

$$\delta(\varepsilon) = \varepsilon\delta_1 + \varepsilon^2\delta_2 + \dots \quad \text{and} \quad \sigma(\varepsilon) = 2 + \varepsilon\sigma_1 + \varepsilon^2\sigma_2 + \dots \tag{5.76}$$

and *seek a 2π-periodic solution* of Eq. (5.54) of the form (5.55). Introducing $S = \sigma s$, (5.76), and (5.55) into Eq. (5.54) and equating the coefficients of each power of ε to zero leads to a sequence of linear problems for $u_1, u_2, \dots$ We then proceed as usual by solving the successive equations for $u_1, u_2, \dots$ and applying solvability conditions.

Analysis of the successive orders

The leading order problem is $O(\varepsilon)$ and is given by

$$4u_1'' + u_1 = RHS \equiv \delta_1 \cos(S). \tag{5.77}$$

The two solutions of the homogeneous problem now are $u_1 = \exp\left(\pm\frac{iS}{2}\right)$ and the solvability condition becomes

$$\int_0^{2\pi} RHS(S) \exp\left(\pm\frac{iS}{2}\right) dS = 0, \tag{5.78}$$

which is identically satisfied for (5.77). Its solution is then

$$u_1 = \left(\alpha \exp\left(\frac{iS}{2}\right) + c.c.\right) - \frac{\delta_1}{6}(\exp(iS) + c.c.), \tag{5.79}$$

where *c.c.* means complex conjugate and α is an unknown complex amplitude. In order to determine α, we need to investigate the higher order problems.

The next problem is $O(\varepsilon^2)$ and is given by the following equation for u_2

$$4u_2'' + u_2 = RHS \equiv -\frac{1}{2}u_1^2 + \delta_2 \cos(S) - 4\sigma_1 u_1''. \tag{5.80}$$

We note that

$$
u_1^2 = (\alpha^2 \exp(iS) + c.c.) + 2\alpha\bar{\alpha} + \left(\frac{\delta_1}{6}\right)^2 [(\exp(2iS) + c.c.) + 1]
$$

$$
- \frac{\delta_1}{6}\left(\bar{\alpha}\exp\left(\frac{iS}{2}\right) + c.c.\right) - \frac{\delta_1}{6}\left(\alpha\exp\left(\frac{3iS}{2}\right) + c.c.\right) \qquad (5.81)
$$

only exhibits higher order harmonics of the basic periodic function $\exp(\pm iS/2)$. Using the solvability condition (5.78) for Eq. (5.80) leads to the condition

$$
\frac{\delta_1}{12}\bar{\alpha} + \sigma_1\alpha = 0. \qquad (5.82)
$$

This equation and its complex conjugate form a homogeneous linear system of two equations for α and $\bar{\alpha}$. It admits a nontrivial solution only if

$$
\delta_1 = \pm 12\sigma_1 \quad \text{and} \quad \alpha \pm \bar{\alpha} = 0. \qquad (5.83)
$$

For a fixed value of the detuning, σ_1, we have two period doubling bifurcations. Note, however, that $|\alpha|$ is arbitrary, meaning that the two bifurcations are vertical at this order of the perturbation analysis. To further progress in our analysis of the bifurcation diagram, we examine another route proposed by (5.82), namely

$$
\boxed{\delta_1 = \sigma_1 = 0.} \qquad (5.84)
$$

In this case, α is still completely arbitrary and we need to investigate the problem for u_3. To this end, we determine u_2 as

$$
u_2 = (\alpha_2 \exp(2iS) + c.c.) + \left(\frac{\alpha^2}{6}\exp(4iS) + c.c.\right)
$$

$$
- \alpha\bar{\alpha} - \frac{\delta_2}{6}(\exp(iS) + c.c.), \qquad (5.85)
$$

where α_2 is a new unknown amplitude.

The equation for u_3 is given by

$$
4u_3'' + u_3 = RHS \equiv \delta_3\cos(S) - u_1u_2 - \frac{1}{6}u_1^3 - 2Au_1' - 4\sigma_2 u_1''. \qquad (5.86)
$$

The solvability condition (5.78) for Eq. (5.86) now leads to the following nonlinear equation for α

$$
\frac{1}{3}\alpha^2\bar{\alpha} - iA\alpha + \frac{\delta_2}{6}\bar{\alpha} + \sigma_2\alpha = 0. \qquad (5.87)
$$

Introducing the decomposition $\alpha = R \exp(i\theta)$ into (5.87), we obtain the following equations for R and θ:

$$R \left[\frac{\delta_2}{6} \cos(2\theta) + \frac{1}{3}R^2 + \sigma_2 \right] = 0, \tag{5.88}$$

$$- \left[\frac{\delta_2}{6} \sin(2\theta) + A \right] R = 0. \tag{5.89}$$

This provides us with the required information, namely the amplitude R of the 2π-periodic solution in the presence of a π-periodic modulation. We may now proceed to the analysis of these solutions. In particular their domain of existence will provide us with the position of the bifurcations.

Bifurcations

The possible solutions of Eqs. (5.88) and (5.89) are (1) the Period 1 solution

$$R = 0 \tag{5.90}$$

and (2) the Period 2 solution $R \neq 0$. Eliminating the trigonometric functions, we obtain the solution (in implicit form)

$$\left(\frac{\delta_2}{6} \right)^2 = \left(\frac{1}{3}R^2 + \sigma_2 \right)^2 + A^2, \tag{5.91}$$

which we rewrite as

$$\frac{1}{9}R^4 + \frac{2\sigma_2}{3}R^2 = \frac{\delta_2^2 - \delta_{PD}^2}{36}, \tag{5.92}$$

where

$$\delta_{PD} = \pm 6\sqrt{A^2 + \sigma_2^2} \tag{5.93}$$

are the period doubling bifurcation points. The bifurcation exhibits hysteresis if $\sigma_2 < 0$ and anticipates a "hard transition" to the Period 2 oscillations as δ_2 surpasses the period doubling bifurcation point. A hard transition means that the laser experiences a quick jump from small to large amplitude oscillations. If $\sigma_2 = 0$, the bifurcation is supercritical but sharp ($R \sim (\delta_2^2 - \delta_{PD}^2)^{1/4}$) while it is parabolic ($R \sim (\delta_2^2 - \delta_{PD}^2)^{1/2}$) if $\sigma_2 > 0$. In this case, the bifurcation transition is called "soft" because the amplitude of the solutions changes gradually as δ_2 passes the period doubling bifurcation point. See Figure 5.8. In the figure, the leading approximation of the maximum of y for the P_1 and P_2 solutions ($y = \delta/3$ and $y = 2\varepsilon R$,

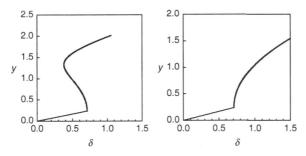

Fig. 5.8 Period doubling bifurcation. $A = 2$ and $\gamma = 10^{-3}$, which implies $\varepsilon^2 = 0.013\,16$. The approximation of the maximum of y for the P_1 solution ($y = \delta/3$) and P_2 solution ($y = 2\varepsilon R$) is represented as a function of δ using Eq. (5.92) for R. Left: $\sigma = 1.9$ ($\sigma_2 = -3.16$). Right: $\sigma = 2.1$ ($\sigma_2 = 3.16$).

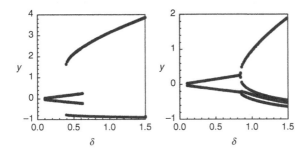

Fig. 5.9 Numerical period doubling bifurcation. The extrema of y are represented as a function of δ. Equation (5.54) is solved numerically with $f(s) = \cos(\sigma s)$, $A = 2$, $\gamma = 10^{-3}$, $\sigma = 1.9$ (left), and $\sigma = 2.1$ (right).

respectively) is represented as a function of δ. The period doubling bifurcation point is located at $\delta_{PD} = 0.71$.

These analytical results are confirmed by numerical simulations as shown in Figure 5.9. In the case of negative (positive) detuning we find the hard transition (the smooth transition) to the P_2 oscillations predicted by the bifurcation analysis. The stability of the periodic solutions could be determined by a two-time perturbation analysis. The P_2 solution near its bifurcation point is shown in Figure 5.10 for the case of positive detuning. It clearly exhibits two maxima and two minima. In the case of negative detuning, the P_2 solution exhibits a large amplitude and only one maximum and one minimum.

In summary, we have described two nonlinear phenomena that result from a weak modulation of a parameter. In the first case, the modulation frequency was close to the relaxation oscillation frequency of the solitary laser and a bistable response for the laser oscillations was possible at large modulation amplitudes. In

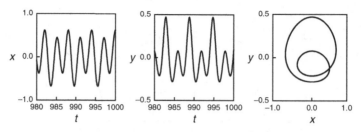

Fig. 5.10 Numerical P_2 solution of Eq. (5.54) with $f(s) = \cos(\sigma s)$ and $A = 2$, $\gamma = 10^{-3}$, $\sigma = 2.1$, and $\delta = 0.85$.

the second case, the modulation frequency was close to twice the relaxation oscillation frequency of the laser (subharmonic modulation), and a period-doubling bifurcation with a hard (or a soft) transition was possible, depending on the sign of the frequency detuning from its resonance value.

5.3 Exercises and problems

5.3.1 Driven Adler's equation

Substitute $\theta = \theta_0(s) + \varepsilon\theta_1(s) + \ldots$ into Eq. (5.2) and equate the coefficients of each power of ε to zero. The equation for θ_0 can be integrated. The equation for θ_1 can be solved after the right hand side is expanded in Fourier series. Verify that the solution for θ_1 becomes unbounded at $\sigma = n^{-1}$ where n is an integer.

5.3.2 Weakly modulated Adler's equation

We wish to compute the expression $(1 - a\sin(\Phi(s)))^{-1}$, where $\Phi(s)$ is defined as the exact solution of Adler's equation for $0 < a < 1$. Introduce the new variable $y = \tan((\Phi + \pi/2)/2)$ and note that the solution of Adler's equation is given by

$$y^2 = \frac{1+a}{1-a}\tan^2\left(\frac{\omega_0}{2}s\right). \tag{5.94}$$

5.3.3 A change of variables in the rate equations

We wish to eliminate the small parameter γ multiplying the right hand side of Eq. (1.8) by changing both the dependent and independent variables. To this end we introduce the new variables

$$s = \sigma t, \quad I = A - 1 + \alpha y, \quad \text{and} \quad D = 1 + \beta x, \tag{5.95}$$

where $(I, D) = (A - 1, 1)$ corresponds to the non-zero intensity steady state. The coefficients σ, α, and β in (5.95) are unknown and will be determined by requiring that (1) γ no longer multiplies a right hand side and that (2) the resulting equations for x and y exhibit a reduced number of parameters. Proceeding in this way, obtain Eqs. (5.38) and (5.39).

5.3.4 Failure of the regular perturbation method

The failure of the regular perturbation method in case of resonance can be simply analyzed using the laser equation in the weak dissipation limit ($\varepsilon = 0$) and with a sinusoidal modulation $f(s) = \sin(s)$

$$u'' = \delta \sin(s) + 1 - \exp(u). \tag{5.96}$$

Seek a regular perturbation expansion solution for u of the form $u = \delta u_1(s) + \delta^2 u_2(s) + \ldots$ and find that u_1 exhibits the secular term $s \cos(s)$.[1] It is called secular as $s \to \infty$ because it increases without bound, thus violating the implicit assumption that successive terms in an asymptotic expansion remain of decreasing size. The regular perturbation expansion is therefore not suitable for s large. Trying a different expansion like $u = \delta^{1/2} u_1(s) + \delta u_2(s) + \ldots$ will only postpone the difficulty to higher order.

---

[1] *Secular* (derived from the Latin *saeculum* for "century") was first used in astronomical applications where the undesired mixed term becomes significant only when time is of the order of a century.

6

Strongly modulated lasers

In the previous chapter, we investigated the case of weakly modulated lasers. We found that a bistable response is possible if the modulation frequency is close to the relaxation oscillation (RO) frequency or to twice the RO frequency of the laser. In this section, we consider stronger modulation amplitudes, which is the case in most experimental studies. A strongly modulated laser may lead to chaos through successive period-doubling bifurcations as we shall see in this chapter.

In the late 1970s and early 1980s, there was a lot of excitement about "deterministic chaos" in all fields of physics, chemistry, and even biology. Deterministic refers to the idea that the future state of a system can be predicted using a mathematical model that does not include random or stochastic influences. Chaos refers to the idea that a system displays extreme sensitivity to initial conditions so that arbitrary small errors in measuring the initial state of the system grow exponentially large and hence practical, long term predictability of the future state of the system is lost. As far as optics was concerned, Kensuke Ikeda suggested in 1979 that an optical ring resonator containing a two-level medium and subject to a delayed feedback could exhibit chaos [113]. His work triggered a lot of experimental research on optical chaos but, as explained in Section 4.2, it took several years before quantitative comparisons were possible. Following a quite different approach, Arecchi *et al.* [136] modulated the losses of a CO_2 laser at a frequency close to the RO frequency and obtained in 1982 a clear period-doubling cascade to chaos (see Figure 6.1). At that time, there was also intense research on chaos in free running (i.e. nonmodulated) lasers, in particular in far-infrared (FIR) lasers. It was motivated by the prediction of the Lorenz–Haken instability but a neat detection of Lorenz type chaos in the NH_3 laser came much later (see Section 11.2). Similarly, there was strong research activity in the former USSR where the concept of laser instabilities was well known from laser physicists. Almost simultaneously with Arecchi *et al.*, Russian scientists Ivanov *et al.* modulated their Nd:YAG laser at a frequency 0.4 times the RO frequency and observed period doubling and chaotic

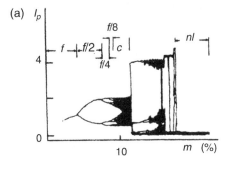

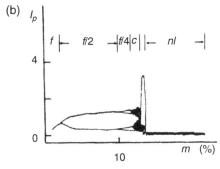

Fig. 6.1 Bifurcation diagram for the CO_2 laser with modulated losses. The peak intensity I_p is represented as a function of the modulation amplitude m. Figures (a) and (b) are obtained by increasing or decreasing m, respectively (f = RO frequency; c = chaos; nl = nonlasing). Adapted Figure 6 with permission from Tredicce *et al.* [141]. Copyright 1986 by the American Physical Society.

emission. They called this irregular behavior "autostochasticity" [137]. Since that time, chaos has been observed in a variety of modulated class B lasers including diode, LNP, $Nd:YVO_4$, Er^{3+}, and Nd^{3+} doped fiber lasers. These experimental results motivated several numerical studies of the laser equations [138, 139, 140]. The excellent agreement between experiments and simulations of rate equations for the modulated CO_2 laser led to the study of many dynamical phenomena using these lasers.

6.1 Generalized bistability

6.1.1 Experiments and simulations

Bifurcation diagrams (BD), where a property of a long time solution (maxima, period) is represented as a function of a control parameter (modulation amplitude or modulation frequency), are useful for exploring how multiperiodic regimes may appear. Typical diagrams showing cascades of period-doubling bifurcations are

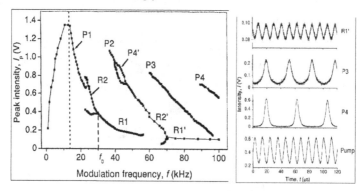

Fig. 6.2 Left: bifurcation diagram of an erbium-doped fiber laser with modulated pump for different values of the modulation frequency. Right: three different regimes are observed for a modulation frequency of 92 kHz. From top to bottom the T, 3T, and 4T time-periodic regimes are represented. The bottom trace is the modulation reference of the pump intensity. Reprinted Figures 3 and 5 with permission from Pisarchik *et al.* [142]. Copyright 2003 IEEE.

shown in Figure 6.1 for increasing or decreasing values of the modulation amplitude. They also exhibit different dynamics (Period 3 and chaos) in the domain ($12\% < m < 14\%$). Such a coexistence of different dynamical regimes for the same values of the parameters is quite common in the modulated CO_2 laser and has been called "generalized bistability (multistability)". The multistability phenomenon has been explored in a series of recent experiments using an erbium-doped fiber laser with a modulated pump (see Figure 6.2).[1] Distinct branches of nT-periodic solutions with their period-doubling bifurcations may be identified. Here the modulation frequency is the control parameter and period-doubling bifurcations are clearly visible, e.g. near 30 kHz and 70 kHz. Attractors with periods T, 3T, and 4T (marked as R1′, P3, and P4 in Figure 6.2) coexist for a modulation frequency of 92 kHz.

A typical numerical BD of the periodic solutions of Eqs. (5.38) and (5.52) is shown in Figure 6.3. The L_2 norm[2] of the periodic solutions is represented as a function of the modulation amplitude δ. The values of the parameters are $\gamma = 2.7 \times 10^{-4}$, $A = 1.03$, $g(s) = \cos(\sigma s)$, and $\sigma = 0.9$. The BD has been obtained by using a continuation method[3] which allows the determination of branches of both stable and unstable solutions. The figure shows the successive branching of

---

[1] This laser most likely operates on many longitudinal modes. This does not significantly modify simple processes such as relaxation oscillations or period-doubling cascades, as shown by several authors.

[2] The L_2 norm of a P-periodic solution $(x(s), y(s))$ is defined as $L_2 \equiv \frac{1}{P} \int_0^P (x^2 + y^2)ds$.

[3] Instead of solving the initial value problem, the continuation method solves a boundary-value problem since each periodic solution satisfies periodic boundary conditions.

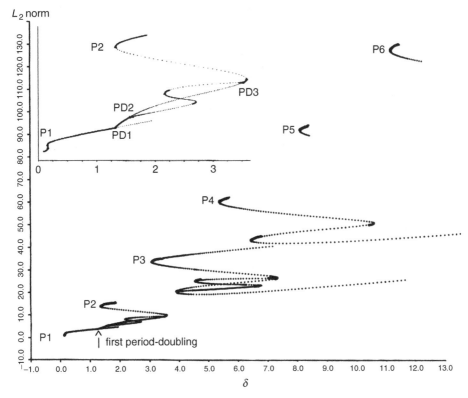

Fig. 6.3 Stable (full lines) and unstable (dots) branches of P_n-periodic solutions ($P_n = n2\pi/\sigma$) appear through successive limit points. In addition, period-doubling bifurcations are possible (PD1–PD3). The region of the first period-doubling bifurcation is enlarged in the left part of the figure. Reprinted from Figures 1 and 2 of Schwartz [139] with permission from Elsevier.

P_n-periodic solutions ($P_n \equiv 2\pi n/\sigma$) emerging from limit points (called primary saddle-node bifurcation points in [139]). These branches of periodic solutions may overlap allowing the coexistence of several regimes for certain range of values of δ. The figure also suggests that the L_2 norm at each limit point is linearly proportional to δ. In addition to the successive branching of P_n-periodic solutions, we note cascades of period-doubling bifurcations. The first period-doubling bifurcation is marked in the figure and is a bifurcation from the P_1 solution. Near this bifurcation, the P_2 solution exhibits a double loop in the phase plane (x, y) and progressively becomes single loop as its amplitude increases.

In summary, the existence of generalized multistability is accompanied by the emergence of isolated subharmonic branches and period-doubling bifurcations. These are key dynamical phenomena for strongly modulated class B lasers. We analyze them in the next two sections.

6.1.2 Branches of subharmonic periodic solutions

In this section, we demonstrate the emergence of branches of subharmonic periodic solutions at limit points and show how to derive analytical expressions for the positions of these points.

Numerically, we note that the limit points $\delta = \delta_n$ for all P_n-periodic solutions move to zero as the damping rate of the laser relaxation oscillations ε^2 goes to zero. This suggests a simple perturbation analysis of the laser equations. To this end, we expand the modulation amplitude δ as

$$\delta = \varepsilon^2 \delta_1 + \dots \tag{6.1}$$

and seek a regular perturbation solution of Eqs. (5.38) and (5.52) of the form

$$x = x_0(s) + \varepsilon^2 x_1(s) + \dots \tag{6.2}$$

$$y = y_0(s) + \varepsilon^2 y_1(s) + \dots \tag{6.3}$$

It is mathematically convenient to introduce the energy of the periodic solution. The first integral (5.42) motivates introducing the energy function $E(x, y)$ defined by

$$E(x, y) \equiv \frac{x^2}{2} + y - \ln(1 + y). \tag{6.4}$$

Differentiating (6.4) with respect to s and using Eqs. (5.38) and (5.52), we obtain the following differential equation for E

$$\frac{dE}{ds} = -\varepsilon^2 x^2 [A + (A - 1)y] - \delta y g(s). \tag{6.5}$$

We solve this equation in the same way as the x and y equations by expanding E in power series of ε^2, i.e.

$$E = E_0(s) + \varepsilon^2 E_1(s) + \dots \tag{6.6}$$

Substituting (6.1), (6.2), (6.3), and (6.6) into Eqs. (5.38), (5.52), and (6.5) leads to a sequence of simpler problems. The leading order problem for x_0, y_0, and E_0 is given by

$$\frac{dx_0}{ds} = -y_0, \tag{6.7}$$

$$\frac{dy_0}{ds} = (1 + y_0)x_0, \tag{6.8}$$

$$\frac{dE_0}{ds} = 0. \tag{6.9}$$

Equations (6.7) and (6.8) are identical to the equations for the conservative system (5.41). This system admits a one-parameter family of period solutions with period changing from 2π to infinite. We denote the P_n-periodic solution ($P_n = 2\pi n/\sigma$) of Eqs. (6.7) and (6.8) by

$$x_0 = X_n(s + \phi_n) \tag{6.10}$$

$$y_0 = Y_n(s + \phi_n), \tag{6.11}$$

where ϕ_n is an arbitrary constant phase. Equation (6.9) simply means that $E_0 = C_n$ is a constant.

To determine ϕ_n, we examine the equation for E_1, namely

$$\frac{dE_1}{ds} = -x_0^2 \left[1 + (A - 1)(1 + y_0) \right] - \delta_1 y_0 g(s). \tag{6.12}$$

We note that the left hand side of Eq. (6.12) admits a constant solution and therefore the right hand side needs to satisfy a solvability condition. This condition requires that the average of the right hand side is zero, i.e.

$$A \int_0^{P_n} X_n^2(s + \phi)ds + \delta_1 \int_0^{P_n} Y_n(s + \phi)g(s)ds = 0 \tag{6.13}$$

where we used the fact that

$$\int_0^{P_n} X_n^2 Y_n ds = - \oint X_n^2 dX_n = 0. \tag{6.14}$$

Setting $\xi = s + \phi_n$, (6.13) can be rewritten as

$$A \int_0^{P_n} X_n^2(\xi)d\xi + \delta_1 \int_0^{P_n} Y_n(\xi)g(\xi - \phi_n)d\xi = 0. \tag{6.15}$$

Without loss of generality, we may define the origin $\xi = 0$ so that $X(\xi)$ is odd and $Y(\xi) = -X'(\xi)$ is even. If $g(s) = \cos(\sigma s)$, then (6.15) reduces to

$$A I_n + \delta_1 \cos(\phi_n) J_n = 0, \tag{6.16}$$

where I_n and J_n are two definite integrals defined by

$$I_n \equiv \int_0^{P_n} X_n^2 d\xi \quad \text{and} \quad J_n \equiv \int_0^{P_n} Y_n(\xi) \cos(\sigma\xi)d\xi. \tag{6.17}$$

Equation (6.16) is now of the form of a steady Adler's equation. A solution of this equation is possible only if $\delta_1 \geq |AI_n/J_n|$ or, equivalently, if

$$\delta \geq \delta_n = \varepsilon^2 |AI_n/J_n|. \tag{6.18}$$

In general, the integrals (6.17) need to be computed numerically but we may analyze the limit n large. In this limit, the period P_n becomes large forcing the P_n-periodic oscillations to be of large amplitude. $X_n(\xi)$ (odd) approaches saw-toothed oscillations of the form

$$X_n = x_n + \xi \ (-x_n \leq \xi < 0)$$

$$= -x_n + \xi \ (0 < \xi < x_n),$$

where $x_n \equiv \pi n/\sigma$. On the other hand,

$$Y_n \simeq -1$$

except near $\xi = 0$ where it pulses and where X_n jumps from x_n to $-x_n$. Using these approximations, we evaluate the two integrals (6.17) as

$$I_n \simeq \int_{-x_n}^{0} (x_n + \xi)^2 d\xi + \int_{0}^{x_n} (-x_n + \xi)^2 d\xi = \frac{2x_n^3}{3}, \tag{6.19}$$

$$J_n \simeq \int_{-x_n}^{x_n} Y_n(\xi)d\xi = -\int_{x_n}^{-x_n} dX_n = 2x_n. \tag{6.20}$$

For (6.20), we have taken into account that the main contribution of the integral J_n is at $\xi = 0$, implying $\cos(\sigma\xi) = 1$. We then used Eq. (6.7) to solve the integral. Using (6.18), the approximation for δ_n then is

$$\delta_n \simeq \sqrt{\frac{\gamma}{A-1}} \frac{Ax_n^2}{3} = \sqrt{\frac{\gamma}{A-1}} \frac{A\pi^2 n^2}{3\sigma^2}. \tag{6.21}$$

Similarly, we evaluate the L_2 norm of the P_n-periodic solution as

$$L_2(n) \equiv \frac{1}{P_n} \int_0^{P_n} (X_n^2 + Y_n^2)d\xi \simeq \frac{1}{2x_n} \int_{-x_n}^{x_n} X_n^2 d\xi = \frac{x_n^2}{3} = \frac{\pi^2 n^2}{3\sigma^2}. \tag{6.22}$$

We conclude that $L_2(n)$ grows linearly with δ_n, which is what Figure 6.3 suggests. The derivation of the leading approximation for (X_n, Y_n) can be substantiated analytically by using the method of matched asymptotic expansions. This analysis is described in [143].

In summary, our analysis reveals the emergence of successive branches of isolated P_n-periodic regimes. The damping rate of the laser relaxation oscillation, proportional to $\varepsilon^2 = \sqrt{\gamma/(A-1)}$, plays an important role in the domain of existence of these solutions. Coexistence of several branches of periodic solutions is more likely to be observed if this damping rate is sufficiently small.

6.2 Map for the strongly modulated laser

In this section, we propose to replace the laser differential equations by equations for a map valid if the oscillations are sufficiently large (see Figure 6.4). We first explore numerically the solutions of the equations in the low dissipation limit. This suggests approximations helping to build the mapping. We then use this mapping to obtain an analytic value of the position of the first period-doubling bifurcation.

6.2.1 Exploring the large oscillation regimes

In the strong modulation limit, $x(s)$ exhibits saw-tooth oscillations while $y(s)$ shows intense pulses separated by intervals where y is close to -1. Specifically, we wish to determine equations for the successive values $x = x_n < 0$ if $y = 0$ as well as the times $s = s_n$ when they appear. For simplicity, we consider the laser

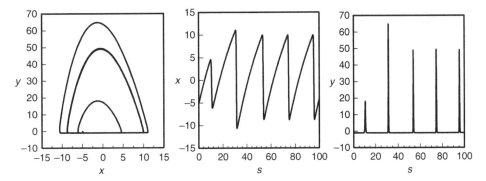

Fig. 6.4 Transient oscillations for the laser with modulated losses ($g(s) = \cos(\sigma s)$). The solution is obtained from Eqs. (5.38) and (5.52). The initial conditions are $(x(0), y(0)) = (-5, 0)$. The values of the parameters are $A = 2$, $\gamma = 10^{-3}$, $\delta = 1.295$, and $\sigma = 0.9$. The laser first oscillates into a Period 1 regime and then goes into its long time Period 2 regime.

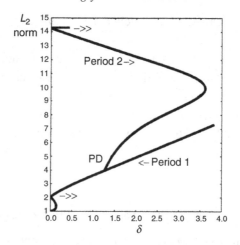

Fig. 6.5 Blow-up of the bifurcation diagram of Eqs. (6.23) and (6.24) for $\sigma = 0.9$ showing the first period-doubling bifurcation. The Period 2 branch emerging at that point turns at a right limit point and connects a left limit point at $\delta = 0$. If ε increases from zero, the limit points at $\delta = 0$ and $L_2 = 2$ and 14 move to the right as suggested by the double arrows. Redrawn from Figure 2 of Schwartz and Erneux [144]. Copyright 1994 Society for Industrial and Applied Mathematics. Reprinted with permission. All rights reserved.

rate equations with loss modulation (5.38), (5.52) in the limit of zero damping ($\varepsilon = 0$). They are given by

$$\frac{dx}{ds} = -y, \tag{6.23}$$

$$\frac{dy}{ds} = (1+y)(1 - \delta \cos(\sigma s)). \tag{6.24}$$

The bifurcation diagram of the Period 1 and Period 2 branches of Eqs. (6.23) and (6.24) obtained by a continuation method is shown in Figure 6.5. The Period 1 and Period 2 left limit points appearing at the L_2 norm, equal to 2 and 14, respectively, are located at $\delta = 0$ because damping is zero ($\varepsilon^2 = 0$ in Eq. (6.18) implies $\delta_n = 0$). The period-doubling bifurcation at $\delta \simeq 1.25$ is new and the periodic orbits in the phase plane before and after the period-doubling bifurcation point are shown in Figure 6.6.

6.2.2 Building the map

We consider Eqs. (6.23) and (6.24) supplemented by the initial conditions

$$x(s_n) = x_n < 0, \quad y(s_n) = 0, \tag{6.25}$$

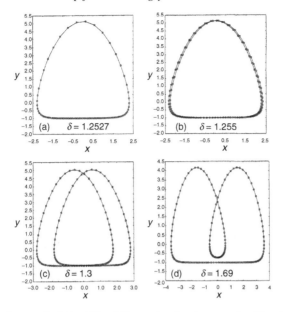

Fig. 6.6 The first period-doubling bifurcation is characterized by the emergence of a double-loop orbit. It progressively becomes single-loop as its amplitude increases. From Figure 4 of Schwartz and Erneux [144]. Copyright 1994 Society for Industrial and Applied Mathematics. Reprinted with permission. All rights reserved.

where $|x_n| >> 1$ is our large parameter that controls the amplitude of the orbit (see Figure 6.7). Furthermore, we assume $\delta = O(1)$ and $\sigma = O(|x_n|^{-1})$ and we build the map by decomposing the evolution into two parts: the slow evolution between the pulses and the fast spiking, which we successively characterize as follows.

Slow evolution

The slow evolution between two successive pulses is characterized by the fact that $y \simeq -1$. In this regime, Eqs. (6.23) and (6.24) reduce to

$$\frac{dx}{ds} = 1, \quad \frac{dy}{ds} = (1 + y)[x - \delta \cos(\sigma s)]. \tag{6.26}$$

Using the initial conditions (6.25) and integrating, the solution of Eq. (6.26) is given by

$$x = x_n + (s - s_n) \quad \text{and} \quad y = -1 + \exp(f(s)), \tag{6.27}$$

where

$$f(s) = x_n(s - s_n) + \frac{1}{2}(s - s_n)^2 - \delta\sigma^{-1}[\sin(\sigma s) - \sin(\sigma s_n)]. \tag{6.28}$$

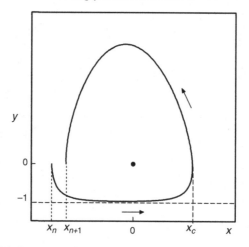

Fig. 6.7 One orbit in the phase plane (x, y). The orbit starts at $(x_n, 0)$ when $s = s_n$, passes through $(x_c, 0)$ when $s = s_c$, and completes its orbit at $(x_{n+1}, 0)$ when $s = s_{n+1}$. If x_n is decreased, the "tear drop" orbit becomes more triangular, spending most of its time near $y = -1$.

Assuming $s - s_n = O(|x_n|) >> 1$, (6.28) simplifies as

$$f(s) = x_n(s - s_n) + \frac{1}{2}(s - s_n)^2. \tag{6.29}$$

Recall that $x_n < 0$, and that the exponential in (6.27) is very small until $f(s_c) = 0$ where it starts growing. s_c is given by

$$s_c = s_n - 2x_n. \tag{6.30}$$

At this time, $x = x_c$ where

$$x_c = -x_n. \tag{6.31}$$

Spiking

Since the pulse is instantaneous, the equations for the fast pulse are the original laser equations where the modulation term is evaluated at $s = s_c = s_{n+1}$. They are given by

$$\frac{dx}{ds} = -y, \tag{6.32}$$

$$\frac{dy}{ds} = (1 + y)\left[x - \delta \cos(\sigma s_{n+1})\right]. \tag{6.33}$$

By dividing the two equations, we obtain a first order differential equation for the trajectories $y = y(x)$. This equation is separable and integration leads to the relation

$$\frac{x^2}{2} - \delta \cos(\sigma s_{n+1})x - y + \ln(1 + y) = C,\qquad(6.34)$$

where C is the constant of integration. This equation must be verified by the starting and end points of the fast pulse defined by $(x_c, 0) = (-x_n, 0)$ and $(x_{n+1}, 0)$, respectively. This leads to the following two conditions for the unknown C and x_{n+1}:

$$\frac{x_n^2}{2} - \delta \cos(\sigma s_{n+1})(-x_n) = C,\qquad(6.35)$$

$$\frac{x_{n+1}^2}{2} - \delta \cos(\sigma s_{n+1})x_{n+1} = C.\qquad(6.36)$$

The solution for x_{n+1} is then obtained by subtracting the two equations and by factorizing $(x_{n+1} + x_n)$. Using (6.31), we eliminate x_c and obtain

$$x_{n+1} = x_n + 2\delta \cos(\sigma s_{n+1}).\qquad(6.37)$$

Equation (6.37) is only the first equation for a map. We need a second equation that describes the period of a complete orbit. Since $s_{n+1} = s_c$ and using (6.30), this equation is

$$s_{n+1} - s_n = -2x_n.\qquad(6.38)$$

This completes the equations for the map which links the amplitude x_{n+1} and time s_{n+1} of the $(n + 1)^{\text{th}}$ pulse to the values for the nth pulse.

6.2.3 The first period doubling bifurcation

We shall now investigate Eqs. (6.37) and (6.38) by seeking the Period 1 and Period 2 solutions. The first period-doubling bifurcation will be given by the existence condition of the Period 2 solution.

The Period 1 solution satisfies the locking condition $s_{n+1} - s_n = 2\pi\sigma^{-1}$ and from Eqs. (6.37) and (6.38) we find

$$x_n = -\pi\sigma^{-1} \quad \text{and} \quad \sigma s_n = \pm\frac{\pi}{2}.\qquad(6.39)$$

The Period 2 single loop solution satisfies the condition $s_{n+1} - s_n = 4\pi\sigma^{-1}$ and from Eqs. (6.37) and (6.38) we have

$$x_n = -n\pi\sigma^{-1} \quad \text{and} \quad \sigma s_n = \pm\frac{\pi}{2}. \tag{6.40}$$

For a Period 2 double loop solution, we need to solve the equations for the map for two successive orbits and use the condition

$$x_{n+2} = x_n \quad \text{and} \quad s_{n+2} = s_n + 4\pi\sigma^{-1}. \tag{6.41}$$

These equations are given by

$$x_1 = x_0 + 2\delta\cos(\sigma s_1), \tag{6.42}$$

$$s_1 - s_0 = -2x_0, \tag{6.43}$$

$$x_0 = x_1 + 2\delta\cos(\sigma s_0), \tag{6.44}$$

$$s_0 - s_1 = -2x_1 - 4\pi\sigma^{-1}. \tag{6.45}$$

We first extract x_0 and x_1 from Eqs. (6.43) and (6.44) and find

$$x_0 = -\frac{1}{2}(s_1 - s_0) \quad \text{and} \quad x_1 = -\frac{1}{2}(s_0 - s_1 + 4\pi\sigma^{-1}). \tag{6.46}$$

The remaining equations then become

$$(s_0 - s_1) + 2\pi\sigma^{-1} + 2\delta\cos(\sigma s_1) = 0, \tag{6.47}$$

$$(s_1 - s_0) - 2\pi\sigma^{-1} + 2\delta\cos(\sigma s_0) = 0. \tag{6.48}$$

We solve these equations by introducing $s_1 - s_0 = 2\pi\sigma^{-1} + \tau$ and rewriting Eqs. (6.47) and (6.48) in terms of τ and s_0

$$-\tau + 2\delta\cos(\sigma s_0 + \sigma\tau) = 0, \tag{6.49}$$

$$\tau + 2\delta\cos(\sigma s_0) = 0. \tag{6.50}$$

We determine an expression for $\sin(\sigma s_0)$ from (6.49), using (6.50). Eliminating s_0 by using a trigonometric identity, we obtain the following equation for τ

$$\delta = \frac{\tau}{|\sin(\sigma\tau)|}\sqrt{\frac{1}{2}(1 + \cos(\sigma\tau))}. \tag{6.51}$$

This equation describes δ as a function of τ. There is a bifurcation point at $\tau = 0$ given by

$$\delta = \delta_{PD} \equiv \sigma^{-1} \tag{6.52}$$

and the bifurcation is supercritical (near $\delta = \delta_{PD}$, $\tau \simeq \pm \sqrt{24\sigma^{-1}(\delta - \delta_{PD})}$, which implies $\delta \geq \delta_{PD}$). The approximation of the period-doubling bifurcation for $\sigma = 0.9$ is $\delta_{PD} \simeq 1.1$, which compares reasonably well with the numerical value ($\delta = 1.25$).

6.3 Dual tone modulation near period-doubling bifurcation

In 1985, Wiesenfeld and McNamara [145, 146] suggested that any system modulated just below the period-doubling bifurcation point with a frequency ω would be very sensitive to additional modulation at the period-doubled frequency $\omega/2$. This phenomenon could be useful for applications where highly sensitive and selective amplification is required. Very quickly after this proposal, Derighetti *et al.* [147] took advantage of this small signal amplification to detect weak input signals on a parametrically modulated "NMR laser". In 1992, experiments on a CO_2 loss-modulated laser, inspired by experiments on noise deamplification in a p-n junction diode, demonstrated that the phase of the additional modulation at $\omega/2$ plays a crucial role and that it was even possible to squeeze noise [148]. The theory of a class B laser subjected to two-tone modulation was later developed together with new experiments on a pump modulated fiber laser [149].

6.3.1 Period-doubling lasers as small-signal detectors

Derighetti *et al.* [147] operated a so-called "NMR laser" which is essentially a ruby NMR experiment in which gain is achieved via microwave pumping near an ESR transition. In their experiment, modulation was achieved via modulation of the NMR linewidth (Q-factor). Derighetti *et al.* concentrated on the first period-doubling bifurcation in their NMR laser whose relaxation frequency ranged between 30 and 80 kHz and the pump quality factor Q was modulated with a frequency $f = 102.7$ Hz, i.e. at about twice the relaxation oscillation frequency. The laser output was sampled at equal time intervals ($T = 1/f$). The discrete output values were then represented as a function of the swept modulation strength a. The resulting bifurcation diagram shows strong low-frequency oscillations for low values of a before the bifurcation point has been reached. Figure 6.8 represents the observation of a period-doubling bifurcation with the superimposition of the amplified input signal. The beat frequency of 1.35 Hz stems from the nonlinear coupling of the first subharmonic of the modulation signal (i.e. $f/2 = 51.35$ Hz)

Strongly modulated lasers

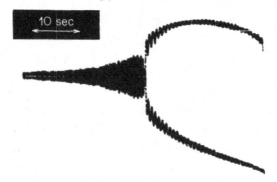

Fig. 6.8 Period-doubling bifurcation for the NMR laser with Q modulation of frequency $f = 102.7$ Hz. The beat of frequency 1.35 Hz indicates the interference with the power line of 50 Hz. Reprinted Figure 1 with permission from Derighetti *et al.* [147]. Copyright 1985 by the American Physical Society.

with the 50 Hz pickup from the powerline. This pickup signal is usually hidden in the thermal noise of the NMR laser. However, near the onset of the bifurcation, it is strongly amplified and clearly visible as illustrated in Figure 6.8.

We first show the connection between their so-called NMR laser and our standard class B laser. Derighetti *et al.* [147] simulated their experiments by considering the following equations

$$\frac{dM_v}{dt'} = -\gamma_\perp(t')M_v + 9C_1 M_z - 9C_2 Q(t')M_v M_z, \qquad (6.53)$$

$$\frac{dM_z}{dt'} = -\gamma_\parallel(M_z - M_e) - C_1 M_v + C_2 Q(t')M_v^2. \qquad (6.54)$$

Here M_z denotes the nuclear magnetization along the direction of the static external field and M_v is the perpendicular magnetization in the rotating frame [150]. The corresponding relaxation rates are $\gamma_\parallel$ and $\gamma_\perp$, respectively. M_e is the pump magnetization and Q is the quality of the coil. The parameters C_1 and C_2 are proportional to the gyromagnetic ratio of the laser-active ^{27}Al nuclei [147]. To simulate the experimental observations, the parametric pump (basic frequency) acts on the coil quality $Q(t')$ and the input signal (period-doubled frequency) changes the perpendicular relaxation rate $\gamma_\perp(t')$. Both parameters become time-periodic functions given by

$$Q(t') = Q\left[1 + a\sin(2\pi f t')\right], \qquad (6.55)$$

$$\gamma_\perp(t') = \gamma_\perp\left[1 + b\sin\left((\pi f + 2\pi\delta)t'\right)\right]. \qquad (6.56)$$

The values of the parameters are [147]: $C_2 = 24.09$ mA^{-1}s^{-1}, $C_1 = -6.9 \times 10^{-3}$ s^{-1}, $M_e = -1.6$ Am^{-1}, $Q = 250$, $\gamma_\perp = 3 \times 10^4$ s^{-1}, $\gamma_\parallel = 10$ s^{-1}, and

$f = 110\,\text{Hz}$. a, b, and δ are three control parameters. Introducing the new variables E, D, and t defined as

$$E \equiv \frac{3C_2Q}{\sqrt{\gamma_\perp \gamma_\parallel}} M_v, \quad D \equiv -\frac{9C_2Q}{\gamma_\perp} M_z, \quad t \equiv \gamma_\perp t' \tag{6.57}$$

into Eqs. (6.53)–(6.56), we find the following equations

$$\frac{dE}{dt} = -r(t)E - \alpha D + q(t)ED, \tag{6.58}$$

$$\frac{dD}{dt} = \gamma \left[A - D + \alpha E - q(t)E^2 \right], \tag{6.59}$$

where $q(t)$ and $r(t)$ are periodic functions of t given by

$$q = 1 + a\sin(\omega t), \tag{6.60}$$

$$r = 1 + b\sin\left(\left(\frac{\omega}{2} + \Delta\right)t\right). \tag{6.61}$$

The definitions and values of the fixed parameters are: $A \equiv -\frac{9C_2Q}{\gamma_\perp} M_e \simeq 2.89$, $\alpha \equiv \frac{3C_1}{\sqrt{\gamma_\perp \gamma_\parallel}} = 3.8 \times 10^{-5}, \gamma = \gamma_\parallel/\gamma_\perp = 3.3 \times 10^{-4}, \omega = 2\pi f \gamma_\perp^{-1} = 2.3 \times 10^{-2}$, $\Delta = 2\pi\delta\gamma_\perp^{-1}$. Neglecting the small α terms, Eqs. (6.58) and (6.59) reduce to the following rate equations of a class B laser

$$\frac{dE}{dt} = E\left[-r(t) + q(t)D\right], \tag{6.62}$$

$$\frac{dD}{dt} = \gamma \left[A - D - q(t)E^2 \right], \tag{6.63}$$

where $q(t)$ and $r(t)$ are given by (6.60) and (6.61), respectively. We compute the relaxation oscillation frequency $\omega_R = \sqrt{2\gamma(A-1)} \simeq 1.2 \times 10^{-2}$ and note that $\omega/\omega_R \simeq 1.9$, i.e. the system is operated at about twice the relaxation oscillation frequency as discussed in Section 5.2.4, meaning that we expect a period-doubling bifurcation for a low value of the parametric pump amplitude a.[4]

6.3.2 Two-tone modulation of a class B laser

Variants of these NMR laser experiments were carried out on real class B lasers (CO_2 [151] and neodymium fiber [149] lasers) with special attention to exact subharmonic modulation. In [149], the laser is pumped with a laser diode and the current of the laser diode is modulated as

---

[4] The factor 2 in the definition of ω_R comes from the fact that we need to rewrite the equation for the field E as an equation for the intensity $I = E^2$ in order to define the RO frequency as in Chapter 1.

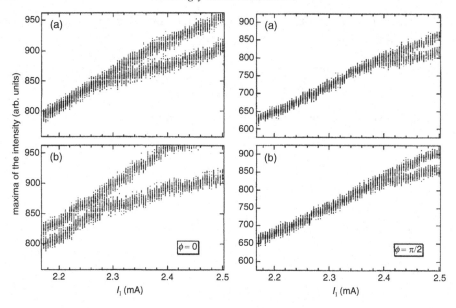

Fig. 6.9 Experimental bifurcation diagram of the two-tone modulated fiber laser. Both diagrams (a) represent the single modulation frequency diagrams ($I_2 = 0$). The figures exhibit a period-doubling bifurcation at $I_1 = 2.3$ mA (left), $I_1 \simeq 2.4$ mA (right). The diagrams (b) represent the effect of the second modulation ($I_2 = 0.06$ mA) for two different phases. Almost no changes are observed for $\phi = \pi/2$. Reprinted Figures 3 and 4 from Newell *et al.* [149]. Copyright 1997 by the American Physical Society.

$$I(t) = I_0 + I_1 \cos(\omega t) + I_2 \cos\left(\frac{\omega}{2}t + \phi\right), \tag{6.64}$$

where $I_0 = 65$ mA ($I_{th} = 34$ mA), I_1 is the bifurcation parameter (2.1 mA $< I_1 < 2.5$ mA), and I_2 is either 0 or 0.06 mA. The bifurcation diagrams representing the maximum of the intensity oscillations as a function of I_1 are recorded for different values of ϕ (see Figure 6.9). Both figures (a) show the bifurcation diagram for the single modulation case ($I_2 = 0$) in the vicinity of the period-doubling bifurcation point at $I_1 \simeq 2.4$ mA. Note that the bifurcation diagrams are slightly different, predominantly due to environmental factors. Consequently, in the experiment, care was taken to first record a reference data set ($I_2 = 0$) then immediately acquire the set in which the perturbation was applied ($I_2 \neq 0$). The environmental factors, ambient temperature, and mechanical drift affect the fiber laser on a very slow time scale (tens of minutes) that does not perturb the bifurcation diagram recording on the millisecond time scale. Comparing diagrams (a) and (b) in Figure 6.9 clearly indicates that the splitting of the Period 1 orbit is maximal at $\phi = 0$ and minimal at $\phi = \pi/2$. The influence of the phase factor ϕ may be anticipated through a simple qualitative analysis. The spectral content of the laser output in the period-doubled regime is made of two spectral components with frequencies ω and $\omega/2$ and a *given* relative phase. Introducing an additional component at $\omega/2$ may interfere constructively or destructively with the $\omega/2$ response of the laser to a single modulation, depending on the phase ϕ.

The change of the bifurcation diagram resulting from the second modulation has been analyzed using the following rate equations which we have shown to be equivalent to our standard rate equations

$$\frac{dE}{dt} = (N - 1)E, \tag{6.65}$$

$$\frac{dN}{dt} = \gamma \left[A(t) - N - N|E|^2 \right], \tag{6.66}$$

where

$$A = A_0 + A_1 \cos(\omega t) + A_2 \cos\left(\frac{\omega}{2}t + \phi\right). \tag{6.67}$$

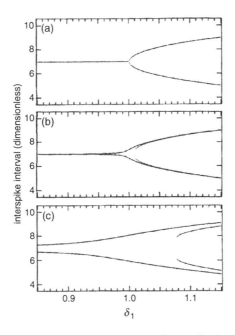

Fig. 6.10 Numerical bifurcation diagrams of the interspike interval $s_{n+1} - s_n$ as a function of δ_1 ($s = \omega_{RO}t$ and $\delta_1 = A_1/(A_0 - 1)$). (a) $A_2 = 0$. A pitchfork bifurcation appears at $\delta_1 = 1$ that creates the Period 2 orbit. (b) $A_2/(A_0 - 1) = 10^{-2}$ and $\phi = 90°$. The pitchfork bifurcation no longer exists and the P_2 bifurcation point is replaced by a limit point located where the two inner branches converge. (c) $A_2/(A_0 - 1) = 2 \times 10^{-3}$ and $\phi = 171°$. At this phase, the unfolding is substantially larger even if the amplitude of the second modulation is smaller than in (b). Reprinted Figure 6 from Newell *et al.* [149]. Copyright 1997 by the American Physical Society.

The values of the fixed parameters appropriate for our experiments are $\gamma = 2.4 \times 10^{-5}$, $A_0 = I_0/I_{th} = 1.93$, and $\omega = 0.9\omega_{RO}$. The bifurcation diagram of the periodic regimes is studied as a function of $\delta_1 \equiv A_1/(A_0 - 1)$ for fixed values of $A_2 << A_1$ and ϕ. It displays the evolution of the interspike interval instead of the intensity observed in most experiments. These two representations are dynamically equivalent since the two quantities are strongly correlated, as demonstrated previously, e.g. by the map; see Section 6.2. Figure 6.10 (a) serves as a reference and Figures 6.10 (b) and (c) clearly demonstrate that the phase of the additional modulation plays a crucial role as observed in the experiments.

To conclude, a strong modulation of a laser with a frequency close to its RO frequency is well described by simple models that capture key nonlinear phenomena such as period-doubling bifurcations. The excellent agreement obtained between analytical, numerical, and experimental results suggests that Class B lasers are good experimental systems for the exploration of generic nonlinear phenomena.

7

Slow passage

What is common to the self-combustion of grain dust in a storage silo and the bursting activity of neurons? In both cases a quick time evolution follows a period of quiescence during which a key parameter is slowly varying. It can be the surrounding temperature in the case of the grain storage silo or the concentration of calcium ions that turns the neuronal activity on and off. These dramatic changes are possible because a slowly varying parameter passes a limit or bifurcation point of a fast dynamical system. But because of the system's inertia close to the bifurcation point, the expected jump or bifurcation transition is delayed. This delay has raised considerable interest not only for lasers but in other areas as well, such as fluid mechanics [152] and chemistry [154–156]. In mechanical engineering, slow passage problems are referred to as "nonstationary processes" [157]. They occur in the start-up and shut-down of engines [158] or in high-rise building elevators when the length of the rope is slowly changing [159]. Although most delay effects are now well understood and illustrated by simple first or second order equations [160], the study of slow passage problems remains a fascinating topic of research for mathematicians [161], biologists [162, 163], and students learning bifurcation theory in the laboratory [164].

Quantitative comparisons between experiments and theory for slow passage problems are always delicate. The evolution equations of a real physical system cannot be reduced to a simple equation if the rate of change is gradually increased, and we often need to take into account the effect of noise present in experiments. The purpose of this chapter is to review some key slow passage problems that have been examined for lasers and optically bistable devices over the last 20 years. Although they often correspond to elementary bifurcations, the experimental investigations of their slow passage properties have generated considerable theoretical discussion.

In this chapter we consider slow passage first through a limit point, illustrating this on a passive optical bistable system, then through a bifurcation point such as we met at the laser threshold. We later investigate passage through a period-doubling bifurcation point, and in the last section we discuss the connection of slow passage with slow–fast dynamics as occurs, for example, in the laser with saturable absorber or the optical parametric oscillator.

7.1 Dynamical hysteresis

The simplest slow passage problem occurs when we investigate a bistable system. Several studies have reported on the enlargement of optically bistable cycles resulting from sweeping the control parameter back and forth [165, 166]. The reference problem is given by the following cubic equation

$$y' = ay - y^3 + \lambda, \tag{7.1}$$

$$\lambda' = \varepsilon, \tag{7.2}$$

where $a = O(1)$ and $\varepsilon \ll 1$. The first equation models a simple hysteretic system driven by some quantity λ, here the field impinging the passive cavity, whose magnitude increases at a constant rate ε.

Typical time evolutions are shown in Figure 7.1 for full ($a = 3$) and nascent ($a = 0$) hysteresis cycles. In both cases, the expected jump transition experiences a delay and the scaling law that relates delay and rate of change ε is indicated in the figures. These scaling laws are determined analytically by investigating the solution of Eqs. (7.1) and (7.2) near the limit point in the limit ε small. The theory predicts that the shift of the switching points in the bistable case scales as the 2/3

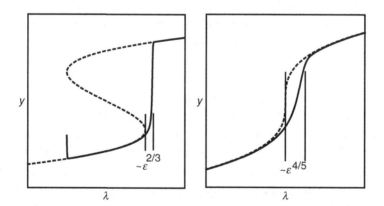

Fig. 7.1 Slow passage through a steady limit point. Solution of Eqs. (7.1) and (7.2) with $a = 3$, $\varepsilon = 0.1$, $y(0) = -1$, and $\lambda(0) = -2$ (left) or with $a = 0$, $\varepsilon = 0.01$, $y(0) = -0.8$, and $\lambda(0) = -0.5$ (right). The broken line is the steady state solution.

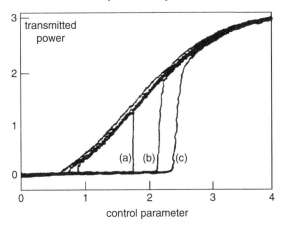

Fig. 7.2 Experimental investigation of the bistable branches by sweeping the voltage applied to the modulator back and forth. Curve (a) obtained at 0.1 Hz gives a good approximation of the static loop. Curves (b) and (c) were obtained at 100 Hz and 500 Hz, respectively. From Figure 1a of Fettouhi *et al.* [168]. With permission of the *European Physical Journal.*

power of the sweeping rate. In the case of nascent hysteresis ($a = 0$), this power increases to 4/5 which is still less than 1. The 2/3 law was verified experimentally in 1990 by using a bistable semiconductor laser [165] and in 1995 by using an injected semiconductor laser [166]. Starting in 1985 with a laser containing a saturable absorber [167], the experimental study of specific optically bistable devices accumulated in the mid 1990s and was a topic of animated discussions.[1] Figure 7.2 shows an experimentally observed bifurcation diagram [168]. The experimental set-up consisted of a 23 m long waveguide Fabry–Pérot cavity filled with $HC^{15}N$ gas at low pressure. The source and the cavity were tuned to the frequency of the $J = 0 \rightarrow 1$ rotational line of $HC^{15}N$, which then behaves as a saturable absorber (purely absorptive bistability). The power transmitted by the cavity was measured as a function of the voltage applied to a PIN diode modulator controlling the input power. The steady state for the field amplitude X in the cavity as a function of input field amplitude Y is given (in implicit form) by

$$Y = X + \frac{2CX}{1 + X^2}, \tag{7.3}$$

where the bistability parameter C is a large parameter (highly nonlinear absorption). See Figure 7.3. The triangular rather than rectangular shape of the hysteresis cycle explains why the determination of the delay as a function of the rate of

---

[1] For example, at the CLEO/Europe–EQEC Meetings at the 5th European Quantum Electronics Conference in Amsterdam, August 28 to 2 September, 1994.

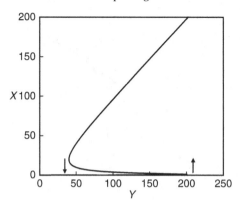

Fig. 7.3 Bistability. The branches of steady states exhibit low and high transmission branches (Eq. (7.3) with $C = 200$). The arrows indicate the switching
positions.

change was best achieved near the right limit point. Near the left limit point, the
observation of a clear fast jump transition is harder. Introducing the scaling law
$Y_J - Y_{LP} = c\varepsilon^\beta$ because the change of X is shorter, where $Y_J - Y_{LP}$ represents
the deviation between the actual jump and limit point values, the experiments indicated that the critical exponent β varies between 0.56 to 0.65 as the pressure is
increased. It thus approaches the theoretical value of $2/3 \simeq 0.67$ at high pressure,
i.e. in the domain where the assumptions used to derive (7.1) are best justified.

7.2 Slow passage through a bifurcation point

The slow passage through the laser threshold was first analyzed in 1984 [169]. The
reference problem describing a slow passage through a steady bifurcation point is
given by the following equations

$$y' = y(\lambda - 1 - y) \tag{7.4}$$

$$\lambda' = \varepsilon \tag{7.5}$$

defined for $y > 0$. The first equation describes, for example, the intensity of the
laser field for a class A laser whose pump parameter is swept (see, e.g., Eq. (1.32)
with $y = |E|^2$, $\lambda - 1 = 2a$, and $b = 1$).

Assuming $y(0) > 0$ and $\lambda(0) < 1$, the passage through the bifurcation point
$\lambda = 1$ exhibits a delay that depends on the rate of change ε. See Figure 7.4. We
arbitrarily define the delay as the deviation $\lambda_j - 1$, where λ_j is the value of λ when
$y = 10^{-2}$. Figure 7.5 represents this delay as a function of the rate of change ε.

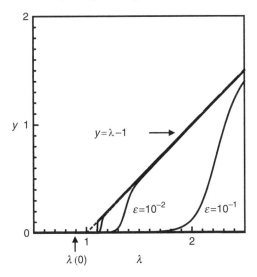

Fig. 7.4 Slow passage through a steady bifurcation point. The numerical solution of Eqs. (7.4) and (7.5) is determined for different rates of change ε. The initial conditions are $y(0) = 10^{-3}$ and $\lambda(0) = 0.9$. From left to right, the different bifurcation transitions correspond to $\varepsilon = 10^{-4}$, 10^{-3}, 10^{-2}, and 10^{-1}.

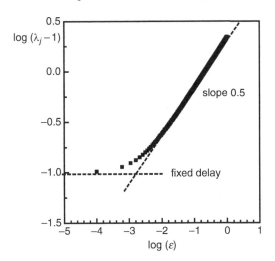

Fig. 7.5 Delay vs. ramp speed. The delay is constant in the limit $\varepsilon \to 0$ and follows a $\sqrt{\varepsilon}$ law for moderately low rates of change. The delay is found numerically by solving Eqs. (7.4) and (7.5). It is defined as the deviation $\lambda_j - 1$, where λ_j is the value of λ when $y = 10^{-2}$.

We note two distinct regimes. For low values of ε ($\varepsilon \leq 10^{-4}$), the delay seems independent of the rate of change ε. For higher values of ε ($\varepsilon > 10^{-2}$), the delay follows a $\sqrt{\varepsilon}$ law. We explain these two different behaviors and show in the next subsection the crucial role played by noise.

7.2.1 The limit $\varepsilon \to 0$ and the role of noise

We first explore the case of very slow passage ($\varepsilon \to 0$) and take advantage of the fact that before the quick jump, y remains very close to zero. Assuming $y \ll 1$, Eqs. (7.4) and (7.5) become

$$y' = y(\lambda - 1) \quad \text{and} \quad \lambda' = \varepsilon. \tag{7.6}$$

Together with the initial conditions $y(0)$ and $\lambda(0)$, the solution of (7.6) is $\lambda = \lambda(0) + \varepsilon t$ and

$$y = y(0) \exp\left[(\lambda(0) - 1)t + \frac{\varepsilon t^2}{2}\right]. \tag{7.7}$$

The expression (7.7) indicates that y is exponentially small until $t = t_j \equiv -2\varepsilon^{-1}(\lambda(0) - 1)$, i.e. when

$$\lambda_j - 1 = 1 - \lambda(0). \tag{7.8}$$

In words, the delay $\lambda_j - 1$ equals the deviation of the initial value of λ from the bifurcation point $\lambda = 1$. From the $\varepsilon = 10^{-4}$ solution in Figure 7.5, we indeed verify that $|\lambda(0) - 1| = \lambda_j - 1 = 10^{-1}$.

The ε small asymptotic result formulated by (7.8) is counter-intuitive and provocative because it implies a delay that is independent of the ramp speed and that the initial conditions control the size of the delay. This problem is resolved by taking into account the fact that noise is always present in experiments and, in particular, in lasers where the lasing action would be impossible without spontaneous emission. There is a considerable literature on the various ways to explore the effect of noise on a delayed bifurcation. A popular way is to add a stochastic term $\xi(t)$ into the right hand side of Eq. (7.4). $\xi(t)$ is often defined as a Gaussian white noise source with zero mean and δ-correlated [70]. The main question is, however, how large noise has to be compared to ε in order to substantially reduce the delay of the bifurcation transition.

This question can be explored by considering Eqs. (7.4) and (7.5) with a small constant term added into the right hand side of Eq. (7.4). The new equation is

$$y' = y(\lambda(0) + \varepsilon t - 1 - y) + \delta, \tag{7.9}$$

where $\delta \ll 1$ needs to be compared to $\varepsilon \ll 1$. Before the jump, $y \ll 1$ and Eq. (7.9) simplifies as

$$y' = y(\lambda(0) + \varepsilon t - 1) + \delta, \tag{7.10}$$

which is linear. The solution of Eq. (7.10) is the sum of two exponentials which can be rewritten as

$$y = y(0) \exp\left[\frac{1}{2\varepsilon}\left(s^2 - s_c^2\right)\right] + \delta \exp\left(\frac{s^2}{2\varepsilon}\right) \int_{-s_c}^{s} \exp\left(-\frac{x^2}{2\varepsilon}\right) dx. \quad (7.11)$$

In (7.11), we have introduced the new time variable

$$s \equiv \varepsilon(t - t_c), \quad (7.12)$$

where $s_c \equiv \varepsilon t_c = (1 - \lambda(0))$. Evaluating the integral in (7.11) for small ε and $s > 0$,[2] we obtain the approximation

$$y \simeq y(0) \exp\left[\frac{1}{2\varepsilon}\left(s^2 - s_c^2\right)\right] + \delta\sqrt{2\pi\varepsilon} \exp\left(\frac{s^2}{2\varepsilon}\right). \quad (7.13)$$

The expression (7.13) now exhibits a competition between two exponentials. If $\delta = 0$, the jump transition occurs at $s = s_c$ and we recover our previous result. On the other hand if δ is exponentially small as

$$\delta = \varepsilon^{-1/2} \exp\left(-\frac{k^2}{2\varepsilon}\right), \quad (7.14)$$

where $k = O(1)$ is a parameter, the jump will occur at $s = k$ if $k < s_c$. Indeed, the first exponential remains small during the interval $0 < s < k$ and (7.13) reduces to

$$y \simeq \sqrt{2\pi} \exp\left[\frac{1}{2\varepsilon}\left(s^2 - k^2\right)\right] \quad (7.15)$$

showing no dependence with respect to the initial conditions.

We conclude that the size of the imperfection needs to be exponentially small as

$$\delta \sim \exp(-1/\varepsilon) \quad (7.16)$$

in order to have a significant effect on the delayed bifurcation transition. Mathematically, the exceptionally long delay comes from the fact that $y = 0$ is a solution of the non-autonomous equation (7.4). This delay remains large provided that δ is exponentially small like (7.16). It will be much smaller if δ is rationally small

---

[2] $\int_{-s_c}^{s} \exp\left(-\frac{x^2}{2\varepsilon}\right) dx = \sqrt{2\varepsilon} \int_{-s_c/\sqrt{2\varepsilon}}^{s/\sqrt{2\varepsilon}} \exp\left(-z^2\right) dz$
$= \sqrt{\frac{\pi\varepsilon}{2}} \left[\text{erf}\left(s/\sqrt{2\varepsilon}\right) + \text{erf}\left(s_c/\sqrt{2\varepsilon}\right)\right]$
$\simeq \sqrt{2\pi\varepsilon}.$

$(\delta \sim \varepsilon^p$ and $p > 0)$. This property is consistent with our observations on the latency time associated with the turn-on of the laser (see Sections 1.3.1 and 1.3.2). The exponential sensitivity confirms that measuring microscopically small initial quantities such as light coming from spontaneous emission in the laser mode is possible through macroscopic quantities (delays).

7.2.2 Moderately small ε

Keeping the same initial conditions but increasing ε, $y(t)$ does not have the time to exponentially relax to the stable zero solution before passing through $\lambda = 1$. This is particularly the case if $\lambda(0)$ is close to the bifurcation point $\lambda = 1$, as it is in the experiments. We analyze this case by introducing the new variables y_1 and s, and parameter λ_1 defined as

$$y = \varepsilon^p y_1, \ s = \varepsilon^r t, \ \lambda(0) - 1 = \varepsilon^n \lambda_1. \tag{7.17}$$

From Eq. (7.4), with $\lambda = \lambda(0) + \varepsilon t$, we obtain

$$\varepsilon^r y_1' = y_1(\varepsilon^n \lambda_1 + \varepsilon^{1-r} s - \varepsilon^p y_1), \tag{7.18}$$

where prime now means differentiation with respect to time s defined in (7.17). Balancing each term in the equation requires that

$$r = n = 1 - r = p. \tag{7.19}$$

The resulting equation for y_1 is now almost the same as the original equation

$$y_1' = y_1(\lambda_1 + s - y_1) \tag{7.20}$$

except that ε has disappeared. This equation can be solved with the initial condition $y_1(0) = \varepsilon^{-1} y(0)$.[3] The important result, however, is the fact that $n = r = 1/2$ which then implies that $\lambda - 1$ is of the form

$$\lambda - 1 = \varepsilon^{1/2}(\lambda_1 + s), \tag{7.21}$$

where both λ_1 and s are $O(1)$ quantities. It is no longer possible to define the delay as the critical value of λ where a sudden exponentially fast jump occurs. Nevertheless, the increase of $\lambda_j - 1$, where λ_j is determined at the same $y = y_j$, is proportional to $\sqrt{\varepsilon}$. This scaling law differs significantly from that associated with passage through a limit point.

---

[3] Physically, assuming a constant intensity is consistent with the fact that in most lasers swept near threshold, the intensity of spontaneous emission, proportional to population inversion, is almost constant.

7.2.3 Experiments

Soon after the delay of the laser bifurcation was theoretically predicted [21], slow passage experiments were investigated in different laboratories with different lasers. The first experimental study was proposed in 1987 by Scharpf [170, 171] who used a commercial argon-ion laser with an acousto-optic modulator (AOM) that slowly changes the cavity losses. By slowly increasing and then decreasing the voltage of the AOM, the losses first decrease, allowing the transition to the lasing mode, and then increase, leading the laser back to its zero intensity regime. See Figure 7.6. The experiments were simulated numerically by assuming a slowly varying pump but, as we shall later show, this is equivalent to a change of the losses for the argon-ion laser.

Consider our laser rate equations (1.7) and (1.8), now supplemented by a slowly varying pump. They are given by

$$\frac{dI}{dt} = I(-1 + D),$$ (7.22)

$$\frac{dD}{dt} = \gamma \left[A(\varepsilon t) - D(1 + I) \right],$$ (7.23)

where I is the intensity of the laser field in the cavity and D is the inversion of population. The pump parameter $A(\varepsilon t)$ is slowly varying with rate ε. It can be a time-periodic function of t or a linear function of t such as

Fig. 7.6 Forward and backward passage through the laser bifurcation. The figure represents the intensity of the laser output as a function of the voltage of the AOM. The forward transition exhibits a much larger delay than the backward transition, allowing a hysteresis cycle (from Figure 3.6A of Scharpf [170]).

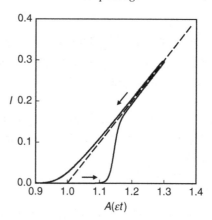

Fig. 7.7 Numerical solution of the laser equations (7.22) and (7.23). The values of the parameters are: $\gamma = 1$, $I(0) = 10^{-3}$, $D(0) = 0.9$. $A = A_0 + \varepsilon t$, where $A_0 = 0.9$ and $\varepsilon = 10^{-3}$ for $0 < t < 400$; $A = A_0 - \varepsilon(t - 400)$, where $A_0 = 1.3$ and $\varepsilon = 10^{-3}$ for $400 < t < 800$. The broken line is the non-zero intensity steady state $I = A - 1$.

$$A(\varepsilon t) = A_0 \pm \varepsilon t. \tag{7.24}$$

We next assume that we start our experiment at or near a stable steady state. For the forward transition ($A = A_0 + \varepsilon t$ with $A_0 < 1$), the initial conditions are

$$I(0) = I_0 << 1 \quad \text{and} \quad D(0) = A_0 \tag{7.25}$$

while for the backward transition ($A = A_0 - \varepsilon t$ with $A_0 > 1$), the initial conditions are

$$I(0) = A_0 - 1 \quad \text{and} \quad D(0) = 1. \tag{7.26}$$

A typical forward and backward slow passage experiment is simulated numerically in Figure 7.7. The values of the parameters are the same as those used in Figure 7.6. For the forward transition, we note a substantial delay followed by a quick jump to the non-zero steady state. For the backward transition, the bifurcation transition is delayed but this delay is much smaller. The broken line represents the non-zero steady state $I = A - 1$. The theory is summarized below and predicts an $O(1)$ delay in the forward transition, followed by a quick transition to the slowly varying solution $I = A - 1 + O(\varepsilon)$, and finally a backward transition near $A = 1$ exhibiting an $O(\varepsilon^{1/2})$ deviation from the laser threshold $A = 1$.

Forward transition

The forward transition can be analyzed in the same way as for our reference problem (7.4) and (7.5) by realizing that (1) if $\varepsilon \to 0$, the intensity I remains exponentially small before its fast jump, and (2) for larger values of ε and starting near the bifurcation point, the dynamical system interacts with the slowly varying bifurcation parameter. In the first case, we find that the delay is independent of the rate of change and only depends on its initial condition ($A_j - 1 = 1 - A_0$) while the delay is proportional to $\sqrt{\varepsilon}$ in the second case ($A_j - 1 \sim \sqrt{\varepsilon}$).

Sharpf *et al.* [170, 171] measured the delay for increasing values of ε. They observed that the delay changes like ε^m where $m = 0.56$ is found using a least-square fit of the data. This critical exponent is close to the theoretical result $m = 0.5$ predicted for moderately small values of ε and starting close to the laser bifurcation. Errors in the experiments were attributed to the inaccurate method used for determining the static threshold and the small number of available data points in the delayed region.

Backward transition

The backward passage problem is more straightforward to analyze. Away from the bifurcation point the solution for the intensity simply follows the slowly varying steady state $I = A(\varepsilon t) - 1 + O(\varepsilon)$. But close to the bifurcation point, a local analysis of the solution is needed. The analysis of the slow passage problem is similar to the analysis of the forward slow passage for moderately small values of ε. The main result is that the intensity deviates from the steady state by an $\sqrt{\varepsilon}$ quantity near the laser bifurcation point $A = 1$.

Changing the losses

In 1993, Bromley *et al.* [172] investigated the slow passage through the laser bifurcation in a CO_2 laser by slowly decreasing and then increasing the losses. In contrast to experiments using the pump parameter, they found that negative hysteresis cycles were possible, i.e. the forward transition occurs at higher intensities than the backward transition. See Figure 7.8. They explain their results by the fact that there exists a competition between the delay of the switch-on and an anticipatory switch-off. Both processes, delay and anticipation, depend on the sweeping rates and the level of noise. The rate equations now are of the form

$$\frac{dI}{dt} = I(-k(\varepsilon t) + D), \tag{7.27}$$

$$\frac{dD}{dt} = \gamma [A - D(1 + I)] \tag{7.28}$$

where

$$k = k_0 \pm \varepsilon t. \tag{7.29}$$

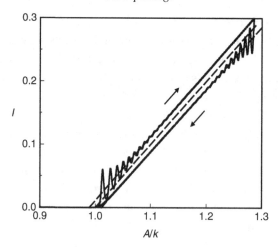

Fig. 7.8 Slow passage through the laser threshold by changing the losses. Numerical solution of Eqs. (7.27)–(7.29) with $\gamma = 10^{-3}$, $k_0 = 1$, $\varepsilon = \mp 10^{-5}$, $A = 0.9$, $\beta = 10^{-6}$, $I(0) = 10^{-3}$, and $D(0) = A$.

Introducing $N = D - k + 1$ into Eqs. (7.27) and (7.28), we find

$$\frac{dI}{dt} = I(-1 + N), \tag{7.30}$$

$$\frac{dN}{dt} = \gamma \left[A - (k-1) - N(1+I) - (k-1)I \right] - k'(\varepsilon t). \tag{7.31}$$

If $k'(\varepsilon t) << \gamma$ and $|k - 1| < 1$, Eqs. (7.30) and (7.31) reduce to the familiar rate equations (7.22) and (7.23) with $A(\varepsilon t) \equiv A_0 \pm \varepsilon t$ and $A_0 \equiv A - (k_0 - 1)$. The conditions on k are verified in the case of the experiments by Scharpf [171] who used an argon-ion laser with $\gamma = O(1)$, $\varepsilon = O(10^{-3})$, and $|k - 1| = 0.2$; see Figure 7.6. On the other hand, if $k'(\varepsilon t) = O(\gamma)$, we may not ignore the effect of the last term in (7.31) and some new phenomena related to the slow evolution of the control parameter could be possible. In Figure 7.8, a negative hysteresis cycle is simulated numerically using Eqs. (7.27) and (7.28) (a term βD with $\beta = 10^{-6}$ has been added to the right hand side of Eq. (7.27) to model the effect of noise). The figure represents intensity vs. A/k rather than k so that forward and backward evolutions can be compared to the one we had with a slowly varying pump. The hysteresis cycle is more rectangular than triangular and seems to follow two slowly varying regimes (upper and lower broken lines). We note that these regimes substantially deviate from the static diagram (middle broken line) by an amount that is larger than ε.

The slowly varying regimes are defined as solutions that only depend on the slow time

$$s \equiv \varepsilon t. \tag{7.32}$$

They are our new reference solutions since the steady state solutions in the static case (namely, $I = 0$ and $I = A - 1$) are no longer solutions of the non-autonomous laser equations. However, in contrast to the static case, only the solution $I = 0$ and $D = A(\varepsilon t)$ is an exact slowly varying solution. To determine whether there is another solution, we shall use a regular perturbation series in ε. Specifically, we seek a solution of (7.27) and (7.28) of the form $I = I_0(s) + \varepsilon I_1(s) + \dots$ and $D = D_0(s) + \varepsilon D_1(s) + \dots$ From Eqs. (7.27) and (7.28) with (7.32) and assuming $\gamma = O(\varepsilon)$, we find that the leading equations for I_0 and D_0 are given by

$$0 = I_0(-k(s) + D_0), \tag{7.33}$$

$$\frac{dD_0}{ds} = \frac{\gamma}{\varepsilon}[A - D_0(1 + I_0)]. \tag{7.34}$$

These equations admit two solutions, namely (1) $I_0 = 0$ and (2)

$$D_0 = k \quad \text{and} \quad I_0 = \frac{A - k}{k} - \varepsilon \gamma^{-1} \frac{k'}{k}. \tag{7.35}$$

The first solution corresponds to the exact slowly varying solution but the second solution is new. The first term in the expression of I_0 matches the non-zero steady state solution $I_0 = A/k - 1$. The second term, proportional to $\varepsilon \gamma^{-1}$, is the deviation due to the sweeping parameter. If $k = k_0 - \varepsilon t$ (forward transition) this correction term is positive, while if $k = k_0 + \varepsilon t$ (backward transition) this correction is negative. The two slowly varying solutions for the forward and backward transitions are shown in Figure 7.8 (upper and lower broken lines).

The decaying oscillations at each jump are the consequence of the small value of γ, allowing relaxation oscillations with a period proportional to $\gamma^{-1/2}$ and decaying on a γ^{-1} time interval. As described in Bromley *et al.* [172], these decaying oscillations are better simulated using a four-level model that introduces more damping than the two-level model. The level of noise and the initial conditions also play a role. Nevertheless, we have explained the negative hysteresis cycle by the presence of substantially different slowly varying attractors as the losses are swept back and forth. The size of this rectangular cycle increases as the ratio ε/γ increases. In the case of slow changes of the pump, this effect is also possible but will require a much larger rate of change of the order of $\varepsilon = O(\sqrt{\gamma})$. But then the change of the pump parameter is comparable to the time scale of the relaxation oscillations and we cannot assume that the pump is a slowly varying parameter.

7.3 Period-doubling bifurcation

Another topic that was and still is a source of theoretical investigations is the slow passage through a period-doubling bifurcation. In 1987, Dangoisse *et al.* [173] investigated experimentally the bifurcation diagram of a modulated CO_2 laser. Several groups had already published on the bifurcation diagram of a modulated laser (see Section 5.2.1) but Dangoisse *et al.* examined the slow passage through the first period-doubling bifurcation. Specifically, the cavity losses were modulated by a T-periodic sine wave and the laser output intensity was sampled at each time T. As a result, T-, $2T$-, and $4T$- periodic regimes will lead to one, two, and four points in the bifurcation diagram, respectively. Figure 7.9 exhibits three slow passage experiments where the modulation amplitude is swept back and forth. As clearly seen in the 300 Hz diagram, the back and forth evolutions are different for the Period 2 branch. As with the steady bifurcation problem, the forward transition experiences a larger delay than the backward transition. Comparing all three diagrams, we note that the width of the forward and backward hysteresis cycle increases with the ramp speed. The experiment was partially motivated by an analysis of the quadratic map with a slowly varying parameter [174] predicting a delayed period-doubling transition proportional to the square root of the ramp

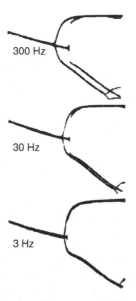

Fig. 7.9 Stroboscopic bifurcation diagram exhibiting the first period-doubling bifurcation. The traces have been obtained by sampling the laser output intensity at each period T. The figure represents both the forward and backward change of the modulation amplitude. Reprinted Figure 20 with permission from Dangoisse *et al.* [173]. Copyright 1987 by the American Physical Society.

speed. Dangoisse *et al.* [173] simulated the laser rate equations and found a good agreement with the experimental diagrams. In dimensionless form (see Section 5.2.1), the evolution equations for the modulated laser are given by

$$x' = -y - \eta x \left[1 + (A - 1)(1 + y) \right] \tag{7.36}$$

$$y' = (1 + y) \left[x - \delta \sin(\sigma t) \right]. \tag{7.37}$$

In these equations, A is the dimensionless pump parameter, $\eta = \omega_{RO}/(A-1)$ represents the damping rate of the relaxation oscillations where $\omega_{RO} = \sqrt{\gamma(A-1)}$ is the relaxation oscillation frequency. $\delta = m/\omega_{RO}$ and $\sigma = \omega_{mod}/\omega_{RO}$ are defined as the modulation amplitude and frequency, respectively. Using the values of the parameters documented in [173] ($\gamma = 2.083 \times 10^{-3}$, $A = 1.05$), we sequentially compute $\omega_{RO} = 1.0205 \times 10^{-2}$, $\omega_{mod} = 2.0944 \times 10^{-2}$, $\sigma = 2.0523$, and $\eta = 0.2041$. The modulation amplitude is changed as

$$\delta = \delta(0) + \varepsilon t, \tag{7.38}$$

where ε ranges from 2.5×10^{-6} to 2.5×10^{-4} for the ramp speeds indicated in Figure 7.9. Figure 7.10 illustrates the forward slow passage through the period-doubling bifurcation point. The T-periodic solution emerges at $\delta = 0$ and the

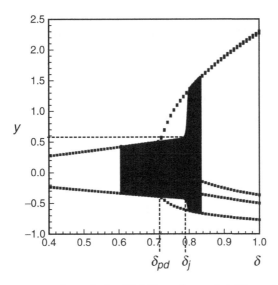

Fig. 7.10 Slow passage through the PD bifurcation point. Numerical solution of Eqs. (7.36)–(7.38) with $\gamma = 2.083 \times 10^{-3}$, $A = 1.05$, $\delta(0) = 0.6$, and $\varepsilon = 3 \times 10^{-5}$. The rapid oscillations jump from the T-periodic to the $2T$-periodic regimes at $\delta = \delta_j$. The dots represent the extrema of the periodic solutions in the static bifurcation diagram ($\varepsilon = 0$).

maximum amplitude follows the straight line $y_M \simeq 0.66\delta$. The static PD bifurcation point is located at $\delta_{pd} \simeq 0.72$. The oscillation jumps at $\delta = \delta_j > \delta_{pd}$ and the delay is defined as the deviation $\delta_j - \delta_{pd}$. We next wish to determine how the delay changes with ε. To this end, we simulate the rate equations for various ε and determine δ_j as the point where $y(t) = 0.66\delta + 0.05$. The best linear fit gives a slope close to 0.45, which is near 0.5. However, we need to be careful when drawing conclusions because the PD transition is dynamically two-dimensional and it is not clear if it can be captured by a one-dimensional theory such as the quadratic map.

7.4 Slow–fast dynamics

Although simple slow passage problems can be identified, we have seen that both theoretical and experimental results need to be carefully examined. There are other important slow passage problems such as the slow passage through a Hopf bifurcation which exhibits both delay and memory effects [175], and the passage through a homoclinic (infinite-period) orbit. Both transitions are possible for lasers subject to a saturable absorber or subject to an injected signal. But these transitions have never been investigated experimentally.

Steady and Hopf slow passage problems are important because they appear as part of more complex dynamical problems such as the polarization dynamics of lasers, the bursting oscillations of optical parametric oscillators (OPOs), and the passage through resonance in a modulated laser. In Chapter 8, we will find that the interpulse regime in an LSA is described by a linear equation for the intensity of the form

$$I' = I(-1 + D(\varepsilon t)),$$

where the population inversion D is slowly varying from negative to positive values. Mathematically, it is a slow passage through the steady bifurcation point $D(\varepsilon t) = 1$ and we know that the intensity will increase when

$$\int_0^t (-1 + D(\varepsilon t'))dt' = 0.$$

Bistable slow–fast cycles appear in OPOs and are explained by identifying a slow passage problem (see Chapter 12) that is not affected by noise. By contrast, noise has a definite effect on the passage through resonance if the modulation frequency is either increased or decreased (see Section 5.2.1).

7.5 Exercise

Analyze Eqs. (7.22) and (7.23) for the forward and backward bifurcation transition. To this end, first determine a slowly varying solution that depends on the slow time εt. In the forward case, it is the stability of the slowly varying solution $I = 0$ with respect to the fast time that determines the delayed bifurcation transition. In the backward transition, the transition from the slowly varying solution $I = A - 1 + O(\varepsilon)$ to $I = O(\varepsilon)$ occurs through a transition layer $|A - 1| = O(\varepsilon^{1/2})$.

Part III

Particular laser systems

8

Laser with a saturable absorber

In a laser with a saturable absorber (LSA), two spatially separated cells are placed in the laser cavity as shown in Figure 8.1. The roles of the two cells are quite different: one of them is pumped so that the atoms have a positive population inversion (active or amplifying medium); the other one is left with a negative population inversion (passive or absorbing medium). As these two media are in general different, they saturate at different power levels. The most interesting case corresponds to the situation where the absorber saturates more easily than the active medium, introducing nonlinear losses inside the cavity. This new nonlinearity is responsible for two phenomena. An LSA may exhibit optical bistability, i.e. two distinct stable steady states may coexist for a range of values of a parameter. It may also produce pulsating intensity oscillations which have been called "passive Q-switching" (PQS) in contrast to "active Q-switching" experiments such as the "gain switching" experiments discussed in Section 1.3.2.

The interest in LSAs varied very much over time, with peaks in the late 1960s for their large intensity pulses, in the mid 1980s for their chaotic outputs, and in the late 1990s for the design of compact microlasers. Historically, physicists trying to explain the irregular intensity pulses delivered by the ruby laser suspected the possible destabilizing role of a saturable absorber. Shimoda [177] proposed that the nonlinear losses generated by a saturable absorber could explain this phenomenon. Self-pulsing was also observed in semiconductor lasers with a degraded section that is acting as an absorber [130]. These first experiments were performed on multimode lasers and require model equations that are mathematically complicated. Experimental and theoretical studies then concentrated on single-mode gas lasers. In 1968, Lisitsyn and Chebotaev [178] obtained clear evidence of optical bistability by introducing a cell containing neon inside the cavity of a He-Ne laser. They observed hysteresis loops for the laser intensity as the losses, the gain, or the cavity detuning were varied. Furthermore, PQS was extensively studied for CO_2 LSAs with the idea that these high intensity pulses could be useful for applications

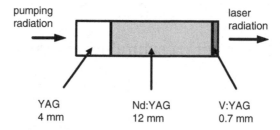

Fig. 8.1 Nd:YAG-V:YAG microchip solid state laser with a saturable absorber. It consists of a sandwich crystal that combines in one piece the cooling undoped part (undoped YAG crystal 4 mm long), the active laser part (YAG crystal doped with Nd^{3+} ions, 12 mm long), and the saturable absorber (YAG crystal doped with V^{3+} ions, 0.7 mm long). The diameter of the crystal is 5 mm. The microchip resonator consists of dielectric mirrors which are deposited directly on the monolith crystal surfaces. Redrawn from Figure 2b of Sulc *et al.* [176].

such as isotope separation. At that time, Brzhazovskii *et al.* [179] carried out a systematic investigation of PQS in CO_2 LSAs. The interest in LSA pulses disappeared when they were outperformed by mode-locking lasers capable of high power pulse production. In the late 1970s, however, the idea of developing an optical computer revived studies on optically bistable systems, including the LSA, but the real new impetus came in the 1980s when it was realized that PQS was a form of spontaneous laser instability that could lead to chaos. Work then concentrated on CO_2 LSAs with the objective of obtaining quantitative comparisons between experiments and theory on chaos [60, 63, 64, 180].

Today, the development of microchip solid state lasers and single mode semiconductor lasers has led to practical applications. PQS microchip lasers deliver extremely short (<1 ns) high-peak-power (>10 kW) pulses of light without the need for an additional external pulse generator. The short pulse widths are useful for high-precision optical ranging with applications in automated production. The high peak output intensities are needed for efficient nonlinear frequency generation or ionization of materials, with applications in micro-surgery and ionization spectroscopy. Self-pulsing semiconductor lasers exhibit a high repetition rate which ranges from hundreds of megahertz to a few gigahertz [33, 181]. They are interesting for telecommunications and for optical data storage using compact disc (CD) or digital versatile disc (DVD) systems [182, 183].

Because of the diversity of LSAs as well as scientific objectives, model equations were derived each time in order to describe particular features of the PQS pulses. Reformulated in dimensionless form, they exhibit similar solutions. In this chapter, we first introduce the new parameters associated with the saturable absorber. We then describe the bistable and pulsating responses. We start with a simple one-variable LSA equation and then introduce progressively more dependent variables as the observed time-dependent regimes become richer.

8.1 LSA parameters

As with the active medium in a laser, (saturable) absorbers may be described by quantum systems with two, three, four, or an infinite number of levels, depending on the nature of the absorber (semiconductor, ion-doped crystal, gas) and the operating wavelength of the laser.

In its simplest version, the absorber is described by a two-level system. Its main dynamical effects are well accounted for by introducing its **absorption coefficient**, its **saturation intensity**, and its **relaxation time**. Specifically, we introduce the saturability a, the absorption parameter $\overline{A}$, and the reduced relaxation rate $\overline{\gamma}$ of the population difference, which should be compared with the active medium parameters A and γ that we previously introduced. It is worthwhile to briefly discuss the range of parameter values which are likely to lead to new phenomena.

- The strength of the interaction of light with the passive medium is measured by the parameter $\overline{A}$ which relates the absorption coefficient to the gain at threshold. In the absence of an absorber, the laser threshold is reached as the (unsaturated) gain compensates the losses ($A = 1$). As the absorber is introduced, the losses are increased and the threshold is shifted proportionally to $\overline{A}$. A strong absorber ($\overline{A} \gg 1$) would block off laser action; a weak one ($\overline{A} \ll 1$) would have no influence. Therefore, interesting values of the absorption are of the order of the gain above threshold, i.e. $\overline{A} \sim A - 1$.
- The saturability coefficient a therefore measures the relative saturation effect of the absorber with respect to the active medium. A large saturability ($a \gg 1$) means that the absorber saturates more easily, i.e. at smaller laser power, than the active medium. On the other hand, a small saturability ($a \ll 1$) implies that the absorption in the passive medium is weak. A bistable steady operation is more likely to appear for large a because the absorption is then strongly nonlinear.
- As for γ, $\overline{\gamma}$ is typically a small parameter for many gaseous absorbers, meaning that the decay of the inversion is slow compared to the decay of the field in the cavity. Often $\overline{\gamma}$ is comparable to γ but if $\overline{\gamma} \gg \gamma$, i.e. a fast absorber as occurs in some solid state absorbers, a simplification of the laser equations is possible. A fast absorber with weak efficiency ($\overline{A} \ll 1$) but large saturability ($a \gg 1$) may exhibit nonlinear behaviors competing in time with the active medium. The fast response compensates for the weaker interaction.

Typical values of the parameters are shown in Table 8.1 for different lasers. We note the small values of γ and $\overline{\gamma}$. They are typically $O(10^{-3})$ small for CO_2 and semiconductor lasers and even smaller for microchip solid state lasers. Solid state

Table 8.1 *Typical values of the parameters for common LSAs.*

Laser	γ	$\overline{\gamma}$	$\overline{A}$	a
$CO_2 + SF_6{}^a$	3×10^{-3}	6.6×10^{-4}	2.78	1768
$Nd^{3+}:YAG + Cr^{4+b}$	1.8×10^{-6}	6.4×10^{-5}	3.96	0.085
semiconductor ^{c}	1.8×10^{-3}	1.4×10^{-3}	4.73	2.16

^{a} [60].
^{b} [184, 185].
^{c} [182, 186].

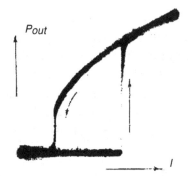

Fig. 8.2 Hysteresis cycles in a He-Ne laser representing the output power as a function of the discharge current. Reprinted Figure 3 with permission from Lisitsyn and Chebotaev [178]. Copyright 1968 American Institute of Physics.

lasers also exhibit a small a compared to CO_2 and semiconductor lasers. The presence of several small parameters suggests the need for asymptotic expansions and opens the possibility of analytic treatments as demonstrated in previous chapters.

8.2 LSA basic phenomena

In this section, we propose an overview of new phenomena arising because of the presence of the absorber, namely, optical bistability and PQS.

8.2.1 Optical bistability

We know from Chapter 1 that the rate equations for a single-mode laser only admit one stable steady state for each value of A. Surprisingly enough, two stable steady states may coexist for the same value of A in LSAs. This phenomenon, called *"optical bistability,"* has been observed for a long time. Figure 8.2 shows the hysteresis cycles obtained in 1968 by Lisitsyn and Chebotaev [178] in a He-Ne laser (0.6328 μm) with an intracavity cell containing Ne at a lower pressure than the amplifying cell. The discharge current in the active medium was used as a control

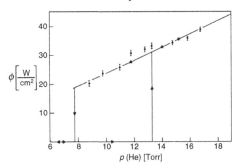

Fig. 8.3 Laser flux as a function of pumping for a CO_2 laser with SF_6 as a saturable absorber. The pumping rate is proportional to the partial pressure of He in the plasma feed mixture. Reprinted Figure 1 from Ruschin and Bauer [187]. With permission from Elsevier.

parameter. They observed that the laser switches on at a discharge current much larger than that for which the laser switches off. In the intermediate domain, the laser may be ON or OFF depending on its previous history. Similar hysteresis phenomena were observed in CO_2 lasers with SF_6, CH_3I, and many other gases (see Figure 8.3).

The minimal rate equation model for an LSA is the class A rate equation (see Section 1.1) where all population variables are adiabatically eliminated. It is given by

$$I' = I\left(-1 + \frac{A}{1+I} - \frac{\overline{A}}{1+aI}\right), \tag{8.1}$$

where the new term $-\overline{A}/(1+aI)$ describes the saturation effect of the absorber. The zero intensity solution admits a bifurcation point at $A = A_{th}$, where

$$A_{th} \equiv 1 + \overline{A}. \tag{8.2}$$

This is the new laser threshold where the gain A compensates the linear losses $1 + \overline{A}$. At that point, we may switch to a non-zero intensity steady state if there exists a positive solution of

$$-1 + \frac{A}{1+I} - \frac{\overline{A}}{1+aI} = 0 \tag{8.3}$$

with $A = A_{th}$. Solving this equation for I leads to the roots $I = 0$ and

$$I = \frac{\overline{A}a - 1 - \overline{A}}{a} > 0. \tag{8.4}$$

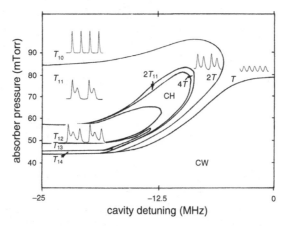

Fig. 8.4 PQS pulse shapes as a function of the absorber pressure and the laser detuning for a $CO_2 + CH_3I$ LSA (from Figure 2 of Dangoisse *et al.* [188]).

The inequality thus requires that

$$\overline{A}a > 1 + \overline{A} = A_{th} \tag{8.5}$$

meaning that the saturation nonlinearity should be larger in the absorber than in the active medium.

8.2.2 Passive Q-switching

PQS is a generic term that covers quite different forms of self-pulsating intensity oscillations in LSAs. Smooth quasi-sinusoidal oscillations (as, for example, in the upper right corner of Figure 8.4) as well as repetitive high intensity spiking (upper left of Figure 8.4) are called PQS. Some PQS lasers even display different forms of bursting and chaos (central part of Figure 8.4). We first describe simple experimental observations on PQS. Richer pulsating activities such as bursting oscillations and chaos in the CO_2 LSA will be considered later (see Section 8.4.1). Recall that large intensity spikes are observed as we turn on a laser from a below- to an above-threshold pump value (see Section 1.3.1). The underlying idea is that, as the laser switches on, it benefits from the accumulated population inversion, i.e. a larger gain than when it is operated in a cw regime. If now a saturable absorber is added to the laser, it will play the role of an optical intracavity switch, hence the name Q-switching. As was already understood in 1966, this switch *"emphasizes the largest amplitude fluctuation occurring at the initiation of the laser oscillation"* [189].

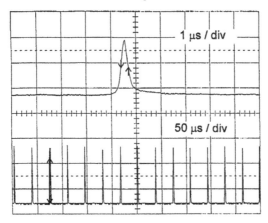

Fig. 8.5 Oscilloscope traces of a train of pulses at 29 kHz from a Nd^{3+}:YAG laser with a Cr^{4+}:YAG saturable absorber. Upper trace is an expanded shape of a single pulse, showing a 300 ns width. Reprinted Figure 1 with permission from Shimony *et al.* [190]. Copyright 1996 IEEE.

As already mentioned, the advent of mode-locked lasers made the PQS technique for pulse production obsolete. But the discovery that LSAs as well as other nonlinear systems may exhibit sustained intensity oscillations triggered a large interest in the investigation of lasers as pure dynamical systems. In this direction, the CO_2/N_2O laser with molecular gases such as SF_6 or CH_3I as saturable absorbers appeared in the 1980s as the simplest and richest optical system displaying several bifurcations. After the craze for optical chaos, the interest in LSAs disappeared until the advent of integrated microchip lasers. The most recent technologies use microchip lasers with two kinds of saturable absorbers. The active medium is most often made of a thin crystal of Nd^{3+}-doped YVO_4 or YAG as the active part, and the saturable absorber is either a Cr^{4+}-doped layer of the same crystal directly deposited on the active part or a semiconductor deposited on the surface of a mirror (SESAM). Typically, a 440 μm long Nd^{3+}-doped YVO_4 with a SESAM mirror delivered 2.6 ns/1.6 kW pulses at 440 kHz. See Figure 8.5. Varying the laser design (cavity length, active medium thickness), it is possible to control both the duration and the frequency of these pulses. For instance, durations as short as 37 ps are possible (see Figure 8.6).

Both the early (*c.* 1966) and the more recent studies (*c.* 2000) were design-oriented. Specifically, they aimed at optimizing the pulsed operation, in terms of pulse duration, energy, peak power, and repetition rate. Here we focus on PQS dynamical properties such as the conditions for its appearance, their frequency, and time evolution. Our objective is to provide simple formulae that make visible the basic physical mechanisms responsible for PQS.

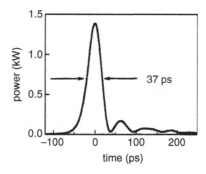

Fig. 8.6 Oscillatory trace of a single-frequency 37 ps Q-switched pulse with a peak power of 1.4 kW and a repetition rate of 160 kHz (from Figure 1 of Spühler *et al.* [191]).

8.3 Rate equations

In this section, we analyze a simple extension of the rate equation (8.1) that takes into account the response of the saturable absorber. Specifically, the rate equations for a two-level laser with a saturable absorber consist of three nonlinear first order differential equations for the intensity of the laser field I, the laser population inversion D, and the absorber state population $\overline{D}$. In dimensionless form, these equations are similar to Eqs. (1.7) and (1.8) for a single-mode laser and we shall not detail their derivation. They are given by

$$\frac{dI}{dt} = (-1 + D + \overline{D})I, \tag{8.6}$$

$$\frac{dD}{dt} = \gamma\,[A - D(1 + I)], \tag{8.7}$$

$$\frac{d\overline{D}}{dt} = \overline{\gamma}\left[-\overline{A} - \overline{D}(1 + aI)\right]. \tag{8.8}$$

In these equations, A and $\overline{A}$ are the pump parameters for the lasing medium and the absorber. γ and $\overline{\gamma}$ are the gain and saturable absorber decay rates normalized to the cavity lifetime. a is the absorber saturability.

The absorber may be fast, like the Cr^{4+} ions in the YAG laser, or slow, like SF_6 in the CO_2 laser. In the first case, a quasistatic approximation may be used to eliminate $\overline{D}$. The resulting two-variable reduction of the LSA rate equations is analyzed in the next subsection and a linear stability analysis of the steady states leads us to two main bifurcations. The three-variable LSA equations (8.6)–(8.8) will then be analyzed but, as we shall demonstrate, the two-variable LSA equations already capture the simple PQS regimes observed with microchip LSAs and even some CO_2 LSAs.

8.3.1 Steady state solutions

We first determine the steady state solutions of Eqs. (8.6)–(8.8). We find a zero intensity solution

$$I = D - A = \overline{D} + \overline{A} = 0 \tag{8.9}$$

and a non-zero intensity solution given by (in parametric form)

$$A = (1 + I)\left(1 + \frac{\overline{A}}{1 + aI}\right), \tag{8.10}$$

$$D = \frac{A}{1 + I} \quad \text{and} \quad \overline{D} = -\frac{\overline{A}}{1 + aI}. \tag{8.11}$$

Note that these steady state solutions depend only on $\overline{A}$ and a. Bistability is possible if there exists a subcritical branch of steady states that folds back at sufficiently large amplitude. Inversely, a supercritical bifurcation at threshold leads to single solution as is the case for the solitary laser. To determine the direction of bifurcation, we analyze Eq. (8.10) for small I. We find a bifurcation point located at $A = A_{th}$, where A_{th} is defined by (8.2), from where a branch of steady states emerges (1) for $A > A_{th}$ if

$$A_{th} - a\overline{A} > 0 \tag{8.12}$$

or (2) for $A < A_{th}$ if

$$A_{th} - a\overline{A} < 0. \tag{8.13}$$

Case 1 (supercritical bifurcation) occurs if a is sufficiently small while Case 2 (subcritical bifurcation) appears if $a > 1$. Condition (8.13) is the necessary condition (8.5) for bistability that we previously derived. In this case, the subcritical branch of steady states folds back at a limit point where $dA/dI = 0$. Using (8.10), we determine dA/dI as

$$\frac{dA}{dI} = 1 + \frac{\overline{A}}{1 + aI} - \frac{(1 + I)\overline{A}a}{(1 + aI)^2} \tag{8.14}$$

and find that the limit point condition $dA/dI = 0$ leads to the following quadratic equation for I

$$a^2 I^2 + 2aI + A_{th} - a\overline{A} = 0. \tag{8.15}$$

Because of (8.13), the last term in Eq. (8.15) is negative and a positive real root of the quadratic equation is always possible. The non-zero intensity steady state

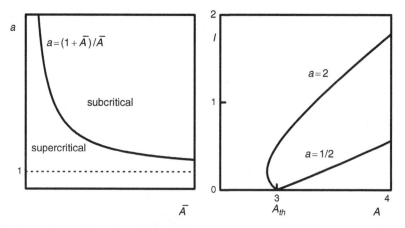

Fig. 8.7 Left: the direction of bifurcation depends on a and $\overline{A}$. If $a < 1$, the steady bifurcation is always supercritical. Right: the branch of non-zero intensity steady states for $a = 1/2$ (supercritical bifurcation) or $a = 2$ (subcritical bifurcation). $\overline{A} = 2$.

solution is represented in Figure 8.7 for the two cases. In the subcritical case, hysteresis is possible. Using Eq. (8.15) with $\overline{A} = a = 2$, we find that the limit point is located at $I_{LP} = 0.207$. However, in order to determine if a bistable response is possible, we need to examine the stability of the coexisting solutions.

8.3.2 Two-variable reduction and PQS

The stability analysis may be performed using the full three-variable model equations (8.6)–(8.8). In this section, we first examine the reduced two-variable model in which the absorber population is adiabatically eliminated. The reduced mode exhibits both bistability and PQS phenomena. The condition for the elimination of $\overline{D}$ by a quasi-steady state approximation is that $\overline{\gamma} \gg \gamma$, which is valid for a fast absorber, such as the Cr^{4+} ions in the Nd^{3+}:YAG laser. As we shall later see, this elimination is also valid if $\overline{\gamma}a \gg \gamma$, which is the case for the $CO_2 + SF_6$ laser. From (8.8), we then find that $\overline{D}$ rapidly changes unless the right hand side is close to zero. After a short initial transition, we expect that

$$\overline{D} \simeq -\frac{\overline{A}}{1 + aI} \tag{8.16}$$

and Eqs. (8.6) and (8.7) reduce to the following equations for I and D

$$\frac{dI}{dt} \simeq \left(-1 - \frac{\overline{A}}{1 + aI} + D\right)I, \tag{8.17}$$

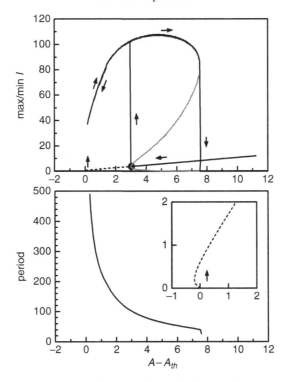

Fig. 8.8 Bifurcation diagram of the steady and pulsating intensity solutions. The solutions satisfy Eqs. (8.17) and (8.18) wih $\overline{A} = 1$, $a = 5$, and $\gamma = 10^{-2}$. Top: extrema of the intensity for the steady and periodic solutions obtained by increasing and then decreasing $A - A_{th}$. The dot marks the Hopf bifurcation point as determined from the linearized theory. Bottom: The period of the stable periodic solutions becomes unbounded as $(A - A_{th})^+ \to 0$. The inset is a blow-up of the branch of steady states shown in the top figure.

$$\frac{dD}{dt} = \gamma \left[A - D(1 + I) \right]. \tag{8.18}$$

These equations are nothing more than the standard rate equations (SRE) where the field losses term has been changed from -1 to $-1 - \overline{A}/(1 + aI)$ and is now a nonlinear function of the intensity. A typical bifurcation diagram of the stable steady and time-periodic solutions is shown in Figure 8.8, and has been obtained numerically by scanning A back and forth. The branch of unstable periodic solutions has been determined numerically by integrating the laser equations backward in time (gray line). For $A < A_{th}$, the zero intensity solution is the only stable steady state. If A is progressively decreased from $A > A_{th}$, the periodic solution branch suddenly stops as the limit-cycle touches the intermediate part of the S-shaped steady state branch ($A = A_{th}$). The period of the oscillations is then infinite. The limit-cycle orbit in the phase plane is called homoclinic because the closed orbit first leaves a saddle-point and then returns to it (homoclinic bifurcation [8]). If $A > A_{th}$, the laser jumps into high amplitude pulsating intensities (PQS). For $A_H < A < A_{PLP}$, where A_H and A_{PLP} denote a Hopf bifurcation point and a limit point of periodic solutions, respectively, the pulsating intensity regime coexists with a stable non-zero intensity steady state. Bistability of distinct steady states is not possible for the values of the parameters of Figure 8.8 because the Hopf bifurcation point has destabilized the upper branch of steady states in the region of coexistence.

Linear stability

The linearized equations for the non-zero intensity steady state are given by

$$\begin{pmatrix} \dfrac{du}{dt} \\ \dfrac{dv}{dt} \end{pmatrix} = \begin{pmatrix} \dfrac{\bar{A}aI}{(1+aI)^2} & I \\ -\gamma D & -\gamma(1+I) \end{pmatrix} \begin{pmatrix} u \\ v \end{pmatrix}. \tag{8.19}$$

The characteristic equation for the growth rate is then of the form

$$\sigma^2 - T\sigma + \Delta = 0, \tag{8.20}$$

where

$$T = \left[\frac{\bar{A}aI}{(1+aI)^2} - \gamma(1+I) \right], \tag{8.21}$$

$$\Delta = \gamma I \left[-\frac{\bar{A}a(1+I)}{(1+aI)^2} + D \right]$$

$$= \gamma I \left[-\frac{\bar{A}a(1+I)}{(1+aI)^2} + 1 + \frac{\bar{A}}{1+aI} \right]. \tag{8.22}$$

Note that $\Delta = \gamma I \frac{dA}{dI}$, where dA/dI is defined by (8.14). We may therefore propose a geometrical condition for stability. A negative Δ implies real roots with opposite signs, meaning instability like a saddle-node. Therefore, the branch $I = I(A)$ with negative slope corresponds to an unstable solution. Determining whether the part with a positive slope is stable or not requires a more refined approach.

Table 8.2 *Low and high intensity Hopf bifurcations.*

	I	ω^2	A
Low intensity	$\gamma/(a\overline{A})$	$\gamma I(A_{th} - \overline{A}a) > 0$	$\simeq A_{th}$
High intensity	$\sqrt{\overline{A}/(\gamma a)}$	γI	$\sqrt{\overline{A}/(\gamma a)}$

A Hopf bifurcation occurs if $\sigma = \pm i\omega$. Substituting this into Eq. (8.20), we obtain the conditions

$$T = 0 \quad \text{and} \quad \Delta > 0. \tag{8.23}$$

The condition $T = 0$ implies

$$\frac{\overline{A}aI}{(1+aI)^2} - \gamma(1+I) = 0. \tag{8.24}$$

For $\gamma = 0$, we find that either $I = 0$ or $I = \infty$ suggesting a small and a large root for $\gamma \neq 0$ small. Assuming $I \ll 1$, we obtain from Eq. (8.24) the approximation

$$\overline{A}aI - \gamma = 0 \quad \text{or} \quad I = \frac{\gamma}{a\overline{A}}. \tag{8.25}$$

On the other hand, assuming I large, we obtain from Eq. (8.24)

$$\frac{\overline{A}aI}{a^2 I^2} - \gamma I = 0 \quad \text{or} \quad I = \sqrt{\frac{\overline{A}}{\gamma a}}. \tag{8.26}$$

We thus find two candidates for a Hopf bifurcation. We still need to verify the second condition in (8.23). For the small intensity Hopf point (8.25) we have

$$\Delta \simeq \gamma I \left(-\overline{A}a + 1 + \overline{A}\right) > 0 \tag{8.27}$$

provided that $-\overline{A}a + 1 + \overline{A} > 0$, which is the condition for a supercritical branch of steady states $(A > A_{th})$. For the high intensity Hopf point (8.26) we have

$$\Delta \simeq \gamma I > 0, \tag{8.28}$$

which is always verified. The properties of the two possible Hopf bifurcation points are summarized in Table 8.2. Note that the low intensity Hopf bifurcation occurs only if the steady bifurcation is supercritical $(A_{th} - \overline{A}a > 0)$. Both Hopf frequencies are $O(\gamma^{1/2})$ small and the high intensity Hopf frequency exactly matches the RO frequency of the solitary laser.

In summary, we have found that sustained oscillations are possible through a Hopf bifurcation. But a Hopf bifurcation only reveals the existence of small amplitude (stable or unstable) solutions and does not explain the strongly pulsating oscillations observed experimentally and numerically. We need a different tool to describe analytically the large amplitude oscillations.

Pulsating solutions

In order to obtain an analytical understanding of the fast pulsating solutions, we consider a method that we already used for the turn-on pulse and for the strongly modulated laser. The PQS regimes are characterized by high intensity pulses separated by long time intervals where the intensity is almost zero. Figure 8.9 shows a

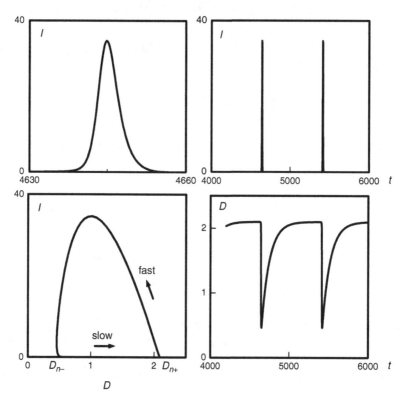

Fig. 8.9 Pulsating oscillations. Between successive pulses, the intensity is almost zero (upper right) and the population inversion slowly increases exponentially (lower right). A single pulse is shown in the upper left figure and the limit-cycle orbit is shown in the phase plane in the lower left figure. The periodic solution has been obtained numerically from Eqs. (8.17) and (8.18) with $A = 2.1$, $\overline{A} = 1$, $a = 5$, and $\gamma = 10^{-2}$.

typical example. Numerical simulations for progressively smaller values of γ indicate that the high intensity pulses occur on an $O(1)$ time interval with an intensity proportional to γ^{-1}. Moreover the interpulse period increases like γ^{-1}. This suggests seeking two separate approximations for the low and high intensity regimes. This analysis is proposed as an exercise in Section 8.5.1. It leads to equations for a map that provide the value of the population inversions at the end of each pulse, $D = D_{n-}$, as well as the interpulse period Δt_{n+1}. These equations are given by

$$(A - 1 - \overline{A})\gamma \Delta t_{n+1} - (D_{n-} - A)(\exp(-\gamma \Delta t_{n+1}) - 1) = 0, \quad (8.29)$$

$$\ln(D_{n+1-}/D_{n+}) - (D_{n+1-} - D_{n+}) = 0, \quad (8.30)$$

where D_{n+} denotes the population at the beginning of the nth pulse and is defined by

$$D_{n+} \equiv (D_{n-} - A) \exp(-\gamma \Delta t_{n+1}) + A. \quad (8.31)$$

A T-periodic limit-cycle solution oscillating between $D = D_-$ and $D = D_+$ therefore satisfies Eqs. (8.29)–(8.31) with $D_{n+1-} = D_{n-}$ or, equivalently, the conditions

$$(A - 1 - \overline{A})\gamma T - (D_- - A)(\exp(-\gamma T) - 1) = 0, \quad (8.32)$$

$$\ln(D_-/D_+) - (D_- - D_+) = 0, \quad (8.33)$$

where D_+ is

$$D_+ = (D_- - A) \exp(-\gamma T) + A. \quad (8.34)$$

Equations (8.32) and (8.33) are transcendental equations which must be solved numerically. However, we note that, numerically, $D_+ > A_{th}$. If $\overline{A}$ is sufficiently large, D_+ will be large and from (8.33), we then note that

$$D_- \simeq D_+ \exp(-D_+) \quad (8.35)$$

is small. Neglecting D_- in (8.32) leads to an implicit solution for the period T as a function of A given by

$$A - A_{th} \simeq \frac{A_{th}(1 - \exp(-\gamma T))}{\gamma T + \exp(-\gamma T) - 1}. \quad (8.36)$$

This approximation is compared to the numerically computed period as a function of the pump parameter in Figure 8.10. The hyperbolic behavior of the period as $A - A_{th} \to 0^+$ is captured analytically by assuming γT large in the expression (8.36). We find that

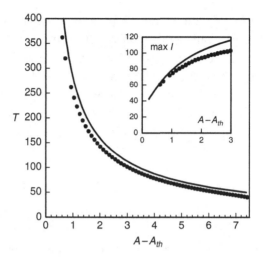

Fig. 8.10 Comparison between numerical (dots) and analytical (full line) approximations of the period as a function of $A - A_{th}$. The inset figure shows the maximum intensity and its approximation. The agreement is better close to threshold. Parameters are $\overline{A} = 1$, $a = 5$, and $\gamma = 10^{-2}$.

$$T \simeq \gamma^{-1} \frac{A_{th}}{A - A_{th}}. \tag{8.37}$$

The maximum intensity is provided by the large-I approximation of the pulse. It is given by

$$I \simeq \gamma^{-1}(D_+ - 1 - \ln(D_+)), \tag{8.38}$$

where $D_+ \simeq (A - A_{th})\gamma T$. In Exercise 8.5.1, we propose to derive frequently used formulae for the repetition rate and pulse width. These expressions require several approximations that may lead to erroneous results if their domain of validity is not respected.

A few remarks are in order. We were successful in describing the PQS train of pulses because the limit-cycle solution could be constructed using two separate approximations. The low intensity approximation satisfies a two-variable linear problem that we were able to solve. The high intensity approximation comes from a nonlinear problem for which we could find a first integral. In order to determine the intensity pulse waveform, a second integration is needed, however. Nevertheless, we were able to describe the bifurcation diagram for the period and the maxima of the intensity. A second point that is worthwhile emphasizing is the fact that parameter a does not appear in the leading approximations of the period or maxima. This is not the case if we apply the same construction technique for the three-variable LSA.

8.3.3 *Three-variable rate equations model*

The two-variable model is deduced from the three-variable one in the limit $\overline{\gamma} \rightarrow \infty$. Therefore both models have the same steady state solutions but their stability properties may be different. In this section we consider the complete three-variable model and analyze the stability of the steady states. We then examine the limit $\overline{\gamma} \rightarrow \infty$ of our results.

Linear stability

The characteristic equation for the three-variable equation is obtained from the following determinant

$$\det \begin{pmatrix} -1 + D + \overline{D} - \sigma & I & I \\ -\gamma D & -\gamma(1 + I) - \sigma & 0 \\ -\overline{\gamma} a \overline{D} & 0 & -\overline{\gamma}(1 + aI) - \sigma \end{pmatrix} = 0, \tag{8.39}$$

which we will evaluate for both the zero and the non-zero intensity solutions.

Stability of the zero intensity solution

For the zero intensity steady state, (8.39) reduces to

$$\det \begin{pmatrix} -1 + A - \overline{A} - \sigma & 0 & 0 \\ -\gamma A & -\gamma - \sigma & 0 \\ \overline{\gamma} a \overline{A} & 0 & -\overline{\gamma} - \sigma \end{pmatrix} = 0 \tag{8.40}$$

and leads to three solutions given by

$$\sigma_1 = A - A_{th}, \quad \sigma_2 = -\gamma, \quad \text{and} \quad \sigma_3 = -\overline{\gamma}. \tag{8.41}$$

The zero intensity solution is thus stable (unstable) if $A < A_{th}$ (if $A > A_{th}$).

Stability of the non-zero intensity solution

For the non-zero intensity steady state (8.10)–(8.11), (8.39) leads to the following third order polynomial in σ

$$\sigma^3 + C_1 \sigma^2 + C_2 \sigma + C_3 = 0. \tag{8.42}$$

The coefficients C_j $(j = 1, 2, 3)$ are defined by

$$C_1 \equiv \gamma(1 + I) + \overline{\gamma}(1 + aI),$$

$$C_2 \equiv I\left(\gamma\frac{A}{1 + I} - a\overline{\gamma}\frac{\overline{A}}{1 + aI}\right) + \gamma\overline{\gamma}(1 + I)(1 + aI),$$

$$C_3 \equiv \gamma\overline{\gamma}I\left(\frac{A(1 + aI)}{1 + I} - \frac{a\overline{A}(1 + I)}{(1 + aI)}\right). \tag{8.43}$$

The necessary and sufficient conditions for all σ having a negative real part are the Routh–Hurwitz conditions [29] given by

$$C_1 > 0, \; C_3 > 0, \quad \text{and} \quad C_1C_2 - C_3 > 0. \tag{8.44}$$

Note that the three conditions imply $C_2 > 0$. The first condition in (8.44) is always satisfied since intensities are positive. In order to analyze the second one, we use (8.14) and rewrite C_3 as

$$C_3 = \gamma\overline{\gamma}I(1 + aI)\left(\frac{A}{1 + I} - \frac{a\overline{A}(1 + I)}{(1 + aI)^2}\right)$$

$$= \gamma\overline{\gamma}I(1 + aI)\frac{dA}{dI}. \tag{8.45}$$

The condition $C_3 > 0$ thus requires that

$$dA/dI > 0 \tag{8.46}$$

as we already found from the two-variable model. The slope of the steady state branch $I = I(A)$ must be positive. The third condition is associated with the Hopf bifurcation condition.

Hopf bifurcation

The conditions for a Hopf bifurcation resulting from the change of stability of the non-zero steady state can be analyzed in the following way. After introducing the decomposition $\sigma = \sigma_r + i\sigma_i$ into the characteristic equation, we seek a solution for σ_r and σ_i assuming $|\sigma_r| \ll 1$ in the vicinity of the Hopf bifurcation. We obtain

$$\sigma_r = \frac{1}{2C_2}(C_3 - C_1 C_2) \quad \text{and} \quad \sigma_i^2 = C_2 + O(\sigma_r). \tag{8.47}$$

Since for stable solutions we have $C_2 > 0$, the Routh–Hurwitz condition $C_1 C_2 - C_3 > 0$ clearly means $\sigma_r < 0$. The Hopf bifurcation occurs if $\sigma_r = 0$ or

$$C_1 C_2 - C_3 = 0. \tag{8.48}$$

The condition (8.48) can be rewritten as

$$\frac{I}{(1 + I)(1 + aI)} \left(\frac{\gamma}{\overline{\gamma}} A - \frac{\overline{\gamma}}{\gamma} a\overline{A} \right) + \gamma(1 + I) + \overline{\gamma}(1 + aI) = 0. \tag{8.49}$$

We next recall that γ and $\overline{\gamma}$ are small dimensionless parameters (see Table 8.1) and explore two limit cases.

Powell and Wolga approximation for the high intensity Hopf bifurcation Assuming that the steady state intensity is not too large, we may neglect the last two terms in (8.49) and obtain the elegant stability condition

$$A < A_{H1} = \left(\frac{\overline{\gamma}}{\gamma} \right)^2 a\overline{A}, \tag{8.50}$$

which was derived by Powell and Wolga in 1971 [59]. The intensity I_H and the frequency ω_H at the Hopf bifurcation are obtained from Eq. (8.10) and $\omega_H^2 = C_2$, respectively. For small γ and $\overline{\gamma}$, ω_H^2 admits the approximation

$$\omega_H^2 = \frac{I_H}{1 + aI_H} \left(\gamma \left(A_{th} + aI_H \right) - a\overline{\gamma}\,\overline{A} \right) > 0. \tag{8.51}$$

However, the approximation (8.50) is asymptotically valid only if $I = O(1)$ and $\overline{\gamma} = O(\gamma)$ as $\gamma \to 0$, which is obviously wrong for the Nd^{3+}:YAG + Cr^{4+} laser but correct for the semiconductor LSAs and only fair for the CO$_2$ + SF$_6$ LSA. We may take into account the high value of a for the CO$_2$ + SF$_6$ laser by assuming $a = O(\overline{\gamma}^{-1})$ and $\gamma = O(\overline{\gamma})$. We then find that the intensity at the Hopf bifurcation point is given by

$$I_H = -\frac{1}{2} + \sqrt{\frac{1}{4} + \frac{\overline{A}}{\gamma a}}, \tag{8.52}$$

which matches the expression of the two-variable approximation if $\gamma a \gg 1$ (see Table 8.2).

Very low intensity Hopf bifurcation There exists another limit of Eq. (8.49) which verifies the scaling $I = O(\gamma)$. This leads to the occurrence of a different Hopf bifurcation in the case of a steady supercritical bifurcation. From Eqs. (8.10) and (8.49), and $\omega_H^2 = C_2$, we find that the Hopf bifurcation location and frequency are

$$I_{H2} = \frac{\gamma + \overline{\gamma}}{\left(\frac{\overline{\gamma}}{\gamma} a \overline{A} - \frac{\gamma}{\overline{\gamma}} A_{th} \right)}, \tag{8.53}$$

$$A_{H2} = A_{th} + I(A_{th} - aA), \tag{8.54}$$

$$\omega_{H2}^2 = I(\gamma A_{th} - a\overline{\gamma}\overline{A} + \gamma\overline{\gamma}) > 0 \tag{8.55}$$

provided that

$$\frac{\overline{\gamma}}{\gamma} a \overline{A} - \frac{\gamma}{\overline{\gamma}} A_{th} > 0. \tag{8.56}$$

Note that the double requirement of a supercritical steady bifurcation ($A_{th} - a\overline{A} > 0$) and the necessary condition (8.56) for a Hopf bifurcation implies the inequality

$$\overline{\gamma} > \gamma. \tag{8.57}$$

The physical meaning of this condition is that a weak absorber ($a < 1$) admits a low intensity Hopf bifurcation provided that the absorber is relaxing faster than the active medium ($\overline{\gamma} > \gamma$). The scaling imposes a very stringent condition for many lasers and only the microchip solid state laser verifies this necessary condition (see Table 8.1).

The limit $\overline{\gamma}/\gamma \to \infty$ of the present approximation correctly matches the low intensity Hopf bifurcation of the two-variable model.

In summary, the linear stability analysis of the non-zero intensity steady state indicates that Hopf bifurcations to sustained oscillations are possible. However, it doesn't anticipate the highly pulsating intensities observed experimentally and numerically. We may use the same technique as for the two-variable model by seeking separate approximations for the interpulse and pulse regimes of the high intensity pulsating oscillations (see Section 8.3.2). The detailed analysis for the three-variable LSA is documented in [192]. A transcendental equation provides the period, and an asymptotic approximation for the pulse intensity allows one to describe the bifurcation diagram. The main difference from the results obtained for the two-variable model is the appearance of the saturability a in these expressions. Szabo and Stein correctly derived the approximation for the pulse intensity in 1965 [193]. The equation for the period was never properly derived, presumably because of the difficulty of obtaining a root of a transcendental equation.

Bifurcation diagrams

The analysis in the previous sections provides us with expressions for the steady states and information on their linear stability properties. In this sub-section, we investigate the bifurcation diagram numerically using A as control parameter.

If the condition (8.13) is satisfied, the steady state bifurcation at $A = A_{th}$ is sub-critical and the branch of steady states that emerges from it is unstable ($C_3 < 0$). This branch of steady states then folds back at a higher intensity but remains unstable ($C_3 > 0$ but $C_1 C_2 - C_3 < 0$) until a Hopf bifurcation point $A = A_H$ is reached. The bifurcation diagram in Figure 8.7 shows that stable pulsating oscillations (PQS) typically appear at $A = A_{th}$ with a large period. This branch of sustained oscillations first coexists with an unstable non-zero intensity steady state and then with a stable steady state if $A > A_H$. But other bifurcation diagrams are possible. We investigate two specific cases that apply to the $CO_2 + SF_6$ laser and to the microchip YAG laser, respectively.

$CO_2 + SF_6$ *laser* The high value of parameter a motivates a rescaling of the LSA equations. Indeed if $\overline{D} = O(1)$, Eq. (8.8) indicates that $\overline{D}$ will decay rapidly until it reaches low $O(a^{-1})$ values. Introducing then $\overline{D} = a^{-1}\overline{F}$, Eqs. (8.6)–(8.8) become

$$\frac{dI}{dt} = (-1 + D + a^{-1}\overline{F})I, \tag{8.58}$$

$$\frac{dD}{dt} = \gamma [A - D(1 + I)], \tag{8.59}$$

$$\frac{d\overline{F}}{dt} = a\overline{\gamma}\left[-\overline{A} - \overline{F}(I + a^{-1})\right]. \tag{8.60}$$

Assuming now $a\overline{\gamma} \gg \gamma$, we may eliminate $\overline{F}$ and reduce the three-variable LSA equations to the two-variable LSA equations (8.17) and (8.18). Figure 8.11 shows the bifurcation diagram of the possible solutions. The unstable periodic solutions have been determined numerically by integrating Eqs. (8.17) and (8.18) backward in time. The bifurcation diagram is similar to the one shown in Figure 8.7 except

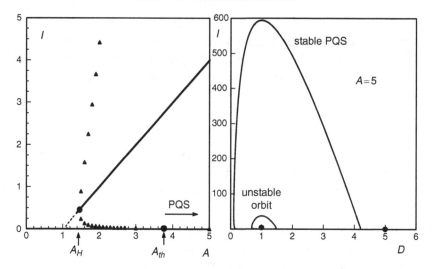

Fig. 8.11 $CO_2 + SF_6$ laser. Numerical solutions of Eqs. (8.17) and (8.18) with the values of the parameters listed in Table 8.1. Left: Bifurcation diagram. Because of the large value of a, the upper branch of the steady state hysteresis curve comes very close to $A = 1$. From $A = A_{th} = 3.78$ to $A \simeq 15$, stable PQS oscillations of large amplitude coexist with a stable steady state (branch of PQS solutions not shown). At $A = A_H \simeq 1.40$, a branch of unstable limit-cycles emerges from a subcritical Hopf bifurcation and its amplitude progressively increases with A. Right: Two periodic orbits coexist in the phase plane if $A > A_{th}$. A stable and large amplitude limit-cycle coexists with an unstable limit-cycle of smaller amplitude, a stable steady state at $(D, I) = (1, 4)$, and an unstable saddle-point at $(D, I) = (5, 0)$.

that $A_H < A_{th}$. This allows bistability between a zero and a non-zero intensity steady state in the domain $A_H < A < A_{th}$.

Nd^{3+}:YAG + Cr^{4+} laser On the other hand, if (8.12) is satisfied, a completely different diagram is obtained (see Figure 8.12). The steady state bifurcation is supercritical and a low intensity Hopf bifurcation at $A = A_{H1}$ appears close to threshold. Detailed comparison of numerical and experimental bifurcation diagrams are difficult because of the large range of values of some of the parameters. The stable coexistence between PQS and a stable non-zero intensity steady state, as suggested by the diagram in Figure 8.11, was observed in 1982 [194] although the bifurcation mechanisms responsible for this coexistence were not understood at that time. Systematic experimental bifurcation studies came later using CO_2 LSAs. They show that PQS always occurs just above the lasing threshold $A = A_{th}$ and that it disappears as the pump parameter is significantly increased.

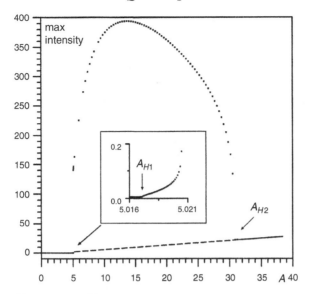

Fig. 8.12 Nd^{3+}:YAG + Cr^{4+} laser. Bifurcation diagram of the periodic solutions, obtained numerically from Eqs. (8.6)–(8.8) with $\gamma = 10^{-2}$, $\overline{\gamma} = 10^{-1}$, $\overline{A} = 4$, and $a = 0.5$. The branch of steady states emerges at $A = A_{th} = 5$ and its intensity changes linearly with A. A branch of stable periodic solutions connects the two Hopf bifurcation points $A = A_{H1}$ and A_{H2}. A_{H1} is slightly larger than A_{th}. The inset in the figure details the Hopf bifurcation vicinity and shows that the branch becomes vertical near $A = 5.02$. The values of γ and $\overline{\gamma}$ have been voluntarily taken larger than the real values (the real values are 1.8×10^{-6} and 6.4×10^{-5}, respectively) to reveal the details of the bifurcation transition. Computations have been done in double precision. If γ and $\overline{\gamma}$ are decreased, the bifurcation diagram near A_{H1} becomes more and more vertical (from Figure 1 of Kozyreff and Erneux [123]).

8.4 PQS in CO$_2$ lasers

The CO$_2$ laser has been one of the most studied LSAs and a great variety of gases have served as saturable absorbers. The first studies were motivated by the possibility of using PQS pulses for isotope separation or as efficient pumps for far-infrared lasers. Although they have lost their main application domain, these lasers have become interesting tools for the experimental study of non-linear responses. In particular, the CO$_2$/N$_2$O LSAs exhibit a large variety of pulse shapes that cannot be described by the simple three-variable model used so far. This has motivated the development of an improved rate equation model that has led to quantitative comparisons between experiments and theory (see Section 8.4.2).

8.4.1 Experiments

Two major observations have marked the progress in the experimental study of the CO_2 pulsating outputs. First, the observation of the Hopf bifurcation transition from damped relaxation oscillations to self-sustained large amplitude oscillations demonstrated that the CO_2 LSA was indeed an optical system for which the theory of dynamical systems was relevant. Second, the transition to chaos via period doubling bifurcations and Shilnikov's saddle-focus dynamics were studied in detail and simulated using rate equations.

Hopf bifurcation transition

The LSA is presumably the first optical system where the Hopf bifurcation transition has been explored. The bifurcation diagram shown in Figure 8.8 indicates a hard transition to sustained oscillations as the pump parameter A is progressively decreased from a large value (subcritical Hopf bifurcation). But the Hopf bifurcation can be supercritical allowing a smoother transition (see Figure 8.13). Experiments were carried out in the late 1970s using a CO_2 laser with low pressure

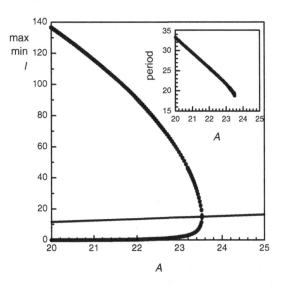

Fig. 8.13 Supercritical Hopf bifurcation. The bifurcation diagram has been obtained numerically from Eqs. (8.6)–(8.8) with $\overline{A} = 4$, $a = 0.5$, $\gamma = 10^{-2}$, and $\overline{\gamma} = 5 \times 10^{-2}$. From the exact Hopf conditions, we determine a Hopf bifurcation point at $A \simeq 23.52$. The period of the oscillations at the Hopf point is $T \simeq 18.78$. As soon as the minimum of the oscillations comes close to zero, i.e. $A \lesssim 22$, the oscillations become strongly pulsating (PQS regime). The inset shows the period of the oscillations, which smoothly increases from T as we progressively decrease A.

CH₃I as a saturable absorber, which was designed for spectroscopic studies. The intensity emitted by such a laser was monitored for different values of A and $\overline{A}$. The damping and the frequency of the oscillations were measured on both sides of a Hopf bifurcation point $A = A_H \simeq 0.09\,\text{W}$. If $A < A_H$, the period of the oscillations was determined by the PQS oscillations. If $A > A_H$, the absorber was subjected to pulsed electric fields that modulated the absorption coefficient through Stark shifting of the energy levels of the absorber. The return to the stable steady state was followed by measurement of the damping and the period of the transient oscillations of the laser output intensity (see Figure 8.14). The figure indicates that (1) the period of the damped oscillations corresponds to the RO oscillations $(T \sim 1/\sqrt{A-1})$ and that it follows the PQS period as A passes through A_H, and (2) the damping time tends to infinity as we approach the Hopf bifurcation point A_H (critical slowing down). Figure 8.14 suggests a smooth transition through a supercritical Hopf bifurcation as the pump parameter is decreased. For other parameter values, a hard transition was more often observed. The onset of sustained RO oscillations through a Hopf bifurcation (called sinusoidal self-modulation) has been further analyzed experimentally and numerically by Tanii *et al.* [62].

Complex oscillations in the CO₂ LSA

Pulse shapes produced by CO_2/N_2O LSAs are much richer than in the YAG lasers. In addition to sinusoidal oscillations and repetitive spikes, pulses with a long tail

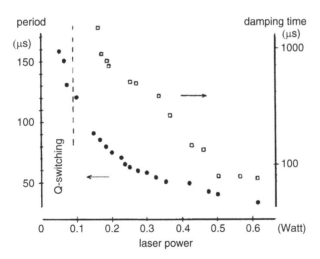

Fig. 8.14 Transients oscillations for a CO₂ laser with a CH₃I absorber. The oscillation period (dots) and damping time (squares) are represented as a function of the laser power in the vicinity of the Hopf bifurcation. Reprinted Figure 2 with permission from Arimondo and Glorieux [195]. Copyright 1978 by the American Institute of Physics.

and sometimes with damped or undamped small oscillations superimposed on plateaus were observed (see Figures 8.4 and 8.15). In the mid 1980s, there was a definite breakthrough for lasers when experiments clarified the role of Hopf and homoclinic bifurcations. Figure 8.4 shows a variety of pulse shapes for a CO_2 laser with CH_3I as a saturable absorber in terms of two parameters (the laser detuning and the absorber pressure). Analyzing the two regimes on the extreme left and right sides of Figure 8.4, we find these two bifurcation mechanisms for the birth of oscillating regimes. Specifically, the low absorber pressure domain corresponds to a steady state and the laser destabilizes through small sinusoidal oscillations for positive detunings (right part of Figure 8.4) suggesting a supercritical Hopf bifurcation. On the large negative detuning side (left part of the diagram), the oscillation period increases and the pulses become stiffer as the laser threshold (high absorber pressure) is approached (homoclinic bifurcation). The central part

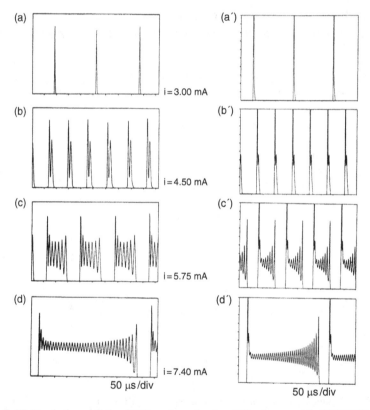

Fig. 8.15 Evolution of the PQS pulses in the CO_2 laser with HCOOH as a saturable absorber. Left: experiments; right: numerical simulations. With kind permission from Springer Science+Business Media (Figure 2 from Tachikawa *et al.* [61]).

of Figure 8.4 demonstrates that there is a continuous transition between pulsating and nearly harmonic oscillations. Chaotic oscillations are also observed in this parameter domain.

Using another absorber leads to similar observations. The series of pulses obtained by Tachikawa *et al.* on CO_2 LSAs [61, 63, 64] are shown in Figure 8.15. They illustrate different forms of bursting oscillations where rapid oscillations are separated by almost static regimes. These oscillations correspond to the central region of the parameter domain in Figure 8.4 but they exhibit longer plateaus.

8.4.2 Bursting oscillations in the CO_2 LSA

The multiplicity of timescales exhibited by the pulsating outputs shown in Figure 8.15 cannot be described mathematically by a simple Hopf or homoclinic bifurcation mechanism. This has motivated the development of a more sophisticated model that we now introduce.

Model equations for a CO_2 LSA

The key ingredient for obtaining a satisfactory description of chaos in these lasers was found by Tachikawa *et al.* [63] who revived the model of Dupré *et al.* [58], introducing a third energy level with slow dynamics. This accounts for the complicated population dynamics due to the transfer between the different vibrational levels involved in the lasing process (pumping and relaxation). From the dynamical system point of view, this third energy level weakly coupled to the previous ones allows the emergence of a new slow time scale. Excellent agreement was then obtained with the experimental observations.

In dimensionless form, the new LSA equations are given by[1]

$$\frac{dI}{dt} = I(U - \overline{U} - 1), \tag{8.61}$$

$$\frac{dU}{dt} = \varepsilon\left[-(I+1)U + W\right], \tag{8.62}$$

$$\frac{dW}{dt} = \varepsilon\left[bU - W + A\right], \tag{8.63}$$

$$\frac{d\overline{U}}{dt} = \overline{\varepsilon}\left[\overline{A} - \overline{U}(1 + aI)\right] \tag{8.64}$$

and only four independent parameters are required. Comparing these equations with Eqs. (8.6)–(8.8) used in our previous study of the LSA, we note that the

---

[1] These equations are the same as the equations studied in [65] with $n = \phi$, $N = \overline{M}$, $B_g f_g = \mathcal{A}$, $B_a = \overline{\mathcal{A}}$, $\frac{l_g}{L} = \zeta$, $\frac{l_a}{L} = \overline{\zeta}$, $k = 2\kappa$, $R_{ij} = \gamma_{ij}$, $r = \overline{\gamma}$, $N^* = M^*$, and $M = N$.

Table 8.3 *Parameters used for the simulations of the CO$_2$ LSA. The two columns for SF$_6$ correspond to different operating conditions. Note that SF$_6$ is highly saturable compared to CH$_3$I (larger values of a).*

	Tachikawa *et al.* [64][a]		Lefranc *et al.* [65][b]
Absorber	SF$_6$	SF$_6$	CH$_3$I
ε	0.076 2	0.076 2	0.137
$\bar{\varepsilon}$	1.664	0.42	1.2
a	14.840 2	91.322	4.17
b	0.989 5	0.989 5	0.85
A	1.312 9	1.260 7	1.4–2.1
$\bar{A}$	2.72	20	2.16

[a] [64].
[b] [65].

variables I, U, and $\overline{U}$ are equivalent to the variables I, D, and $\overline{D}$, respectively. Moreover, the relaxation rates ε and $\bar{\varepsilon}$ correspond to γ and $\overline{\gamma}$, respectively. The main difference from our previous model equations comes from the fact that the population inversion in the active medium is not directly controlled by the pump A but is mediated through a new population variable W.

Bursting oscillations exhibiting fast and slow evolutions were found experimentally by Tachikawa *et al.* [63, 64] and they were simulated numerically using Eqs. (8.61)–(8.64). Excellent quantitative agreement was obtained using the values of the parameters listed in Table 8.3 (the first two columns correspond to the data for Figures 1c' and 1d' in [64]).

We have found numerically that the bursting response remains unchanged if we make a quasi-steady state approximation for the variable $\overline{U}$. Mathematically, this is justified if $\bar{\varepsilon}\overline{A}$ is sufficiently large. From Eq. (8.64), $\overline{U}$ is then given by

$$\overline{U} = \frac{\overline{A}}{1+aI} \tag{8.65}$$

and Eqs. (8.61)–(8.64), reduce to the following three equations

$$\frac{dI}{dt} = I\left(U - \frac{\overline{A}}{1+aI} - 1\right), \tag{8.66}$$

$$\frac{dU}{dt} = \varepsilon\left[-(I+1)U + W\right], \tag{8.67}$$

$$\frac{dW}{dt} = \varepsilon\left[bU - W + A\right]. \tag{8.68}$$

We next analyze the bifurcation diagram for these equations.

Steady states, stability, and bifurcation diagram

Equations (8.66)–(8.68) admit the zero intensity steady state

$$I = 0, \; U = W = \frac{A}{1-b} \tag{8.69}$$

and the non-zero intensity steady state (in implicit form)

$$A = (1 - b + I)\left(1 + \frac{\overline{A}}{1 + aI}\right) \tag{8.70}$$

$$U = 1 + \frac{\overline{A}}{1 + aI}, \; W = (1 + I)\left(1 + \frac{\overline{A}}{1 + aI}\right). \tag{8.71}$$

The laser threshold appears at $A = A_{th} \equiv (1 - b)(1 + \overline{A})$ and requires that $b < 1$. Expanding (8.70) for small I indicates that $A = A_{th}$ is a bifurcation point of the zero intensity steady state. The bifurcation is supercritical (subcritical) if

$$1 + \overline{A} - a\overline{A}(1 - b) > 0 \; (< 0). \tag{8.72}$$

We may analyze the stability of the steady states in the same way as we did for Eqs. (8.6)–(8.8). For the zero intensity solution, we find that the steady state is stable if $A < A_{th}$. For the non-zero intensity steady state, the characteristic equation takes the form

$$\sigma^3 - T_1\sigma^2 + T_2\sigma - T_3 = 0 \tag{8.73}$$

where the coefficients are defined by

$$T_1 = \frac{\overline{A}aI}{(1 + aI)^2} - \varepsilon(2 + I) \tag{8.74}$$

$$T_2 = -\varepsilon(2 + I)\frac{\overline{A}aI}{(1 + aI)^2} + \varepsilon I\left(1 + \frac{\overline{A}}{1 + aI}\right) + \varepsilon^2(1 + I - b) \tag{8.75}$$

$$T_3 = -\varepsilon^2 I\left[-(1 - b + I)\frac{\overline{A}a}{(1 + aI)^2} + 1 + \frac{\overline{A}}{1 + aI}\right]. \tag{8.76}$$

A Hopf bifurcation is possible if $T_1 T_2 - T_3 = 0$ and $T_2 > 0$ but needs to be explored numerically. In the limit $\varepsilon \to 0$, this condition simplifies and provides a simple expression for the Hopf bifurcation point given by

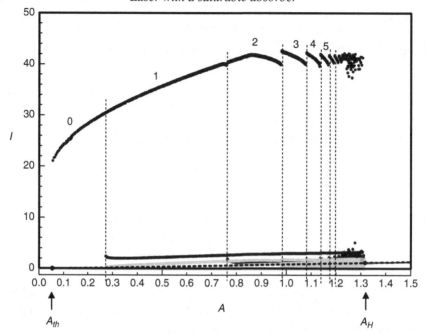

Fig. 8.16 Bifurcation diagram of the Tachikawa *et al.* LSA equations (8.66)–(8.68). The values of the parameters are $\varepsilon = 0.076, \overline{A} = 2.72, a = 14.84$, and $b = 0.989$. Dark and gray points correspond to the maxima and minima, respectively. The broken line is the non-zero intensity steady state. The labels $1, 2, \ldots 5$ denote the number of low-amplitude oscillations.

$$I_H = \frac{1}{a}\left[\sqrt{\overline{A}(2a-1)} - 1\right] > 0 \tag{8.77}$$

$$A_H = \left(1 - b + \frac{1}{a}\left[\sqrt{\overline{A}(2a-1)} - 1\right]\right)\left(1 + \sqrt{\frac{\overline{A}}{2a-1}}\right). \tag{8.78}$$

The bifurcation diagram of the stable solutions of Eqs. (8.66)–(8.68) is shown in Figure 8.16. The values of the fixed parameters ε, $\overline{A}$, a, and b are given by the first column of Table 8.3. From left to right, we note the emergence of successive periodic solutions with a progressively increasing number of extrema. The bifurcation diagram starts with a homoclinic orbit located at or close to the laser steady bifurcation point at $A = A_{th}$. This orbit leads to large-period oscillations reminiscent of the PQS oscillations of our previous LSA models. Close to $A = 0.27$, the high-amplitude pulsating oscillations exhibit one extra low-amplitude oscillation. A second, extra low-amplitude oscillation appears close to $A = 0.76$. This process continues until $A = 1.2$ where the oscillations are no longer periodic. All oscillatory activities stop close to the Hopf bifurcation point at $A_H = 1.31$. The

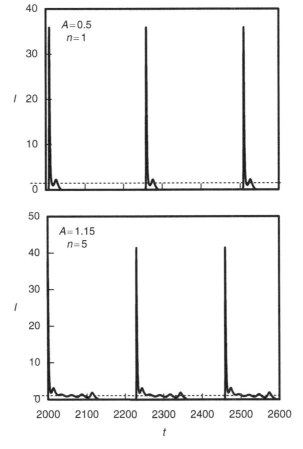

Fig. 8.17 Periodic solutions of the Tachikawa *et al.* LSA equations (8.66)–(8.68) for two specific values of A and the parameters listed in Figure 8.16. Specific domains of values of A admit periodic solutions with n low-amplitude extrema. The broken line indicates a plateau where these oscillations appear during a progressively increasing interval of time as A is increased.

non-zero intensity steady state then becomes the only stable solution. Figure 8.17 shows two typical periodic solutions corresponding to this diagram.

Bursting oscillations

The new dynamical phenomenon revealed by the model equations (8.66)–(8.68) is obviously the onset of the low-amplitude oscillations. They do not correspond to a well-known bursting mechanism based on a hysteresis cycle in a slow–fast phase plane. Nevertheless, it is possible to identify a slowly varying process that explains how the laser low-amplitude oscillations appear. As noted in Figure 8.17, these oscillations remain close to a non-zero intensity plateau (broken line in the figure). Analyzing these oscillations in more detail indicates that U oscillates close to 1 while W is slowly decaying exponentially. From Eqs. (8.66)–(8.68) with the new variables $U = 1 + \sqrt{\varepsilon}u$ and $s = \sqrt{\varepsilon}t$, we obtain

$$\frac{dI}{ds} = I\left(u - \frac{\overline{A}}{\sqrt{\varepsilon}(1 + aI)}\right), \tag{8.79}$$

$$\frac{du}{ds} = -(I + 1)(1 + \sqrt{\varepsilon}u) + W, \tag{8.80}$$

$$\frac{dW}{ds} = \sqrt{\varepsilon}\left[b(1 + \sqrt{\varepsilon}u) - W + A\right]. \tag{8.81}$$

From Eq. (8.81), we note that W is a slowly decaying function of $\sqrt{\varepsilon}s$ approaching $W = A + b$. We then examine the remaining equations for I and u with W treated as a slowly varying parameter. By analyzing the stability of the non-zero intensity steady state, we find that it is stable (unstable) if $W > W_H$ ($W < W_H$), where $W_H \simeq 2.43$ denotes a Hopf bifurcation point ($I_H \simeq 1.10$ and $U_H \simeq 1.16$). As W progressively decreases, we slowly pass a Hopf bifurcation. See Figure 8.18. In summary, we have shown that the emergence of the low-amplitude oscillations results from the passage through a subcritical Hopf bifurcation point. A deeper analysis is necessary, however, to find out how the number of low-amplitude oscillations is related to the control parameter A.

Comparison with experimental results

In the first experimental investigations of laser dynamical instabilities, a single parameter was progressively changed and reasonable agreement with theory was obtained by fitting parameters. The model developed in the previous section has excellent predictive possibilities and reproduces accurately the observations on CO_2 LSAs with absorbers as different as CH_3I and SF_6. This then allowed investigations where at least two parameters were changed. As a result, a more global understanding of dynamical phenomena became possible through the identification of particular points in parameter space or subtle changes in the waveforms.[2] By the end of the 1980s, researchers armed with a collection of new theoretical tools (phase portraits, Poincaré maps, one-dimensional maps) were ready to explore the chaotic dynamics of the LSA. Chaos in CO_2 lasers with various saturable absorbers (CH_3I, SF_6) was observed almost simultaneously in several laboratories [188, 196, 197]. The transition to chaos was later analyzed by Lefranc *et al.* [65] in great detail using Eqs. (8.61)–(8.64) for the simulations. Experimental chaotic regimes have been shown to display the typical features of homoclinic chaos.

---

[2] A similar approach has been followed more recently for the optically injected laser (see Section 9.4).

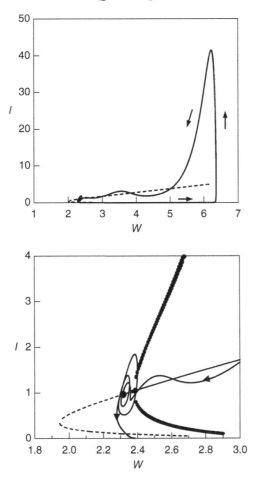

Fig. 8.18 Top: limit-cycle solution in the phase plane (W, I) for $A = 1.15$. It has been obtained numerically from Eqs. (8.66)–(8.68) with the values of the parameters listed in Figure 8.16. The broken line represents the branch of steady states of the reduced two-variable equations. Bottom: blow-up. The parabolic full and broken lines correspond to the stable and unstable parts of the steady states, respectively. The Hopf bifurcation is subcritical and the dots are the extrema of a branch of unstable periodic solutions. The limit-cycle solution passes the Hopf bifurcation, then spirals out before jumping to a nearly zero intensity regime.

Bifurcation diagrams, temporal sequences and first return maps can be obtained numerically, and give us precious indications of the possible routes to chaos. In experiments where only one or sometimes two variables are accessible, trajectories may be reconstructed from single-variable measurements using the time delayed method. This method relies on the fact that a pseudo-trajectory reconstructed from time-shifted values of a single variable such as $\{I(t), I(t + \tau), I(t + 2\tau) \ldots\}$ has the same topological properties as the real trajectory. This technique has been

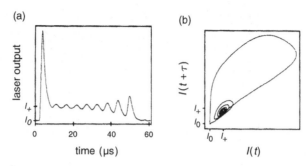

Fig. 8.19 (a) PQS regime for a $CO_2 + SF_6$ LSA and (b) reconstructed phase space trajectory with $\tau = 0.6$ μs (from Figure 4 of de Tomasi *et al.* [196]).

extensively used for the analysis of chaotic signals, through either dimensional or topological analyses.[3] Figure 8.19 shows an example for a CO_2 laser with SF_6 as a saturable absorber. It clearly illustrates the saddle-focus character of I_+. However it is misleading for the exact role played by I_0 because of the low resolution near the origin.

In spite of their simplicity, the model equations (8.61)–(8.64) reproduce in detail the evolution of the dynamical regimes observed in the pulsating CO_2 LSAs. Because comparative studies between experiments and simulations are possible, the LSA as well as other laser devices are used to investigate new ideas in the nonlinear dynamics community. In the 1990s interest shifted significantly to the control of chaos in low-dimensional systems. Several control schemes were used to stabilize chaotic lasers, which is an important achievement from a practical as well as a fundamental perspective (see [200] for a recent review).

8.5 Exercises and problems

8.5.1 Asymptotic analysis of the pulsating solutions

As pulsating intensity oscillations develop in time, they exhibit two distinct regimes that we call the interpulse regime and the pulse. During the interpulse regime, the intensity is very small and the increase of D is slow. On the other hand, the pulse part is characterized by a large intensity and a rapid change of both I and D.

(1) Derive simplified equations for the interpulse and pulse regimes. Hint: the reduced equations for the interpulse regime are linear and can be solved exactly using the initial conditions $D(0) = D_{1-}$ and $I(0) = I_0 \ll 1$ (see

---

[3] The LSA was the first optical system for which topological methods were tested [198]. By unfolding the topological organization of the unstable periodic orbits embedded inside the chaotic attractor, Papoff *et al.* [199] demonstrated that the chaotic output of a CO_2 LSA could be explained by low-dimensional dynamics.

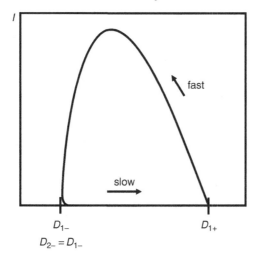

Fig. 8.20 Analytical description of the limit-cycle in the phase plane. It consists of a slow increase from D_- to D_+ with an almost zero intensity, followed by a quick change from D_+ to D_{2-} with a high intensity.

Figure 8.20). This solution provides the interpulse period T as well as $D = D_{1+}$, the point where the fast pulse emerges. The expressions for both T and D_{1+} are independent of I_0, in first approximation as $\gamma \to 0$. The reduced equations for the pulse are nonlinear but can be integrated in the phase plane leading to an expression for $I = I(D)$. The limit-cycle trajectory satisfies the conditions $I(D_{1+}) = I(D_{2-}) = 0$ which then provides an equation for D_{2-} as a function of D_{1+}. Finally, we require that $D_{2-} = D_{1-}$ to close the orbit.

Answer: the interpulse period is the non-zero root of the transcendental equation

$$(A - A_{th})\gamma T - (D_{1-} - A)(\exp(-\gamma T) - 1) = 0. \tag{8.82}$$

The maximum value of $D = D_{1+}$ is given by

$$D_{1+} = A + (D_{1-} - A)\exp(-\gamma T) \tag{8.83}$$

and the relation between D_{1+} and D_{1-} is

$$\ln(D_{1-}/D_{1+}) - (D_{1-} - D_{1+}) = 0. \tag{8.84}$$

(2) Analyze these equations for $A - A_{th} \to 0$ (close to threshold) assuming that $\gamma T \to \infty$ in this limit. Show that the repetition rate, defined as $f \equiv T^{-1}$, admits the approximation

$$f = \gamma \frac{A - A_{th}}{A_{th}} \frac{1}{\Delta D}, \tag{8.85}$$

where $\Delta D \equiv D_{1+} - D_{1-}$ is defined as the gain reduction ($D_{1+} = A_{th}$, in first approximation as $A - A_{th} \to 0$).

(3) Consider the additional limit $\overline{A} \to 0$ and show that $\Delta D = 2\overline{A}$, in first approximation as $\overline{A} \to 0$. The resulting expression for f is used in the engineering literature as a guide for improving the repetition rate (see Eq. (7) in [191]).

8.5.2 Dimensionless formulation

Spühler *et al.* [191] investigated microchip lasers with semiconductor saturable absorbers. They formulated the following rate equations for the laser power $P(t)$, the intensity gain coefficient per cavity round trip $g(t)$, and the intensity saturable loss coefficient per cavity round-trip time $q(t)$

$$T_R \frac{dP}{dt} = (g - q - l)P,$$

$$\frac{dg}{dt} = -\frac{g - g_0}{\tau_L} - \frac{gP}{E_L},$$

$$\frac{dq}{dt} = -\frac{q - q_0}{\tau_A} - \frac{qP}{E_A},$$

where T_R is the cavity round-trip time, E_L is the saturation energy of the gain, τ_L is the upper-state lifetime of the gain medium, E_A is the saturation energy of the absorber, and τ_A is the relaxation time of the absorber. l denotes the total nonsaturable loss coefficient per round trip. $g = g_0$ and $q = q_0$ are the equilibrium values of g and q for $P = 0$. The values of the fixed parameters are

$$T_R = 2.61 \text{ ps}, \; l = 14\%, \; \tau_L = 50 \,\mu\text{s},$$

$$\tau_A = 200 \,\text{ps}, \; E_L/E_A = 10^3, \quad \text{and} \quad q_0 = 5\%.$$

The pump parameter is defined as $r \equiv g_0/g_c$, where $g_c \equiv l + q_0$ is the laser threshold.

Reformulate these equations in the dimensionless form (8.6)–(8.8) and discuss the values of the dimensionless parameters. Note that $\overline{\gamma}a/\gamma$ is large, which motivates the adiabatic elimination of $\overline{D}$. Moreover $\overline{A}$ is small.

8.5.3 Symmetric pulse and pulse width

Derive approximate analytical expressions for the pulse shape and its width on the basis of the analysis proposed in Exercise 8.5.1.

Within the approximations used above, the equations for the pulse are

$$I' = I(-1 + D), \quad D' = -\gamma DI, \quad \text{and} \quad D(-\infty) = D_{1+},$$

where $D_{1+} \to A_{th} = 1 + \overline{A}$ if $A - A_{th} \to 0$. The time variable is $s = t - t_p$, where t_p is defined as the time at which the pulse is maximum.

Solve these equations in the limit $\overline{A} \to 0$. Introducing $\varepsilon = \overline{A}$, the boundary condition suggests the new variable $X = (D - 1)/\varepsilon$. The leading-order equations for $\varepsilon \to 0$ then become

$$I' = \varepsilon XI, \quad \varepsilon X' = -\gamma I, \quad \text{and} \quad X(-\infty) = 1.$$

Show that the solution of these equations is

$$X = -\tanh\left(\frac{\varepsilon s}{2}\right) \quad \text{and} \quad I = \frac{\varepsilon^2}{2\gamma}\operatorname{sech}^2\left(\frac{\varepsilon s}{2}\right).$$

The expression for I represents a symmetric pulse. Its width at half maximum is

$$\Delta s = \frac{4ar\cosh(\sqrt{2})}{\varepsilon} \simeq \frac{3.52}{\varepsilon} = \frac{3.53}{\overline{A}}.$$

This expression for the pulse width is used in the engineering literature as a guide (see Eq. (9) in [191]).

8.5.4 The LSA with two-photon absorber

A GaAs plate inserted inside the cavity of a Nd^{3+}:YAG laser acts as a saturable absorber because two-photon absorption of the 1.06 μm photons dominates at high irradiance and is intensity-dependent. Gu *et al.* [201] numerically simulated the pulsating oscillations by using the following equations for the photon density Φ and the population inversion N

$$\frac{d\Phi}{dt} = \frac{1}{\tau_R}\left[\left(2\sigma l N - \gamma - 2\alpha l_q\right)\Phi - B\Phi^2\right],$$

$$\frac{dN}{dt} = -c\sigma N\Phi - \frac{N}{\tau} + P(N_T - N).$$

They considered the following values of the parameters: $\tau_R = 2.7$ ns, $\sigma = 7.6 \times 10^{-19}$ cm^2, $l = 10$ mm, $\gamma = 0.02$, $\alpha = 1.1$ cm^{-1}, $l_q = 0.628$ mm, $B = 7.88 \times 10^{-17}$ cm^{2}W^{-1}, $\tau = 230$ μs, $N_T = 1.5 \times 10^{20}$ cm^{-3}, $c = 3 \times 10^8$ ms^{-1}, and $P = 50$ s^{-1}.

(1) Reformulate these equations in dimensionless form and determine the number of independent parameters.
(2) Determine the steady state solutions and their linear stability properties.

9

Optically injected semiconductor lasers

There is a recurrent need for high power, frequency-stable lasers in applications as diverse as laser radar, remote sensing, gravitational wave interferometry, and nonlinear optics. This need is often satisfied by using a low power, frequency-stable laser followed by a chain of amplifiers, but a preferred approach is to injection-lock a high power (slave) laser to a lower power, frequency-stable (master) laser. Other advantages of the injection-locking technique are the possibility of ensuring single-mode operation, eliminating mode partition noise, mode hopping, preventing spurious feedback effects, and synchronizing one or more free-running lasers to the same pump. As explained in Section 3.4, the main benefits of optical injection occur when the frequencies of both lasers are close together and for sufficiently large injected power. The slave laser then gets the spectral properties of the master one in terms of frequency and linewidth [202, 203, 204]. Stover and Steier [77] did the first optical injection experiment in 1966 using gas lasers. The first optical injection experiment using semiconductor lasers (SLs) came much later and was done by Kobayashi and Kimura in 1980 [205]. At that time, it was not clear that SLs would be useful lasers, but the performance of these lasers has dramatically improved during the last 30 years, providing reliable devices for a large variety of applications. Optical injection is used to reduce noise (frequency noise [206, 207, 208], mode partition noise [209], or intensity noise [210]), to generate microwave signals [205, 211, 212], or to produce chaotic outputs for secure communication [116, 213, 214, 215].

Besides phase-locking, optically injected SLs admit a rich variety of dynamical responses that have motivated quantitative comparisons between theory and experiments [216]. Most comparisons have been performed by comparing experimental data with numerical simulations of rate equations. Here, preference is given to asymptotic studies which provide analytical information of physical significance. How good is Adler's theory of phase-locking for a SL? What is the effect

213

of the linewidth enhancement factor α? Can we describe the separate effects of the detuning and the RO frequency of the solitary laser?

9.1 Semiconductor lasers

Before we introduce the rate equations, we need to stress three important features of a SL that have an immediate impact on the way we examine the experimental observations. First, SLs are highly sensitive to noise because of their particular geometries and their small size. SLs are produced in two geometries, namely edge-emitting lasers (EELs) and vertical cavity surface emitting lasers (VCSELs). See Figure 9.1. For edge-emitting lasers, light propagates in a guided mode parallel to different layers of semiconductor, and the lateral facets are coated with mirrors. In VCSELs, Bragg mirrors are epitaxially grown on both sides of a very thin active layer and light freely propagates perpendicular to the layers. Table 9.1 summarizes the main properties of these two lasers. The circular geometry of VCSELs leads to several advantages compared to EELs. The output beam of the fundamental mode has a circular cross section, a small divergence (between $10°$ and $20°$), and no astigmatism. As a result, the laser beam can be focused or coupled into an optical fiber more easily. The threshold current of a VCSEL can be orders of magnitude smaller than that of an EEL, implying higher efficiency. Finally, due to the short cavity, VCSELs are inherently single longitudinal mode lasers.

Because of their geometries and/or guided character of propagation, both EELs and VCSELs efficiently collect spontaneous emission in contrast to open

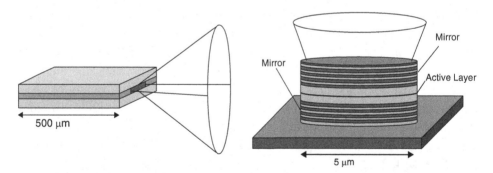

Fig. 9.1 In an EEL (left), the cavity is oriented parallel to the wafer surface while in a VCSEL (right), the orientation of the cavity is perpendicular to the surface (vertical). Despite their success and implementation in various applications, EELs have some disadvantages. For example, the shape of the output beam is elliptical and has astigmatism, which is not appropriate for efficient coupling into an optical fiber. By contrast, the output beam of a VCSEL has a circular cross section, a small divergence, and no astigmatism, allowing easier coupling to an optical fiber.

Table 9.1 *Cavity length and shape, dimensions, reflectivity of the mirrors, and beam shape of EELs and VCSELs.*

laser	cavity length	cavity shape	dimensions	reflectivity	beam
EEL	$\sim 500 \ \mu m$	rectangular	$\sim (50 \times 5) \ \mu m$	$\sim 30\%$	elliptical
VCSEL	$\sim 10 \ \mu m$	cylindrical	diameter $5 \ \mu m$	$\sim 99.9\%$	circular

Fabry–Pérot cavities used for gas or solid state lasers. Unavoidable mechanical or thermal fluctuations then induce large frequency variations for the short cavity SLs. For example, a length change of 1 μm in a 500 μm long diode laser operating at a wavelength of 1 μm produces a frequency shift as large as 60 MHz while the same length change causes a shift of only 3 kHz in a 1 m long CO_2 laser operating at a wavelength of 10.6 μm. Any quantitative comparison between theory and experiments for a SL will need a theory that takes into account these laser fluctuations.

A second feature of SLs is the experimental difficulty of investigating time series. The intrinsic time scales of SLs are extremely fast, of the order of tens of picoseconds or smaller. It is impossible to have measurements of representative time series of the intensity of a diode laser, except when a streak camera is used, but then only a good resolution on a short time interval or a poor resolution on a long time interval are possible. The preferred method of investigation for SLs relies on spectral analysis. We can measure the radiofrequency spectrum of the laser intensity with a radiofrequency spectrum analyzer and the optical spectrum of the laser with a monochromator. Physically, they provide the squared modulus of the Fourier transforms of the intensity and the electrical field, respectively. Such spectra are time averaged quantities and must be interpreted carefully. But combining optical and power spectra, we may identify specific bifurcations and determine regions in parameter space where a particular response is possible. A drawback of this approach is the identification of possible coexisting attractors with the same spectral content or transient attractors, leading to incorrect spectral signatures. Sometimes numerical simulations of the equations that describe the system dynamics must be used to actually interpret the spectra. In such cases, we must use equations that have been tested before and have good predictive power, and must also be certain of the specific values of the parameters that appear in these equations.

Finally, SLs differ from other lasers in that the free carriers which control the optical gain of the system also have a strong effect on the refractive index, giving it the characteristics of a detuned oscillator. This phenomenon is parametrized by a single parameter, the α parameter, which is defined as the ratio of the real and imaginary parts of the pump-induced susceptibility. In 1967, Lax [217] and Haug and Haken [218] separately noted that the linewidth of the laser field is enhanced by a factor $1 + \alpha^2$. But in 1982 it was found by Henry [40] that α is of the order of 5, which explained the origin of the large linewidth. Since then, the α parameter has been known as the linewidth enhancement factor or Henry's α.

9.2 Injection-locking

As described in Section 1.7, emission in SLs results from electron–hole recombinations between energy bands rather than discrete energy levels. Fortunately, the dynamical response of a single-mode SL can still be described in terms of two rate equations [33, 34, 44]. A typical experimental set-up is shown in Figure 9.2. If we consider the injection of a monochromatic optical field, $\mathcal{E}_{in}(\tau) = E_{in} \exp(i(\omega_{in}\tau + \phi))$, Eq. (1.94) for the carrier density remains unchanged but Eq. (1.93) for the amplitude of the field admits an additional term. The new equation is given by

$$\frac{dE}{d\tau} = \frac{\Gamma G_N}{2}(1 + i\alpha)nE + \kappa E_{in} \exp(i\nu\tau), \tag{9.1}$$

where $\nu \equiv \omega_{in} - \omega_0$ is defined as the detuning frequency between the injected signal and the solitary laser, and κ is the injection rate. (The phase ϕ can be eliminated by simply changing the origin of time.) We reformulate the dimensionless equations using t and Z defined in (1.95) and Y given by

$$Y \equiv \sqrt{\frac{\tau_s G_N}{2}} E \exp(-i\nu\tau). \tag{9.2}$$

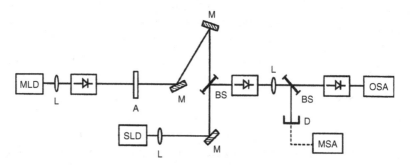

Fig. 9.2 The output field from a master laser diode (MLD) of frequency ω_{in} is injected into a slave laser diode (SLD) exhibiting a field with frequency ω_0 close to ω_{in}. An optical isolator guarantees that the injection is unidirectional and that no feedback light goes back to the master laser. L: lens; M: mirror; A: variable attenuator; BS: beam splitter; D: fast photodiode; OSA: optical spectrum analyzer; MSA: RF/microwave spectrum analyzer. The solid lines are optical paths and the dashed line is an electrical signal path. Reprinted Figure 1 from Simpson [219] with permission from Elsevier.

In terms of the new variables, Eqs. (1.94) and (9.1) become

$$\frac{dY}{dt} = (1 + i\alpha)YZ - i\Omega Y + \eta, \tag{9.3}$$

$$T\frac{dZ}{dt} = P - Z - (1 + 2Z)|Y|^2, \tag{9.4}$$

where the new parameters

$$\eta \equiv \sqrt{\frac{\tau_s G_N}{2}}\tau_p \kappa E_{in} \quad \text{and} \quad \Omega \equiv \nu\tau_p \tag{9.5}$$

are our control parameters. η is proportional to the injection rate and Ω is proportional to the detuning between the frequency of the master laser and the frequency of the free-running slave laser. It will be convenient to introduce the amplitude and the phase of the field Y. Inserting $Y = R\exp(i\psi)$ into (9.3) and (9.4) leads to the following equations for R, ψ, and Z

$$\frac{dR}{dt} = ZR + \eta\cos(\psi), \tag{9.6}$$

$$\frac{d\psi}{dt} = -\Omega + \alpha Z - \frac{\eta}{R}\sin(\psi), \tag{9.7}$$

$$T\frac{dZ}{dt} = P - Z - (1 + 2Z)R^2. \tag{9.8}$$

These equations are now in a convenient form for analysis.

9.3 Adler's equation

Adler's equation is expected to be a valid description of the master–slave locking phenomenon for low values of both the injection rate η and the detuning Ω. The asymptotic analysis is not so obvious as in Section 3.4 for a class A laser and we examine the problem in detail. Specifically, we consider η as a small parameter and expand Ω in power series of η as

$$\Omega = \eta\Omega_1 + \eta^2\Omega_2 + \ldots \tag{9.9}$$

We next seek a solution of Eqs. (9.6)–(9.8) of the form

$$R = R_0(t, \tau) + \eta R_1(t, \tau) + \ldots \tag{9.10}$$

$$\psi = \psi_0(t, \tau) + \eta \psi_1(t, \tau) + \ldots \tag{9.11}$$

$$Z = \eta Z_1(t, \tau) + \ldots, \tag{9.12}$$

where $\tau \equiv \eta t$ is defined as a slow time variable. This slow time variable is suggested by the right hand side of Eq. (9.7) which is $O(\eta)$ small since both Ω and Z are proportional to η. Introducing (9.10)–(9.12) into Eqs. (9.6)–(9.8), we obtain a sequence of problems to solve. From the first two problems we find that

$$R = P^{1/2} + \left[\eta \frac{1 + 2P}{2P} \cos(\psi_0) + \text{EDT} \right] + O(\eta^2) \tag{9.13}$$

$$Z = \eta \left[-P^{-1/2} \cos(\psi_0) + \text{EDT} \right] + O(\eta^2), \tag{9.14}$$

where $\psi_0 = \psi_0(\tau)$ is a function of τ and EDT means exponentially decaying functions of t that we do not need to describe. Solvability of the ψ_1 equation with respect to the fast time t then leads to an equation for ψ_0. In terms of the original time variable and parameters, this equation is given by

$$\frac{d\psi}{dt} = -\Omega - \eta P^{-1/2} (\alpha \cos(\psi) + \sin(\psi)) \tag{9.15}$$

$$= -\Omega - \eta_{loc} \sin(\psi + \psi_{loc}), \tag{9.16}$$

where

$$\eta_{loc} \equiv \sqrt{\frac{1 + \alpha^2}{P}} \eta \quad \text{and} \quad \psi_{loc} \equiv \arctan(\alpha). \tag{9.17}$$

Equation (9.16) is Adler's equation for an optically injected SL. It differs from the one derived in Section 3.4 by the α factor. Steady state locking is possible only if $\eta_{loc} > |\Omega|$. Equivalently, the *locking range* is given by

$$\Delta\Omega = 2\sqrt{(1 + \alpha^2) \frac{\eta^2}{P}} \tag{9.18}$$

and increases with α. The expression (9.18) is documented in many textbooks (see, for example, Siegman [20]). Adler's equation (9.15) admits an exact solution and offers the possibility of analyzing the behavior of the slave laser both in the locking region and outside the locking region.

Outside the locking range, the phase ψ is an unbounded periodic function of τ although R and Z given by (9.13) and (9.14) are bounded periodic functions of τ. They are called *four-wave mixing (FWM)* regimes [220] because of their typical optical spectra: the optical field emitted by the laser exhibits a strong component at frequency $\omega_1 \simeq \omega_0$ where ω_0 is the solitary laser frequency, and two weaker components, one at the injection frequency $\omega_2 = \omega_{in}$ and one at the conjugate frequency $2\omega_1 - \omega_2$.[1] Another property of the FWM pulsating oscillations is the *frequency pulling* phenomenon, which is best observed far from the locking region. Recall Adler's frequency (3.12). For Eq. (9.16), it is given by

$$\Omega_{out} = \sqrt{\Omega^2 - P^2/(1+\alpha^2)}. \tag{9.19}$$

Well outside the locking region ($|\Omega|$ sufficiently large), (9.19) can be expanded as

$$\Omega_{out} \simeq -\Omega\left[1 - \frac{P^2}{2(1+\alpha^2)\Omega^2}\right], \tag{9.20}$$

where the minus sign in front of Ω comes from Eq. (9.16), indicating $d\psi/d\tau = -\Omega$ as $|\Omega| \to \infty$. Using (9.2) for E and $\psi = \Omega_{out}t$, the optical field $E\exp(i\omega_0\tau)$ has the form

$$E\exp(i\omega_0\tau) \sim Y\exp[i(\omega_0 + \nu)\tau]$$
$$= R\exp[i(\omega_{out}\tau)], \tag{9.21}$$

where τ is the original time and $\omega_{out} \equiv \omega_0 + \nu + \Omega_{out}\tau_p$. With the expression (9.20) and noting that $\tau_p\Omega = \nu$, the optical frequency becomes

$$\omega_{out} \simeq \omega_0 + \nu - \tau_p\Omega + \tau_p\frac{P^2}{2(1+\alpha^2)\Omega}$$
$$= \omega_0 + \frac{\tau_p^2 P^2}{2(1+\alpha^2)\nu} \tag{9.22}$$

meaning that it is *pulled* from ω_0.

The derivation of Adler's equations is valid for low injection rate and low detuning. It ignores the time scales of the laser intensity given by the relaxation oscillation frequency $\omega_R = O(T^{-1/2})$ and the damping rate $\xi = O(T^{-1})$ defined by (1.99). If Eq. (9.16) is a valid description of the long time behavior of the laser, the evolution of the laser phase ψ should be slower than the decay of the laser relaxation oscillations, which is a function of $T^{-1}t$. This then implies the inequality

$$\eta_{loc} \equiv \sqrt{\frac{1+\alpha^2}{P}}\,\eta << T^{-1}. \tag{9.23}$$

---

[1] The FWM pulsating intensity oscillations typically appear outside the locking region but they may coexist with a stable locked state near the locking boundary [216].

This condition considerably restricts the domain of injection rate since (9.23) requires that

$$\eta \ll \sqrt{\frac{P}{1+\alpha^2}\frac{1}{T}}. \qquad (9.24)$$

This condition is valid for all lasers subject to an injected signal but is more restrictive for a SL because the upper bound decreases with α. In Section 9.6.3, we concentrate on the case $\eta = O(T^{-1})$ and show that a series of new bifurcations appear which cannot be described by Adler's equation. The interaction between Adler locking dynamics and the laser relaxation oscillations is investigated in Section 9.7.

9.4 Experiments and numerical simulations

From low to moderate injection powers, three distinct long time phenomena are observed depending on the detuning. At a fixed injection rate, steady state locking occurs if the detuning is sufficiently small. Outside the locking region, the FWM pulsating intensities are observed and they appear as soon as the injection rate is increased. Stable locking occurs predominantly at negative detunings relative to the frequency of the free-running laser. But as the detuning is progressively increased from negative to positive values, the locked steady state may transfer its stability to sustained relaxation oscillations (ROs). This transition is a Hopf bifurcation.

Numerical simulations of the laser rate equations and experiments indicate that the domain of stable locking is limited in size and that different dynamical outputs are possible. Because of the large number of these regimes, it was important to draw a map of the injected slave behavior in parameter space defined by the injected power and the detuning. The mapping of the locking area was presented by Mogensen in 1985 [221]. Then multiple regimes were mapped out, such as relaxation regime [222], chaos [223], and wave-mixing [223]. More recently, accurate injection maps were obtained [224–227]. In Figure 9.3, full lines represent stability boundaries obtained numerically and dots correspond to experiments. Only the cross-hatched region for negative detuning corresponds to the region of stable locking. The experiments were done using a distributed feedback (DFB) semiconductor laser operating at 1.557 μm, which is strictly single-mode throughout the entire range of injection values. All the fixed parameters have been estimated experimentally [228]. They are given by $\alpha = 2.6$, $P = 0.22$, and $T = 125$. The parameters Λ and Δ correspond to the injection rate and detuning normalized by the RO frequency, respectively. They are defined as $\Lambda = \eta/(\sqrt{P}\omega_R)$, $\Delta = \Omega/\omega_R$, where $\omega_R = \sqrt{2P/T}$. The lines emerging from the origin $(\Lambda, \Delta) = (0,0)$ are lines above which steady state locking is possible as we keep the detuning fixed and

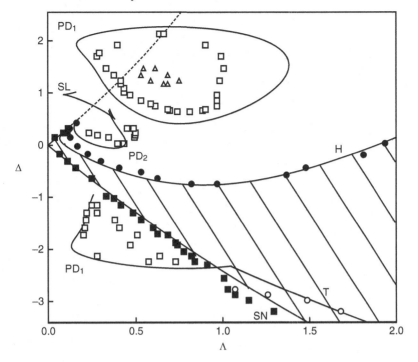

Fig. 9.3 Experimental and numerical stability diagram. Only the stable boundaries are shown, i.e. those corresponding to stable solutions. Full lines were obtained numerically. Experimental data are denoted by dots, squares, and triangles: • Hopf, ■ SN or steady locking, ○ Torus, □ first period doubling, and △ second period doubling. Adapted Figure 1 with permission from Wieczorek *et al.* [228]. Copyright 2002 by the American Physical Society.

progressively increase the injection rate (SN or saddle-node bifurcation). The curve denoted by H is the line of Hopf bifurcations where a stable steady state loses its stability and time-periodic solutions appear. The lines PD_1 and PD_2 mark period-doubling bifurcations. Two PD_1 lines appear symmetrically for low injection for Δ close to $\Delta = \pm 2$. The line T means a line of torus bifurcations from periodic to quasi-periodic oscillations. The two lines SL (saddle-node of limit cycles) that appear for positive detuning are lines of limit points where a periodic solution appears.

We wish to investigate how these different bifurcations appear and how they are related to the physical parameters. As in all dynamical system problems, we first concentrate on the steady state solutions and analyze their linear stability properties. From Figure 9.3, we note that the domain of stable steady states is bounded by the SN curve, above which locking is possible, and by the Hopf curve, above which oscillatory regimes appear. Both stability boundaries can be derived analytically and admit interesting approximations.

9.5 Stability of the steady states

9.5.1 Multiple steady states

The steady state solutions satisfy the conditions $dR/dt = d\psi/dt = dZ/dt = 0$ and from Eqs. (9.6)–(9.8), we obtain

$$ZR + \eta \cos(\psi) = 0, \tag{9.25}$$

$$-\Omega + \alpha Z - \frac{\eta}{R} \sin(\psi) = 0, \tag{9.26}$$

$$P - Z - (1 + 2Z)R^2 = 0. \tag{9.27}$$

Eliminating the trigonometric functions in Eqs. (9.25) and (9.26) leads to

$$\eta^2 = R^2 \left(Z^2 + (\alpha Z - \Omega)^2 \right). \tag{9.28}$$

Using then Eq. (9.27), we find an expression for $R^2(Z)$ given by

$$R^2 = \frac{P - Z}{1 + 2Z} > 0. \tag{9.29}$$

Finally, substituting (9.29) into Eq. (9.28) gives $\eta^2(Z)$ as

$$\eta^2 = \frac{P - Z}{1 + 2Z} \left(Z^2 + (\alpha Z - \Omega)^2 \right). \tag{9.30}$$

The expressions (9.29) and (9.30) provide a parametric solution for the intensity of the laser field $I \equiv R^2$ as a function of the injection rate η (Z is the parameter and $-1/2 < Z \le P$).

We first consider the case $\Omega = 0$ and $P > 0$. Equation (9.30) simplifies as

$$\eta^2 = \left(1 + \alpha^2\right) \frac{P - Z}{1 + 2Z} Z^2 \tag{9.31}$$

and implies two solutions for $\eta^2 = 0$, namely

$$(1)\ Z = P \quad \text{and} \quad (2)\ Z = 0. \tag{9.32}$$

They correspond to the steady state solutions of the free-running laser with $R^2 = 0$ and $R^2 = P$, respectively. From these two points a Z-shaped branch of solutions emerges in the Z vs. η diagram (see Figure 9.4, broken lines). The hysteresis of the steady state curve does not necessarily mean *bistability*, i.e. two stable steady states coexisting for the same value of η. In order to claim for bistability, we need

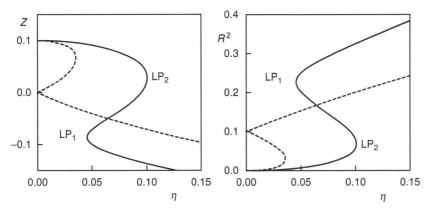

Fig. 9.4 Z-shaped branch of steady states for the carrier density $Z = Z(\eta)$, and S-shaped branch of steady states for the intensity $R^2 = R^2(\eta)$. These are obtained using (9.29) and (9.30) with $\alpha = 3$ and $P = 0.1$. The broken and full lines correspond to $\Omega = 0$ and $\Omega = -0.3$, respectively.

to investigate the stability properties of the two steady states. In general, the middle and lower branches of the $R^2 = R^2(\eta)$ steady state curve are unstable. For lower values of the pump parameter ($P = O(\varepsilon)$), the lower branch can be either fully stable or partially stable (see Exercise 9.8.1).

We next consider the case $|\Omega| \neq 0$. As $|\Omega|$ progressively increases from zero, the Z-shaped curve unfolds (see Figure 9.4, solid lines) and only $Z = P$ is a possible solution for $\eta = 0$. The curve exhibits two limit points (LP_1 and LP_2) which we would like to determine. They satisfy the geometrical condition $d\eta^2/dZ = 0$. Using (9.30), we find

$$\frac{d\eta^2}{dZ} = \frac{F(\Omega, Z)}{(1 + 2Z)^2} = 0, \tag{9.33}$$

where the numerator is defined by

$$F \equiv -(\alpha Z - \Omega)^2(1 + 2P) + 2\alpha(\alpha Z - \Omega)(P - Z)(1 + 2Z)$$
$$+ 2Z(P - Z)(1 + 2Z) - (1 + 2P)Z^2. \tag{9.34}$$

The condition (9.33) implies $F = 0$, which gives a quadratic equation for $\alpha Z - \Omega$. We solve this equation for $\alpha Z - \Omega$ at fixed values of Z and then extract Ω. Having $\Omega(Z)$, and with $\eta^2(Z)$ given by (9.30), we have a parametric solution for the limit points. Figure 9.5 shows the limit point curves LP_1 and LP_2 in the η vs. Ω parameter space. Hysteresis is possible for both Ω positive and Ω negative.

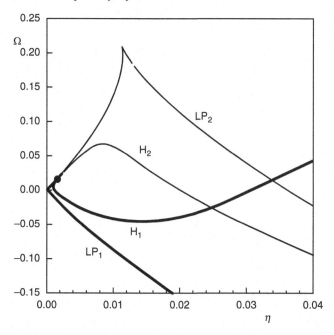

Fig. 9.5 Steady and Hopf stability boundaries. The parameters are $\alpha = 3$, $P = 10^{-1}$, and $\varepsilon \equiv T^{-1} = 8 \times 10^{-3}$. The dot marks the crossing of the SN and Hopf bifurcation lines. The curves LP_1 and H_1 correspond to SN and Hopf bifurcation points to stable solutions, respectively. The curves LP_2 and H_2 refer to SN and Hopf bifurcation points to unstable solutions, respectively.

The region bounded by the LP_1 and H_1 lines corresponds to the cross-hatched region in Figure 9.3.

9.5.2 Linear stability analysis

The starting point of all bifurcation studies is the linear stability analysis of the steady states. For the laser subject to injection, two stability boundaries are of particular interest. First, the transition to locking is marked by a limit point of the steady states. Second, the transition from steady to time-periodic intensities appears through a Hopf bifurcation. These two stability boundaries can be determined analytically.

The linearized equations for the deviations (u, v, w) from the steady state (R, ψ, Z) are given by

$$\frac{d}{dt} \begin{pmatrix} u \\ v \\ w \end{pmatrix} = L \begin{pmatrix} u \\ v \\ w \end{pmatrix}, \tag{9.35}$$

where

$$
L \equiv \begin{pmatrix} Z & -(\alpha Z - \Omega) R & R \\ (\alpha Z - \Omega) R^{-1} & Z & \alpha \\ -2R(1 + 2Z)\varepsilon & 0 & -(1 + 2R^2)\varepsilon \end{pmatrix} \tag{9.36}
$$

is the Jacobian matrix. We have eliminated $\cos(\psi)$ and $\sin(\psi)$ by using Eqs. (9.25) and (9.26). The small parameter ε is defined as

$$
\varepsilon \equiv T^{-1} \tag{9.37}
$$

and is useful if we look for approximations of the Hopf bifurcation point. We now seek a solution of Eq. (9.35) of the form $u = c_1 \exp(\sigma t)$, $v = c_2 \exp(\sigma t)$, and $w = c_3 \exp(\sigma t)$. The condition for a nontrivial solution leads to the following characteristic equation for σ

$$
\sigma^3 + C_1 \sigma^2 + C_2 \sigma + C_3 = 0, \tag{9.38}
$$

where the coefficients are defined as

$$
C_1 \equiv -2Z + \varepsilon \frac{1 + 2P}{1 + 2Z}, \tag{9.39}
$$

$$
C_2 \equiv -2\varepsilon Z \frac{1 + 2P}{1 + 2Z} + 2\varepsilon(P - Z) + Z^2 + (\alpha Z - \Omega)^2, \tag{9.40}
$$

$$
C_3 \equiv -2\varepsilon(P - Z)(\alpha(\alpha Z - \Omega) + Z) + \varepsilon \frac{1 + 2P}{1 + 2Z}\left(Z^2 + (\alpha Z - \Omega)^2\right). \tag{9.41}
$$

We have used (9.29) in order to express R^2 in terms of Z. As for the laser with a saturable absorber, the stability of the steady states can be determined by the Routh–Hurwitz conditions [29] given by

$$
C_1 > 0, \quad C_3 > 0, \quad C_1 C_2 - C_3 = 0. \tag{9.42}
$$

These conditions lead to the steady and Hopf bifurcation boundaries as we shall now describe.

9.5.3 Limit point or saddle-node bifurcation

By comparing (9.41) and (9.33), we note that

$$
C_3 = -\varepsilon(1 + 2Z)\frac{d\eta^2}{dZ}, \tag{9.43}
$$

where $\eta^2 = \eta^2(Z)$ is the implicit steady state solution (9.30). Since $Z > -1/2$, the sign of C_3 is directly related to the direction of branching $d\eta^2/dZ$. A limit point verifies the condition $d\eta^2/dZ = 0$ and thus implies

$$C_3 = 0, \tag{9.44}$$

which also means a zero eigenvalue of Eq. (9.38). The condition (9.44), or equivalently Eq. (9.33), is the equation for the limit points.

9.5.4 Hopf bifurcation

A Hopf bifurcation point satisfies the two conditions

$$C_1 C_2 - C_3 = 0 \tag{9.45}$$

and

$$C_2 > 0. \tag{9.46}$$

These conditions are found by substituting $\sigma = i\mu$ into Eq. (9.38) and separating the real and imaginary parts. The real part gives (9.45) and the imaginary part gives $\mu^2 = C_2$, which implies (9.46). The first condition leads to the following quadratic equation for $\alpha Z - \Omega$:

$$(\alpha Z - \Omega)^2 Z - \varepsilon \alpha (\alpha Z - \Omega)(P - Z) - \varepsilon \left(\varepsilon(P - Z) + 2Z^2 \right) \frac{1 + 2P}{1 + 2Z}$$

$$+ Z^3 + \varepsilon Z(P - Z) + \varepsilon^2 Z \frac{(1 + 2P)^2}{(1 + 2Z)^2}$$

$$= 0. \tag{9.47}$$

Solving this equation for $\alpha Z - \Omega$ and then extracting Ω leads to zero, one, or two real roots. Together with (9.30), we have the Hopf bifurcation point in the parametric form $\Omega(Z)$ and $\eta(Z)$. Figure 9.5 shows a typical stability diagram with the Hopf bifurcation lines.

9.5.5 Approximations of SN and Hopf bifurcation points

Because $\varepsilon = O(10^{-3})$ is small and the detuning Ω is proportional to the relaxation oscillation frequency $\omega_R = \sqrt{2P\varepsilon} = O(\varepsilon^{1/2})$, it is reasonable to look for approximations of the SN and Hopf bifurcation lines [229]. From (9.34) with $Z = O(\varepsilon^{1/2})$ and $\Omega = O(\varepsilon^{1/2})$, we find

$$\eta_{SN} = \pm \sqrt{\frac{P}{1 + \alpha^2}} \Omega, \tag{9.48}$$

which matches Adler's estimate but becomes inaccurate for larger η.

Inspecting the Hopf bifurcation condition (9.47) for small ε suggests two different scalings of the parameters. First, we may assume $\Omega = O(\varepsilon^{1/2})$ and find that $Z = O(\varepsilon^{1/2})$, which implies $\eta = O(\varepsilon^{1/2})$. From (9.47), the leading order problem is $O(\varepsilon^{3/2})$, given by

$$(\alpha Z - \Omega)^2 Z - \varepsilon \alpha (\alpha Z - \Omega) P + Z^3 + \varepsilon Z P = 0, \tag{9.49}$$

which must be solved for $\alpha Z - \Omega$. From (9.30), we then determine η^2 as

$$\eta^2 = P\left(Z^2 + (\alpha Z - \Omega)^2\right). \tag{9.50}$$

Using (9.49) and (9.50), we determine Z as

$$Z = \frac{-\varepsilon \alpha P \Omega}{\eta^2 P^{-1} + \varepsilon P(1 - \alpha^2)}. \tag{9.51}$$

We then eliminate Z in (9.50) and obtain $\Omega = \Omega_H(\eta)$ as

$$\Omega_H = \pm \frac{\eta}{\sqrt{P}} \frac{\eta^2 P^{-1} + \varepsilon P(1 - \alpha^2)}{\sqrt{\varepsilon^2 \alpha^2 P^2 + (\eta^2 P^{-1} + \varepsilon P)^2}}. \tag{9.52}$$

The expressions (9.52) provide a good approximation of the two Hopf bifurcation lines (H_1 and H_2 in Figure 9.5) except in the vicinity of $\Omega = 0$ where a different approximation with different scaling laws is needed (see Eq. (9.58)). Note that there exist two Hopf bifurcation points if $|\Omega|$ is sufficiently small. This can be seen by looking for the zeros of (9.52): $\Omega = 0$ at $\eta = 0$ and at

$$\eta = P\sqrt{\varepsilon(\alpha^2 - 1)} \tag{9.53}$$

if $\alpha > 1$. This second Hopf bifurcation at high injection rate is responsible for the stability recovery of the steady state. This was ignored in early studies of the optically injected SL, which mainly concentrated on low injection regimes. The frequency of the Hopf bifurcation is defined by $\omega_H^2 = C_2 > 0$. Using (9.40) with $\Omega = O(\varepsilon^{1/2})$ and $Z = O(\varepsilon^{1/2})$, and then (9.50), we find the elegant approximation

$$\omega_H = \sqrt{2P\varepsilon + \eta^2}. \tag{9.54}$$

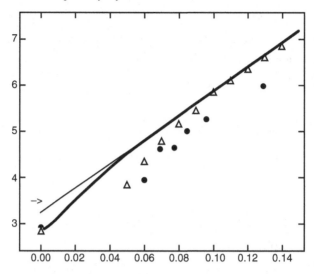

Fig. 9.6 Resonance frequency (in GHz) as a function of the injection rate r. r is proportional to η, and $r = 0.008$ at the first Hopf bifurcation point. Solid line corresponds to the numerically estimated Hopf frequency from the linearized theory. Triangles are computed frequencies from the full rate equations. Dots are the experimental data. The mode calculations used experimentally determined parameters for the slave laser. Reprinted Figure 1 with permission from Simpson *et al.* [230]. Copyright 1994 by the American Institute of Physics.

The expression (9.54) indicates that ω_H is close to the relaxation oscillation frequency $\omega_R = \sqrt{2P\varepsilon}$ for small injection rates but approaches the straight line

$$\omega_H = \eta \qquad (9.55)$$

for large injection rates. In Figure 9.6, experimental data for the main resonance frequency as a function of the injection rate are compared with theoretical predictions [231]. The monotonically increasing full line is the Hopf frequency obtained numerically from $\omega_H = \sqrt{C_2}$. The dots are the experimental data for the main resonance frequency. The numerically computed frequencies using the full nonlinear rate equations are shown by triangles. The expression (9.54) suggests that the Hopf bifurcation frequency approaches a straight line for sufficiently large injection rate (this line is indicated by an arrow in Figure 9.6). This limit is clearly followed by the experimental and numerically computed frequencies. The possibility of reaching higher frequencies by operating the laser close to the second Hopf bifurcation was successfully used to produce a source for photonic microwave transmission. Microwave frequencies over six times the RO frequency were generated [232].

Another distinct approximation of the Hopf bifurcation appears if we concentrate on the small detuning regime. Assuming $\Omega = O(\varepsilon)$ instead of $O(\varepsilon^{1/2})$ we find that Z needs to be scaled as $Z = O(\varepsilon)$, implying $\eta = O(\varepsilon)$. From (9.47), the leading approximation is now $O(\varepsilon^2)$ and given by

$$\varepsilon P \left[-\alpha(\alpha Z - \Omega) - \varepsilon(1 + 2P) + Z \right] = 0. \tag{9.56}$$

The solution for Z then is

$$Z = \frac{\varepsilon(1 + 2P) - \alpha\Omega}{1 - \alpha^2} \tag{9.57}$$

and using (9.50), we find

$$\eta_H = \frac{\sqrt{P}}{|\alpha^2 - 1|} \left[(1 + \alpha^2)(1 + 2P)\varepsilon^2 - 4\varepsilon\alpha\Omega(1 + 2P) + (1 + \alpha^2)\Omega^2 \right]^{1/2}. \tag{9.58}$$

If $\alpha = 0$, we find the expression derived for a solid state laser (see Exercise 9.8.4). In the (η, Ω) stability diagram, the Hopf bifurcation line (9.58) crosses the SN bifurcation line (9.48) ($\Omega > 0$) at a critical point called a fold-Hopf bifurcation point (see Exercise 9.8.2). From $\eta_{SN} = \eta_H$, we find

$$\Omega_{FH} = \frac{\varepsilon(1 + 2P)}{2} \frac{(1 + \alpha^2)}{\alpha}, \tag{9.59}$$

and then using (9.48)

$$\eta_{FH} = \frac{\varepsilon(1 + 2P)}{2} \frac{\sqrt{P(\alpha^2 + 1)}}{\alpha}. \tag{9.60}$$

This point is shown by a black dot in Figure 9.5. Such degenerate Hopf bifurcation points are called organizing centers because several bifurcation lines may emerge from this point (see Section 9.6.3). Note that the expressions (9.59) and (9.60) are the product of the damping of the RO oscillations (namely, $\Gamma_{RO} = \varepsilon(1 + 2P)/2$) and a nonlinear function of α. If $\alpha \to 0$, this point moves to infinity leading to a dynamically more stable laser.

9.6 Nonlinear studies

As is largely the case for engineers and applied scientists, a model is often considered as a numerical model. The difficulty with this approach is that computation limits insight because of an inability to pose questions properly. We cannot ignore the possibilities offered by our PCs but we also need to think about the objectives of our research. To this end, asymptotic approaches based on the natural

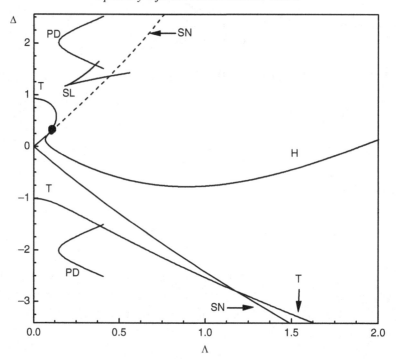

Fig. 9.7 Analytical bifurcation lines. H, T, PD, and SN denote the Hopf, torus, period-doubling, and saddle-node bifurcation lines, respectively. The stability diagram compares quantitatively with the experimental and numerical diagram of Figure 9.3.

values of the parameters smoothly complement detailed simulations by emphasizing particular features of our model equations. From an applied mathematical point of view the SL rate equations offer challenging (singular) limits requiring us to adapt known techniques to our laser equations. In this section, we summarize some of the most pertinent results and propose a stability diagram in Figure 9.7 solely based on analytical results using the same values of the parameters as the experimental-numerical diagram in Figure 9.3.

9.6.1 Formulation

The simplest limit to analyze is the limit of small injection rate. In this limit, we expect that the free running laser is weakly perturbed by the injected signal. From Chapter 1, we know that the free running laser exhibits $2\pi/\omega_R$ ROs that slowly decay in time. This suggests the introduction of the laser time (5.36) as our basic time scale. We next proceed as in Section 5.2.1 and remove the large T parameter multiplying dZ/dt in Eq. (9.8) by introducing the following dependent variables y and z

$$R = \sqrt{P}(1 + y) \quad \text{and} \quad Z = \omega_R z. \tag{9.61}$$

Inserting (9.61) into Eqs. (9.6)–(9.8) and simplifying leads to the following equations for y, z, and ψ

$$y' = (1 + y)z + \Lambda \cos(\psi), \tag{9.62}$$

$$\psi' = -\Delta + \alpha z - \frac{\Lambda}{1 + y} \sin(\psi), \tag{9.63}$$

$$z' = -\frac{1}{2}(2y + y^2) - \frac{\xi}{1 + 2P} z \left(1 + 2P(1 + y)^2\right), \tag{9.64}$$

where

$$\Lambda \equiv \frac{\eta}{\sqrt{P}\omega_R}, \quad \Delta \equiv \frac{\Omega}{\omega_R}, \quad \text{and} \quad \xi \equiv \frac{1 + 2P}{2P}\omega_R = \frac{1 + 2P}{\sqrt{2PT}}. \tag{9.65}$$

Prime means differentiation with respect to time s. The RO frequency now equals 1 in the time s variable and the large T parameter appears through the small parameter ξ. The new control parameters Λ and Δ are properly scaled by ω_R. The low injection limit implies the limit $\Lambda \to 0$ but the solution of Eqs. (9.62)–(9.64) depends on Δ.

9.6.2 Δ is arbitrary

The first step of our analysis is to determine a solution in power series of Λ assuming the small damping rate $\xi = O(\Lambda)$. The analysis is long and tedious but reveals simple results. We find that x is of the form

$$z \simeq \Lambda a \exp(i(s + \phi)) + c.c., \tag{9.66}$$

where a and ϕ denote slowly varying functions of $\tau = \Lambda^2 s$. In terms of the original time s, the amplitude $a(s)$ satisfies a linear ordinary differential equation given by

$$a' = \frac{a}{2}\left[-\xi + \frac{\alpha \Lambda^2}{\Delta(1 - \Delta^2)}\right]. \tag{9.67}$$

Equation (9.67) implies that the steady state $a = 0$ is stable either if

$$\Delta(1 - \Delta^2) < 0 \tag{9.68}$$

or if

$$\Delta(1 - \Delta^2) > 0 \quad \text{and} \quad \Lambda < \Lambda_T \equiv \sqrt{\frac{\xi \Delta(1 - \Delta^2)}{\alpha}}. \tag{9.69}$$

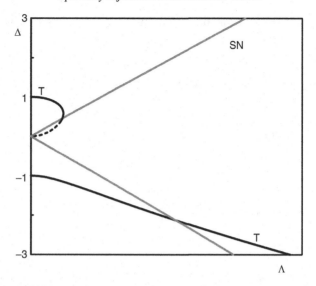

Fig. 9.8 Torus bifurcation line. Outside the locking region, the FWM pulsating intensity regime may change stability through a torus bifurcation as $\Lambda > \Lambda_T$. For negative detuning, the torus bifurcation line emerges at $\Delta = -1$ and then crosses the SN bifurcation line for large negative detunings. Stable steady and periodic solutions may then coexist in the domain bounded by the SN and T lines. For positive detuning, the torus bifurcation line emerges at $\Delta = 1$ and connects $\Delta = 0$. However a finer analysis near $\Delta = 0$ indicates that the T line ends at the SN line (see Figure 9.9).

The critical point Λ_T represents a torus bifurcation point to quasi-periodic oscillations. See Figure 9.8. A second result of our analysis is that the expansion of the solution becomes nonuniform or singular at and near the critical points $\Delta = 0, \pm 1, \ldots$ meaning points of resonance where the solution locks into a specific time-periodic regime. Each case needs to be examined independently by an appropriate new expansion of the solution. We summarize the main results.

9.6.3 $|\Delta|$ is small

If $|\Delta| = O(\Lambda)$, the expansion of the solution obtained assuming Δ arbitrary is singular and a refined analysis near $\Delta = \Lambda = 0$ is necessary. The leading order solution is now given by

$$z \simeq \Lambda^{1/2} a \exp(i(s + \phi)) + c.c., \tag{9.70}$$

$$\psi = \psi_0 - \Lambda^{1/2} \alpha a \exp(i(s + \phi)) + c.c., \tag{9.71}$$

where a, ϕ, and ψ_0 are functions of the slow time $\tau = \Lambda s$. In terms of the original time, the amplitude $a(s)$ and phase $\psi_0(s)$ satisfy two coupled equations given by

$$a' = -\frac{a}{2}\left[\xi + \Lambda\left(\alpha\sin(\psi_0) + \cos(\psi_0)\right)\right], \tag{9.72}$$

$$\psi_0' = -\Delta - \Lambda\left(\alpha\cos(\psi_0) + \sin(\psi_0)\right). \tag{9.73}$$

Equation (9.73) is equivalent to Adler's equation (9.15). However its steady or time-dependent solution will affect the stability of the laser steady state. First, Eqs. (9.72) and (9.73) admit a steady state solution $a = 0$ and $\psi_0 = \psi_s$, where ψ_s satisfies the condition

$$\Delta + \Lambda\left(\alpha\cos(\psi_s) + \sin(\psi_s)\right) = 0. \tag{9.74}$$

From (9.74), we determine the steady state limit point given by

$$\Lambda_{LP} = \frac{|\Delta|}{\sqrt{1 + \alpha^2}}, \tag{9.75}$$

which is a well-known approximation of the steady state locking point [229]. Second, the steady state exhibits a Hopf bifurcation point given by $\psi_0 = \psi_H$ at $\Lambda = \Lambda_H$, where ψ_H and Λ_H satisfy the two conditions

$$\xi + \Lambda_H\left(\alpha\sin(\psi_H) + \cos(\psi_H)\right) = 0 \tag{9.76}$$

$$\Delta + \Lambda_H\left(\alpha\cos(\psi_H) + \sin(\psi_H)\right) = 0. \tag{9.77}$$

Eliminating the trigonometric functions, we find

$$\Lambda_H = \frac{1}{|\alpha^2 - 1|}\sqrt{\left[(\alpha\xi - \Delta)^2 + (\Delta\alpha - \xi)^2\right]}, \tag{9.78}$$

which matches the approximation of the Hopf bifurcation point for small $|\Delta|$ [229]. Third, there exists a quasi-periodic solution characterized by an unbounded phase ψ_0 satisfying Eq. (9.73). This equation as well as Eq. (9.72) can be solved exactly. We note that ψ_0' is P-periodic, where P is defined by

$$P \equiv \frac{4\pi}{\sqrt{\Delta^2 - \Lambda^2(1 + \alpha^2)}} \tag{9.79}$$

and $\psi_0(s + P) = \psi_0(s) + 4\pi$. Integrating then Eq. (9.72) from $s = 0$ to $s = P$ shows that $a = 0$ is stable either if

$$\Delta < \Delta^* \equiv \frac{\xi}{2\alpha}(1 + \alpha^2) \tag{9.80}$$

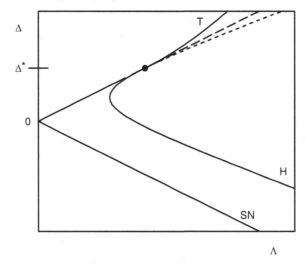

Fig. 9.9 Organizing center. The three curves of saddle-node (SN), Hopf (H), and torus (T) bifurcations cross at the same point, called the organizing center. The SN and H curves are unstable as the detuning is increased past this point, i.e. they correspond to unstable solutions.

or if

$$\Delta > \Delta^* \quad \text{but} \quad \Lambda < \Lambda_{QP} \equiv \sqrt{\frac{\xi}{\alpha} \left[\Delta - \frac{\xi}{4\alpha}(1 + \alpha^2) \right]}. \tag{9.81}$$

The critical point $\Lambda = \Lambda_{QP}$ is a quasi-periodic bifurcation point. As Δ is progressively increased from zero, this bifurcation point appears at $\Delta = \Delta^*$ and $\Lambda = \Lambda^*$, where $\Lambda^* = \Lambda_{QP}(\Delta^*)$. Note that Λ_{QP} matches the expression (9.69) as Δ further increases (i.e. the limit Δ large of (9.81) is identical to the limit Δ small of (9.69)). A uniform expression for the quasi-periodic bifurcation can be derived using (9.69) and (9.81) and is given by

$$\Lambda_{QP} = \sqrt{\frac{\xi}{\alpha} \left[\Delta(1 - \Delta^2) - \frac{\xi}{4\alpha}(1 + \alpha^2) \right]}. \tag{9.82}$$

Figure 9.9 shows the three bifurcation lines in the Δ vs. Λ diagram. They either emerge or change stability at the fold-Hopf bifurcation point (9.59) and (9.60) ($\Delta^* = \Omega_{FH}/\omega_R$ and $\Lambda^* = \eta_{FH}/(\sqrt{P}\omega_R)$; dot in Figure 9.9).

9.7 A third order Adler's equation

Asymptotic methods are used to determine specific solutions of our laser equations, for example by seeking a periodic solution near a Hopf bifurcation point.

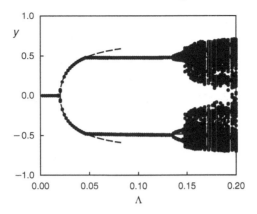

Fig. 9.10 Bifurcation of the third order Adler equation (9.88). $\Delta = 0, \xi = 0.1$, $\alpha = 5$. The extrema of y ($y = -\alpha^{-1}\psi'' + O(\xi)$) are determined as a function of Λ.

But asymptotic theories are also powerful tools for simplifying the original equations. Taking account of the large values of T and redefining the time variable as the time of the relaxation oscillations allowed us to highlight the laser conservative equations subject to weak damping. A further simplification is possible if we consider α as a large parameter. We know that relatively large values of α contribute to laser dynamical instabilities and we are interested in the limit α large. As we shall demonstrate, the laser equations are then reduced to the equations for a weakly damped harmonic oscillator subject to a strong feedback of the phase.

Specifically, we consider Eqs. (9.62)–(9.64) and examine the limit α large

$$\Lambda = O(\alpha^{-1}), \quad y = O(\alpha^{-1}), \quad \text{and} \quad z = O(\alpha^{-1}).$$

From Eqs. (9.62)–(9.64), we then find that the leading order problem is given by

$$y' = z + \Lambda \cos(\psi), \tag{9.83}$$

$$\psi' = -\Delta + \alpha z, \tag{9.84}$$

$$z' = -y - \xi z. \tag{9.85}$$

Expressing y and z as functions of ψ using (9.84) and (9.85), we derive from Eq. (9.83) a third order equation for ψ only. Specifically, we sequentially find

$$z = \alpha^{-1}(\psi' + \Delta), \tag{9.86}$$

$$y = -\alpha^{-1}\psi'' - \xi\alpha^{-1}(\psi' + \Delta), \tag{9.87}$$

and

$$\psi''' + \xi\psi'' + \psi' + \Delta + \alpha\Lambda\cos(\psi) = 0. \tag{9.88}$$

The first three terms in (9.88), namely $\psi''' + \xi\psi'' + \psi' = 0$, correspond to the equation of a damped linear oscillator. The last three terms in (9.88), namely $\psi' + \Delta + \alpha\Lambda\cos(\psi) = 0$, represent the Adler equation. Equation (9.88) is the simplest equation that describes the interaction between the free laser oscillations and the injected field. It is also a convenient equation for analysis [233].

We consider the case $\Delta = 0$. The steady state solutions are $\psi = \pm\pi/2$. The linear stability analysis indicates that $\psi = -\pi/2$ is always unstable as a saddle while $\psi = \pi/2$ undergoes a Hopf bifurcation at $\Lambda = \alpha^{-1}\xi$. We may progress further with the nonlinear analysis by considering the limit $\xi \to 0$ and assuming $\Lambda = O(\xi)$ small. Specifically, we assume

$$\Lambda = \xi\Lambda_1 + \xi^2\Lambda_2 + \dots \tag{9.89}$$

and seek a solution of the form

$$\psi = \psi_0(s) + \xi\psi_1(s) + \dots \tag{9.90}$$

Introducing (9.89) and (9.90) into Eq. (9.88), we equate to zero the coefficients of each power of ξ. The first two problems are given by

$$\psi_0''' + \psi_0' = 0 \tag{9.91}$$

$$\psi_1''' + \psi_1' = -\psi_0'' - \alpha\Lambda_1\cos(\psi_0). \tag{9.92}$$

The periodic solution of Eq. (9.91) is

$$\psi_0 = A\sin(s) + B, \tag{9.93}$$

where A and B are two unknown amplitudes. Inserting (9.93) into the right hand side of Eq. (9.92) and expanding $\cos(\psi_0)$ in Fourier modes, we apply two solvability conditions. They are given by

$$J_0(A)\cos(B) = 0, \tag{9.94}$$

$$-A + 2\alpha\Lambda_1 J_1(A)\sin(B) = 0, \tag{9.95}$$

where $J_0(A)$ and $J_1(A)$ are the Bessel functions of order zero and one, respectively. These equations admit two nontrivial solutions given by (in implicit form)

$$(1): B = \pi/2, \quad \Lambda_1 = \frac{A}{2\alpha J_1(A)} \tag{9.96}$$

and

$$(2) : J_0(A) = 0, \ \Lambda_1 = \frac{A}{2\alpha J_1(A) \sin(B)}. \tag{9.97}$$

The first solution corresponds to the Hopf bifurcation branch emerging at $\Lambda_1 = \alpha^{-1}$. The second solution implies that A equals a root of the Bessel function $J_0(A)$ (the first root is $A_1 \simeq 2.4$). The equation for B then indicates that Λ_1 is proportional to $1/\sin(B)$, implying an isolated C-shaped branch of solutions. For $A = A_1$, $J_1(A_1) > 0$ and the branch emerges from a limit point at $\Lambda_1 = A_1/(2\alpha J_1(A_1))$, where $B = \pi/2$. The upper and lower parts of the branch are bounded by the lines $B = 0$ and $B = \pi$.

In summary, our analysis has allowed us to predict the coexistence of a Hopf bifurcation branch and isolated branches of periodic solutions.

9.8 Exercises and problems

9.8.1 Hopf bifurcation close to the laser threshold

Investigate the possibility of a Hopf bifurcation for low values of the pump parameter P.

Solution: Equation (9.30) for the steady state requires that $Z \leq P$. Assuming then $Z = O(P)$ small, (9.30) reduces to

$$\eta^2 \simeq (P - Z)\Omega^2 \tag{9.98}$$

and the Hopf bifurcation condition (9.47) simplifies as

$$\Omega^2 Z + \varepsilon \alpha \Omega (P - Z) - \varepsilon^2 P + 2\varepsilon^2 Z \simeq 0. \tag{9.99}$$

Solving Eq. (9.99) for Z, we obtain

$$Z = \frac{\varepsilon^2 - \varepsilon \alpha \Omega}{\Omega^2 - \varepsilon \alpha \Omega + 2\varepsilon^2} P \tag{9.100}$$

and from (9.98)

$$\eta^2 = \frac{\Omega^2 + \varepsilon^2}{\Omega^2 - \varepsilon \alpha \Omega + 2\varepsilon^2} P \Omega^2 \geq 0. \tag{9.101}$$

The expression (9.101) tells us that a Hopf bifurcation point emerges from $\eta = 0$ if the inequality in (9.101) is verified. From the second Hopf condition $\omega^2 = C_2 > 0$, we also find that the Hopf bifurcation frequency is $\omega \simeq |\Omega|$, in first approximation. This means that the Hopf bifurcation is initiated by the detuning rather than the RO frequency as is the case for $P = O(1)$.

9.8.2 Fold-Hopf bifurcation

A fold-Hopf bifurcation (or Gavrilov–Guckenheimer or zero-pair or Hopf-SN bifurcation) appears when a single Hopf bifurcation line crosses the SN bifurcation line in the injection vs. detuning stability diagram. It is characterized by one zero eigenvalue and a pair of purely imaginary eigenvalues. The conditions for such a point are given by

$$C_1 = C_3 = 0 \quad \text{and} \quad C_2 > 0. \tag{9.102}$$

Investigate these conditions in the limit $\varepsilon \to 0$.

Solution: using (9.39), we obtain from $C_1 = 0$ a quadratic equation for Z

$$4Z^2 + 2Z - \varepsilon(1 + 2P) = 0. \tag{9.103}$$

After solving Eq. (9.103) for Z, we use (9.41) and determine from $C_3 = 0$ a quadratic equation for $\Omega - \alpha Z$ given by

$$-2(P - Z)(\alpha(\alpha Z - \Omega) + Z) + \frac{1 + 2P}{1 + 2Z}\left(Z^2 + (\alpha Z - \Omega)^2\right) = 0. \tag{9.104}$$

If $\varepsilon \to 0$, $Z \to \frac{\varepsilon}{2}(1 + 2P)$ from (9.103). Using (9.104) and (9.30), we obtain the expressions (9.59) and (9.60) for the small ε fold-Hopf bifurcation point.

9.8.3 Bogdanov–Takens bifurcation

A Bogdanov–Takens (or double zero) bifurcation appears if

$$C_2 = C_3 = 0. \tag{9.105}$$

The additional condition

$$C_1 > 0 \tag{9.106}$$

guarantees that the third eigenvalue is real and negative. Show that these conditions cannot be satisfied in the limit $\varepsilon \to 0$.

Solution: using (9.40) and (9.41), condition (9.105) implies

$$-2\varepsilon Z \frac{1 + 2P}{1 + 2Z} + 2\varepsilon(P - Z) + Z^2 + (\alpha Z - \Omega)^2 = 0 \tag{9.107}$$

and

$$-2(P - Z)(\alpha(\alpha Z - \Omega) + Z) + \frac{1 + 2P}{1 + 2Z}\left(Z^2 + (\alpha Z - \Omega)^2\right) = 0. \tag{9.108}$$

If $\varepsilon \to 0$, Eq. (9.107) requires that

$$Z^2 + (\alpha Z - \Omega)^2 = 0,$$

which is equivalent to $\eta = 0$. This condition cannot be satisfied unless $Z = \Omega = 0$. A different conclusion is possible if we assume $P = O(\varepsilon)$.

9.8.4 Injected solid state laser

In 2005, Valling *et al.* [226, 234] studied the stability diagram of a solid state Nd:YVO$_4$ laser. The interesting feature of the solid state laser is the fact that α should be zero. Best agreement between experimental and numerical maps is actually obtained by introducing an effective α factor ($\alpha = 0.35$). The α factor was later measured using pump modulation experiments, providing the result $\alpha = 0.25 \pm 0.13$ [235]. The authors considered the following rate equations (we neglect the gain saturation terms because $\gamma_p/\gamma_c \sim 10^{-5}$ and use the fact that $\gamma_n \simeq \gamma_s \tilde{J}$)

$$\frac{da}{dt} = \left[\frac{1}{2}(1 - i\alpha)\gamma_c(n - 1) + i\Omega \right] a + \kappa \tag{9.109}$$

$$\frac{dn}{dt} = \gamma_s \left[1 - n + \tilde{J}(1 - |a|^2) + \tilde{J}|a|^2(1 - n) \right], \tag{9.110}$$

where $\Omega \equiv 2\pi(\nu_{ML} - \nu_{SL})$ and $\tilde{J} \equiv (J - J_{th})/J_{th}$. By introducing the new variables

$$Y = \sqrt{\frac{\tilde{J}}{2}}a, \ Z = \frac{n - 1}{2}, \ s = \gamma_c t, \tag{9.111}$$

show that Eqs. (9.109) and (9.110) can be rewritten as Eqs. (9.3) and (9.4), where

$$T \equiv \gamma_c/\gamma_s, \ P \equiv \frac{\tilde{J}}{2}, \ \eta \equiv \sqrt{\frac{\tilde{J}}{2}}\frac{\kappa}{\gamma_c},$$

$$\alpha \to -\alpha, \quad \text{and} \quad \Omega \to -\frac{\Omega}{\gamma_c}.$$

Using the values of the parameters in [226], we determine $T = 1.53 \times 10^6$ and $P = 1.25$. Taking into account the large value of T ($\varepsilon = T^{-1} << 1$), determine the leading approximations of the Hopf and SN stability boundaries for the case $\alpha = 0$.

Solution: We first consider the Hopf bifurcation condition (9.47) with $\alpha = 0$. Assuming $\Omega = O(\varepsilon)$ and $Z = O(\varepsilon)$, we find

$$Z = \varepsilon(1 + 2P) \tag{9.112}$$

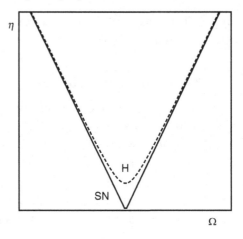

Fig. 9.11 Stability diagram for $\alpha = 0$. The Hopf bifurcation line emerges at a finite injection rate at $\Omega = 0$ and approaches the SN bifurcation boundaries as $|\Omega| >> \varepsilon$ increases from zero.

and then, using (9.30) with $\alpha = 0$, we obtain

$$\eta = \sqrt{P\left[\varepsilon^2(1 + 2P)^2 + \Omega^2\right]}. \tag{9.113}$$

Similarly, from the SN bifurcation condition, we find

$$Z = 0 \tag{9.114}$$

in first approximation and, using (9.30), we find

$$\eta = \pm|\Omega|. \tag{9.115}$$

See Figure 9.11.

10

Delayed feedback dynamics

In the study of the laser subject to an electrical feedback (see Section 4.1), we assumed that the response time of the feedback was instantaneous. The time to sense information and react to it was neglected because it was much smaller than any time scale of the CO_2 laser. This is, however, not the case for semiconductor lasers (SLs) exhibiting a very short photon lifetime inside the cavity ($\sim 10^{-12}$ s) and optical feedback response times 10^3 times larger.

In this chapter, we consider a variety of systems in which the dynamics are greatly affected by the response time of the feedback. We first concentrate on the so-called low frequency fluctuations or LFF observed with SLs, because they have been the topic of many investigations in the last 30 years. We first describe the LFF from an experimental point of view and then interpret the phenomenon in terms of numerical bifurcation diagrams. In the second part of this chapter, we show how optical feedback may also be used to improve the sensitivity of imaging systems. The last section is dedicated to optoelectronic feedback systems for which pulsating instabilities appear as a possible source of high frequency (microwave) electrical signals.

10.1 History

Optical feedback (OFB) cannot be fully avoided in experiments. Any optical element placed in front of a laser, such as a detector or even an antireflection coated lens, back-scatters part of the laser beam. This retroreflected light may re-enter the laser cavity and interfere with the field already existing inside the cavity, leading to a serious perturbation of the laser output. Soon after the development of the first laser, it was discovered that even a small amount of back-reflected radiation could strongly alter the response of a laser and that the intensity of the laser emission was modulated *"through one cycle as the distance between the laser and the reflector was varied by one half-wavelength"* [236]. If the reflecting target is moving, the frequency of the radiation reinjected inside the laser is Doppler shifted, producing a temporal modulation of the laser emission.

The extreme sensitivity of the laser to feedback permitted the determination of tiny effects such as the refractive index variations through laser reflection from a plasma [237]. It was also useful for more common practical applications such as Doppler velocimetry and range-finding. Doppler velocimetry is applied for instance to blood flow and pressure checks in human tissue (laser dopplerometry). Range-finding techniques using OFB allow imaging and profiling of surface microstructures to 20 nm depth resolution [238]. OFB is also used in extended-cavity SLs to improve their frequency stability and spectral purity or to increase their tunability.

Systematic experimental investigations of the influence of OFB started in the early 1970s and were particularly relevant for SLs. In contrast to most other lasers emitting a low divergence beam, SL divergence is large. The diffraction limit in a standard laser[1] is typically one milliradian or better for most modern lasers and it reaches a fraction of a radian for edge-emitting lasers [239]. As the SL is emitting over a large angle, it will efficiently collect light backscattered from many directions (reciprocity principle). So even small amounts of feedback will add up, making the SL very sensitive to feedback from any backscattering object. Even tiny amounts of OFB (less than 0.01%) can cause the laser to enter a state of erratic pulsating instabilities and irregular chaotic transitions.

Besides practical applications which most often rely on stable operation, the interest in studying OFB in SLs clearly arises from the rich phenomenology observed, ranging from multistability, bursting, intermittency, irregular intensity drops (LFF), and transition to developed chaos (coherence collapse). An analytical understanding of the physical mechanisms responsible for such complex behaviors is, however, still missing. In particular, the origin of the LFF (stochastic, deterministic, or both) has been under debate since their very first observations and yet this puzzling problem has not been fully solved.

Practically speaking, the dynamical regimes observed as the feedback strength is progressively increased from zero are quite different for short feedback response times (OFB from an optical fiber tip in telecommunications applications) and for large feedback response times (OFB due to reflection from a faraway target in a laser range-finder). If the external cavity formed by the retroreflecting target and the laser is of the order of 1 m, noise peaks appear at GHz frequencies and are referred to as "high frequency noise". In addition, "low frequency" noise dominates at frequencies less than 100 MHz and appears to be proportional to the external cavity length (see [240] for a recent review).

---

[1] $\sim \lambda / \pi w_0$, where w_0 is the minimum beam waist of the Gaussian laser beam.

Key to this effect is the coherence of the reflected light.[2] Modeling OFB therefore requires introducing the electrical field as a dynamical variable instead of the field intensity. Experiments indicate that an important parameter is the laser–mirror–laser round-trip time, rather than the single dephasing caused by the feedback loop. The necessity of explicitly introducing this round-trip time in the laser rate equations significantly increases the complexity of the problem because they are now delay differential equations (DDEs). A widely used system of rate equations was formulated by Lang and Kobayashi (LK) [243] in 1980 in an effort to provide a simplified but effective analysis of a SL optically coupled with a distant reflector. Thanks to intensive computer simulations, many observed phenomena were successfully reproduced using the LK equations.

10.1.1 Low frequency fluctuations

Because of OFB, the laser exhibits pulsating intensity outputs which result from a combination of effects involving delay and relaxation oscillations (RO) that sometimes fall in the same range of times. More quantitatively, the typical time scales of the semiconductor laser are the photon lifetime $\tau_p \sim 1$ ps, the carrier lifetime $\tau_c \sim 1$ ns, and the laser–mirror–laser round-trip time $\tau \equiv 2L/c \sim 1$ ns $= 10^{-9}$ s (L is the laser–mirror distance and c is the speed of light[3]). The normalized delay $\theta \equiv \tau/\tau_p$ is a large quantity like 10^3 and the relative decay rate of the carrier $\Gamma \equiv \tau_p/\tau_c$ is small like 10^{-3}. A small Γ means that the solitary laser is weakly stable and exhibits damped RO oscillations (see Chapter 1). A large delay generally generates Hopf bifurcation instabilities [245] and this is enhanced by the fact that the laser–mirror–laser round-trip time is of the order of the RO period which is typically ~ 1 ns. As a consequence these lasers are particularly sensitive to GHz signals, threatening reliable performance at the high-speed transmission rates which are now common. In addition, the linewidth enhancement factor α – unique to SLs – introduces a strong coupling between the amplitude and the phase of the laser field, which may be another source of instabilities. A recent and excellent review of the experimental literature on OFB is presented by Gavrielides and Sukow in [240]. Here, we emphasize some basic properties of the LFF.

---

[2] Incoherent OFB may be useful as demonstrated by Houlihan *et al.* [241] and Lu *et al.* [238] (see [242] for a recent application). For coherent OFB, as considered in this chapter, the coherence length of the laser emission must exceed twice the feedback distance.

[3] If $L = 15$ cm and $c = 3 \times 10^8$ m s^{-1}, we compute $\tau = 2L/c = 10^{-9}$s. $L = 15$ cm is typical for a laboratory experiment. Distances in telecom devices are smaller but the ratio $\theta = \tau/\tau_p$ remains large compared to 1, except if L decreases below 1 mm.

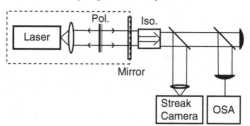

Fig. 10.1 Experimental set-up. A temperature-stabilized laser diode is subject to delayed optical feedback from a semitransparent dielectric mirror. The laser beam is collimated using an aspheric lens, and feedback strength is controlled with a polarizer (Pol.). The optical isolator (Iso.) shields this external cavity configuration from eventual perturbations from the detection branch. The light is analyzed using a single-shot streak camera and the optical spectrum is monitored with a grating spectrometer (OSA). Figure 1 adapted with permission from Heil *et al.* [244]. Copyright 2001 by the American Physical Society.

Typical experiments in the laboratory consider the case of an external mirror located at 0.1 to 1 m from the laser [247]; see Figure 10.1. The round-trip frequency of light $\nu_{EC} = \tau^{-1}$ corresponds to the intermode spacing of the "external cavity" formed by one facet of the laser diode and the external mirror. It is then some hundreds of MHz and is substantially lower than the GHz range of the RO frequency ν_{RO}. If the laser is operated close to the solitary laser threshold and the feedback is comparable with the laser facet's reflectivity (i.e. a few percent), the output intensity exhibits irregular drop-offs (LFF), a behavior characterized by at least two distinct time scales. Figure 10.2 shows an example of LFF recorded under these conditions. Figure 10.2 (top) shows irregular fluctuations of the laser intensity on a time scale of microseconds, which is very slow compared with the round-trip time and the RO period. Figure 10.2 (bottom) shows the same dynamics on faster time scales indicating that indeed there is a faster dynamics in the frequency range of the ROs ($\nu_{average} \simeq \nu_{RO}$) underlying the slow dynamics. Note in Figure 10.2 (top) the irregular intensity drops.

However, in many practical applications such as fiber couplers or compact discs, the external cavity is only a few millimeters long. The ratio of the two basic frequencies ν_{EC} and ν_{RO} is reversed and a different laser response is possible [244]. See Figure 10.3. Note that the intensity output is more regular than the one shown in Figure 10.2. The laser intensity shows a periodic emission of regular pulse packages separated by short intervals of very low intensity. The dynamics on the short time scale are now dominated by the delay time.

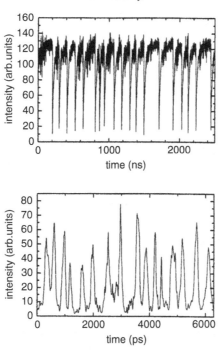

Fig. 10.2 Intensity time series recorded for a laser operating close to its threshold. Top: oscilloscope single-shot measurement, bandwith 1 GHz. Bottom: streak camera single-shot measurement, bandwidth more than 50 GHz (from Figure 1 of Heil *et al.* [246]).

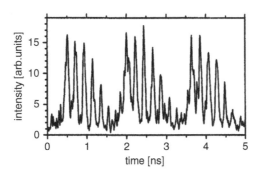

Fig. 10.3 Streak camera measurements of the intensity time series of a laser operating in the short cavity regime. The injection current is $I = 1.15I_{th,sol}$. The external cavity is 3.2 cm long corresponding to $\nu_{EC} = 4.7$ GHz. Reprinted Figure 2 with permission from Heil *et al.* [244]. Copyright 2001 by the American Physical Society.

Mathematically, we consider the idealized case of a single-mode laser subject to a *weak* optical feedback so that multiple reflection can be ignored. With the injection of the delayed optical field $Y(t - \theta)$, the laser equations for the amplitude of the field Y and the carrier density Z are given by [44, 243]

$$\frac{dY}{dt} = (1 + i\alpha)ZY + \eta \exp(-i\Omega_0\theta)Y(t - \theta), \tag{10.1}$$

$$T\frac{dZ}{dt} = P - Z - (1 + 2Z)|Y|^2, \tag{10.2}$$

where $t = t'/\tau_p$. P is the pump parameter above threshold ($P = O(1)$), $T \equiv \tau_c/\tau_p$ and $\theta \equiv \tau_L/\tau_p$ are ratios of times, η is the feedback rate ($\eta << 1$) and $\Omega_0\theta \equiv \omega_0\tau$ is a phase called the feedback phase. Ω_0 is the solitary (stand-alone) laser resonance frequency. The delay θ appears in the delayed field amplitude $Y(t - \theta)$ and in the phase factor $\exp(-i\Omega_0\theta)$. Varying the position of the external mirror over one-half an optical wavelength (250–750 nm) corresponds to a variation in the phase $\Omega_0\theta$ of 2π.

Equations (10.1) and (10.2) are known as the (dimensionless) Lang and Kobayashi (LK) equations. These equations have been extensively studied both analytically and numerically. Computer simulations have shown that they correctly describe the dominant effects observed experimentally. These include the occurrence of mode hopping [248, 249], LFF [250, 251, 252], the onset of coherence collapse [248, 253], and coexisting time-periodic attractors. This motivates further analytical investigations of the solutions of these equations.

10.1.2 ECM solutions

The LK equations admit simple solutions known as external cavity modes (ECMs). These modes are the building blocks for all analytical or numerical bifurcation studies developed here. A basic ECM solution of Eqs. (10.1) and (10.2) is the single-frequency solution

$$Y = A \exp\left(i(\Omega - \Omega_0)t\right) \quad \text{and} \quad Z = B, \tag{10.3}$$

where A, Ω, and B are constants. Substituting (10.3) into Eqs. (10.1) and (10.2) leads to three equations for A, B, and $\Delta \equiv \Omega\theta$ given by

$$B = -\eta \cos(\Delta), \tag{10.4}$$

$$\Delta - \Omega_0\theta = -\eta\theta\left(\alpha \cos(\Delta) + \sin(\Delta)\right), \tag{10.5}$$

$$A^2 = \frac{P + \eta \cos(\Delta)}{1 - 2\eta \cos(\Delta)} \geq 0. \tag{10.6}$$

Δ is called the ECM frequency.[4] It satisfies the transcendental equation (10.5) and the implicit solution is

---

[4] This definition of Δ may seem unnatural. In fact the accumulated phase shift rules the constructive/destructive character of OFB and the present definition of Δ allows for simpler expressions of, e.g., Eqs. (10.4)–(10.7).

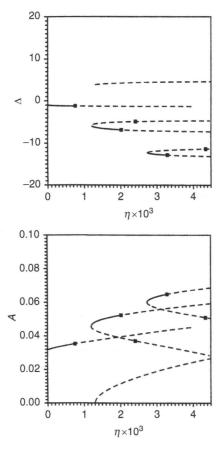

Fig. 10.4 Bifurcation diagram of the ECMs. The upper and lower figures represent the ECM frequency Δ and amplitude A, respectively. The values of the parameters are $P = 10^{-3}$, $T = \theta = 10^{3}$, $\alpha = 4$, and $\Omega_0\theta = -1$. Full and broken lines represent stable and unstable solutions, respectively. Black squares denote Hopf bifurcation points which were obtained numerically.

$$\eta\theta = -\frac{\Delta - \Omega_0\theta}{\alpha\cos(\Delta) + \sin(\Delta)}. \tag{10.7}$$

In the case of weak feedback ($\eta << 1$), Eq. (10.7) reduces to $A^2 = P + \eta\cos(\Delta)$, which opens the possibility of negative pumping if $\cos\Delta > 0$. This means that, in agreement with intuitive thinking, feedback lowers the effective laser threshold if the reflected wave constructively interferes with the intracavity wave, resulting in an effective loss reduction by an amount $O(\eta)$.

By continuously changing Δ from negative to positive values, we determine $\eta\theta$ from Eq. (10.7) and find several branches of solutions. See Figure 10.4. In the very weak feedback limit ($\eta \to 0$), there is only one ECM mode and this can be useful for imaging techniques (see Section 10.2). Except for the first mode that appears

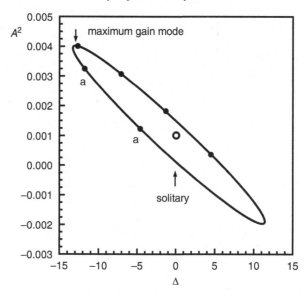

Fig. 10.5 ECM fixed points located on the ECM ellipse. The values of the parameters are $P = 10^{-3}$, $\Omega_0 \theta = -1$, $\eta = 0.003$, and $\alpha = 4$. The labels a denote
antimodes which are dynamically unstable like a saddle-point. Other ECMs are
stable or unstable with respect to the ROs. The maximum of A^2 on the ellipse
is referred to as the maximum gain mode. An ECM located close to this point
has the best chance of being stable. The open circle refers to the solitary laser
solution.

at $\eta = 0$, all other modes emerge by pairs from limit points and their number progressively increases with η. As expected, increasing the feedback rate allows more
and more modes to oscillate because the feedback reduces the effective losses for
these modes in the external cavity.

10.1.3 ECM ellipse, maximum gain mode, and LFF

The ECMs result from the interference between the laser field and the delayed
field returning from the external cavity. If the laser is replaced by a passive resonator, the resulting eigenfrequencies will have constant spacing and
show alternative constructive (even multiples of π) and destructive (odd multiples of π) interference. Since the semiconductor laser exhibits gain as well as
phase–amplitude coupling (the α parameter), the fixed points will have more
complicated properties. A popular way to represent the ECMs at a fixed feedback rate is with a diagram showing the power A^2 vs. the frequency shift Δ (see
Figure 10.5). The so-called "ellipse" is obtained by eliminating the trigonometric
functions in Eqs. (10.4) and (10.5). This leads to the following relation between
Ω and B.

$$(\Omega - \Omega_0 - \alpha B)^2 + B^2 = \eta^2. \qquad (10.8)$$

Using then the expression (10.6), we may determine the intensity A^2 of each mode. As found in the numerical simulations (see Figure 10.4), only a finite number of ECM points are possible,[5] and they are shown by dots on the ellipse. The points labeled "a" are called antimodes and are associated with destructive interference. The other ECMs fall into two categories: a stable fixed point near the maximum power, and points subject to undamped ROs caused by Hopf bifurcations. Finally we observe that the ellipticity of the ellipse is determined by the value of α. A standard class B laser (with $\alpha = 0$) does not display the rich dynamics of the SC laser with feedback.

We next introduce two particular ECM points. The "maximum gain mode" is defined by the condition $\Delta = 0 \mod 2\pi$. From the expression (10.4), we then note that this maximum gain mode corresponds to the maximum possible suppression of carriers. The system is then benefiting maximally from the feedback and the resulting power is

$$A^2 = \frac{P + \eta}{1 - 2\eta}. \qquad (10.9)$$

This state is also referred to as the "minimum threshold state," which can be understood in the following way. Without feedback, the system is characterized by the field intensity $A^2 = P$ and the threshold current is simply $P = 0$. With feedback, many ECMs may exist, each with its own effective threshold current. But for the maximum gain mode, we note from (10.9) that the effective threshold is reduced maximally to $P = -\eta$.

The second ECM fixed point is the "minimum linewidth mode," defined by the condition

$$\Delta = \Omega_0 \theta = -\arctan(\alpha), \qquad (10.10)$$

which implies $\Delta \simeq -\pi/2 \mod 2\pi$ for α sufficiently large. From Eq. (10.3), we note that there is no frequency change since $\Omega = \Omega_0$. It was long believed that this mode is the most stable mode because it has the minimum linewidth. But this conclusion was derived by an analysis of the phase equation considered as decoupled from the population equation. In other words, it ignores the stability of the ECMs with respect to the ROs.

We may analyze the possibility of a change of stability of ECMs due to a Hopf bifurcation by taking advantage of the large value of T and assuming a low feedback rate $\eta = O(T^{-1})$.[6] With $\alpha \neq 0$, the large T approximation of the first Hopf bifurcation is given by [254]

---

[5] A graphical solution of the problem is obtained by plotting the ellipse (10.8) together with the sine-wave curve (10.4) in the (Δ, B) plane. ECM solutions correspond to the intersections of these two curves.

[6] The asymptotic analysis is done in Section 10.2.1 for the special case $\alpha = 0$.

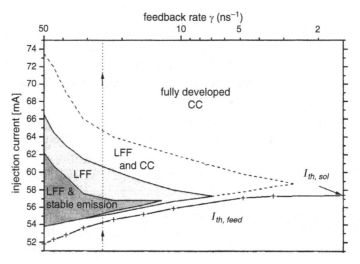

Fig. 10.6 Dynamical responses of a semiconductor laser subject to optical feedback in feedback (γ)–current (I) space. The LFF regime is depicted in light gray; the dark gray region embedded in the LFF regime corresponds to the region of coexistence of the stable emission state and the LFF state. The unshaded region encompassed by the dashed line is the transition region between the LFF regime and the fully developed coherence collapse regime (CC). Reprinted Figure 2 with permission from Heil *et al.* [256]. Copyright 1998 by the American Physical Society.

$$\eta_H = -\frac{1+2P}{2T \sin^2\left(\dfrac{\omega_R \theta}{2}\right)(\cos(\Delta) + \alpha \sin(\Delta))}, \qquad (10.11)$$

where $\omega_R \equiv \sqrt{2P/T}$ is the RO frequency. In the case of the maximum gain mode $(\Delta = 0 \bmod 2\pi)$, $\eta_H < 0$, meaning stability for low η. In the case of the minimum linewidth $(\Delta = -\arctan(\alpha) \bmod 2\pi)$, $\eta_H > 0$ if $\alpha > 1$, meaning instability [255].

The conclusions of this asymptotic analysis are limited to small feedback rates but lead to the idea that LFF may coexist with a stable emission on the maximum gain mode occurring for a wide parameter range. This was later verified experimentally [256, 257] and further explored numerically [258].

10.1.4 LFF experimental results

Experiments reported on a systematic investigation of the possible dynamical regimes by progressively increasing the pump current from below threshold through the whole accessible range. This procedure was repeated for several values of the feedback strength ranging over three orders of magnitude. See Figure 10.6 for the transition region. Increasing the pump current along the vertical dotted line in Figure 10.6 ($\gamma = 25\,\mathrm{ns}^{-1}$),[7] all possible regimes appearing in the (γ, I)

---

[7] $\eta = \gamma \tau_P$.

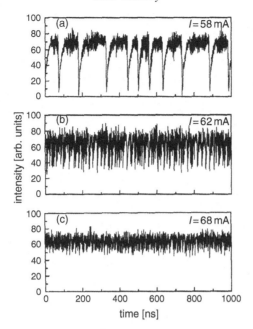

Fig. 10.7 Intensity time series of a SL subject to optical feedback and three different pump currents I. The optical feedback amounts to $\gamma \approx 25\,\mathrm{ns}^{-1}$, which corresponds to the vertical line in Figure 10.6. We note the progressive transition from LFF to CC as the current is increased. Reprinted Figure 3 with permission from Heil *et al.* [257]. Copyright 1999 by the American Physical Society.

diagram have been observed. After passing the feedback-reduced laser threshold ($I_{th,feed}$) the laser emission is stable on a single longitudinal diode mode and several ECM modes. The mode-beating of the ECMs appears as sharp peaks in the power spectrum. The LFF regime is reached by increasing the injection current by approximately 1 mA above threshold. There, a dominant low frequency contribution builds up in the power spectrum; the ECM beatings broaden significantly. The emission of the laser is still dominated by one longitudinal diode mode. Figure 10.7 (a) shows a time series of the dynamical behavior within the LFF regime. With increasing injection current, the time intervals between dropouts decrease. Finally a continuous transition to a fully developed coherence collapse (CC) takes place, accompanied by a broad power spectrum and completely irregular intensity time series. This dynamical behavior is illustrated in Figures 10.7 (b) and (c) showing a time series of the transition regime and the fully developed coherence collapse regime, respectively. Moreover, there exists a large region within the LFF domain where discrete transitions between LFF and stable emission on a single ECM occur. An example of such a behavior is presented in Figure 10.8.

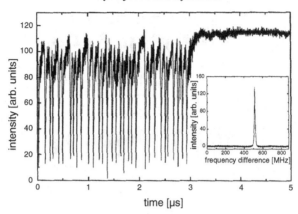

Fig. 10.8 Intensity time series for a SL subject to OFB with $I = 60$ mA and $\gamma = 45\,\text{ns}^{-1}$. The inset shows the optical spectrum of the stable emission state. Reprinted Figure 4 with permission from Heil *et al.* [257]. Copyright 1999 by the American Physical Society.

This figure shows the intensity time series of a transition from the LFF state to a stable emission state appearing at 3 µs. We note that the intensity stabilizes on a higher level than the LFF. Accordingly, the transition to stable emission is characterized by a sudden increase in the time-averaged recorded intensity. Second, the power spectrum is completely flat; no frequency component remains. Third, the stable emission occurs on a single ECM, resolved by the scanning Fabry–Pérot interferometer. The inset of Figure 10.8 shows the optical spectrum of the stable emission state, showing a single sharp peak. All these checks support the interpretation of the behavior shown in Figure 10.8 as due to the coexistence of a stable state with LFF.

10.1.5 Numerical simulations and bridges

Optical sources pulsating with high frequencies of several tens of gigahertz are required for a number of signal processing applications. By the end of the 1990s, the group of Bernd Sartorius from the Heinrich-Hertz-Institut (HHI) in Berlin had started to be interested in generating tunable self-pulsations (SPs) with frequencies above 20 GHz in laser devices. It was later discovered that Tager and Elenkrig (1993) [259] and Tager and Petermann (1994) [260] already were concerned with this problem. Using the single-mode LK equations, they analyzed the possibility of a Hopf bifurcation to a high frequency mode-antimode beating (MB) regime. They found that a short external cavity[8] and a high feedback rate were necessary

---

[8] A simple rule of thumb suggests that a short cavity favors high frequency MB oscillations because of a larger intermode spacing. This is confirmed by mathematical analysis.

for this type of output. But the authors didn't give any clue as to the stability of such high frequency SPs. Could a stable mode and an unstable mode combine and produce a stable two-mode regime? Starting in 1999, a series of workshops was organized at the Weierstrass Institute for Applied Analysis and Stochastics (WIAS) with the aim of attracting mathematicians and engineers and discussing these issues. In 2000, an asymptotic analysis of the LK equations based on the limit T large showed that the high frequency MB regimes belong to branches that are connecting isolated ECM branches (bridges) [261, 262]. Therefore, the LK equations may exhibit two types of Hopf bifurcations, namely, the bifurcation to RO oscillations or the bifurcation to MB regimes. How these bifurcations interact in parameter space was carefully investigated in [263]. In 2002, Sieber [264] proposed a detailed bifurcation analysis of the traveling wave laser equations, emphasizing the domains of parameters where the high frequency pulsations are possible. To achieve the required high feedback, Bauer *et al.* [265] from the HHI have attached to the passive short external cavity an active amplifier section. The carriers in the amplifier introduce an additional degree of freedom leading to a stabilization of the MB regime [266] as well as a complex dynamics including chaos [267]. High frequency dynamical regimes of passive feedback lasers were reported by Ushakov *et al.* in 2004 [268] using an integrated distributed feedback device that allows the control of the feedback phase.

We first summarize numerical results. The bifurcation diagram of the maxima and minima of $|Y|$ obtained by integrating the LK equations in time for gradually increasing (or decreasing) values of η is shown in Figure 10.9 top (same values of the parameters as in Figure 10.4). The figure shows successive stable ECM branches, each undergoing a Hopf bifurcation. The same diagram now obtained by a continuation method (BIFTOOL) for the steady and time-periodic solutions is shown in the bottom figure. Only the maxima are shown. The figure reveals that bridges do connect two Hopf bifurcation points belonging to distinct ECMs. What is the nature of these bridges?

It can be shown by an asymptotic analysis of the LK equations based on the limit T large that these bridges correspond to solutions combining two single ECMs of the form [261]

$$Y = A_1 \exp\left(i(\Omega_1 - \Omega_0)t\right) + A_2 \exp\left(i(\Omega_2 - \Omega_0)t\right) + O(T^{-1}). \qquad (10.12)$$

In contrast to a single ECM for which the intensity $I = A^2$ is constant, the intensity of the two-ECM solution (10.12) exhibits time-periodic intensity oscillations of the form

$$I = |A_1|^2 + |A_2|^2 + 2|A_1||A_2| \cos\left((\Delta_1 - \Delta_2)\theta^{-1}t + \phi\right), \qquad (10.13)$$

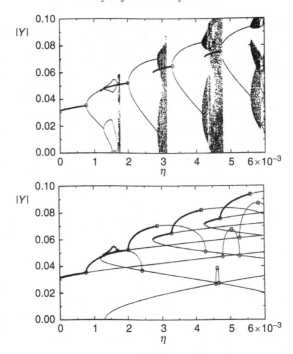

Fig. 10.9 Top: bifurcation diagram of the stable solutions obtained by integration. Bottom: bifurcation diagram of the stable and unstable steady and periodic
solutions obtained by a continuation method. Reprinted Figure 1 with permission
from Pieroux *et al.* [262]. Copyright 2001 by the American Physical Society.

where ϕ is a constant phase. The period of the oscillations is the mode-beating
period

$$P_{MB} = 2\pi\theta \left|\Delta_1 - \Delta_2\right|^{-1}, \tag{10.14}$$

which is clearly proportional to the delay θ.

Numerical bifurcation studies suggest that bridges are either unstable or are
partially stable [261, 262, 269]. However, stable bridges are possible if α is sufficiently low ($\alpha \leq 1$) [269]. For an arbitrary value of $\alpha > 1$, a stable bridge may
change its stability at a torus bifurcation point as we increase the feedback rate. The
torus bifurcation leads to quasi-periodic oscillations with two distinct frequencies.
The first and second frequencies are the bridge intensity frequency which coincides with the MB frequency P_{MB}^{-1}, and the usual RO frequency $\omega_R = \sqrt{2P/T}$
[262, 269]. This confirms our initial intuition of a simple interaction between the
RO and MB dynamics. The different ECM modes are interconnected through
bridges but this was not predicted by our first analysis of the steady states (see
Figure 10.4).

10.2 Imaging using OFB

Few people need to inspect an object in a glass of milk. If, however, an imaging technique can see through milk, it can probably image objects effectively through other diffusing media such as blood or even a suspension of silica powder in a polishing workshop. It then becomes an effective inspection tool in applications as diverse as manufacturing inspection, medical imaging of living tissues, even tasks requiring undersea visibility.

Current options for imaging through diffusing media include techniques such as time-resolved holography, optical coherence tomography, and scanning confocal microscopy. Each has its benefits and limitations. With confocal techniques, for example, researchers can avoid some problems intrinsic to tomography, such as the requirement to solve an inverse problem. There are, however, issues related to limited imaging sensitivity and the complexity and cost of required equipment.

In 1999 Frédéric Stoeckel at the Laboratoire de Spectrométrie Physique of the Université Joseph Fourier de Grenoble had the idea of taking advantage of optical feedback using Nd^{3+}:YAG microchip lasers [270, 271, 272]. Together with his colleague Eric Lacot, they developed a new technique called LaROFI for laser relaxation oscillation frequency imaging [273]; see Figure 10.10. The technique relies on the resonant sensitivity of a short-cavity laser to optical feedback produced by ballistic photons retrodiffused from the medium. The method produces two- and three-dimensional imaging in turbid media that is similar to heterodyne scanning confocal microscopy, but resolves some of the limitations just discussed.

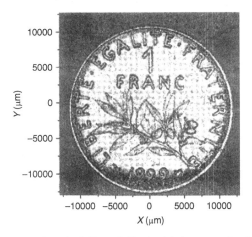

Fig. 10.10 Two-dimensional (262 × 262 pixels) image of a French one-franc coin using the laser relaxation oscillation frequency imaging technique. The coin is immersed in a glass of milk (milk depth 1 cm). The pixel dimensions are 100 μm × 100 μm (from Figure 3 of Lacot *et al.* [271]).

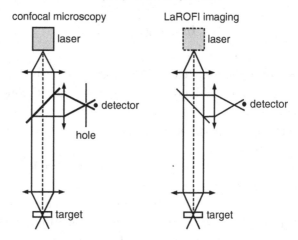

Fig. 10.11 In confocal microscopy, spatial filtering is controlled by the size of the hole, and the quality of the detection depends on the detector. In LFI and LaROFI imaging, spatial filtering is achieved by selection of one laser mode, and the quality of the detection depends on the laser.

One important advantage of the LaROFI method is that the laser source is also the detector. In addition to its optical amplification duties, it provides self-aligned spatial and temporal coherent detection (acts as both a spatial and a temporal filter). See Figure 10.11. Another novelty of the LaROFI technique is that the frequency of the intensity relaxation oscillations is measured together with the intensity of the laser field. This provides 100 times higher sensitivity compared to previous techniques based on external cavity frequency measurements (laser feedback interferometry or LFI). The main objective of the next two subsections is to explain why this is the case using the LK equations. Our analysis provides approximations of the laser intensity and laser relaxation oscillations that compare quantitatively with the experimental data.

10.2.1 Stability analysis

The LK equations (10.1) and (10.2) describe the response of a single-mode laser subject to optical feedback from a distant mirror. Introducing the amplitude R and the phase ϕ of the field $Y = R \exp(i\phi)$, these equations with $\alpha = 0$ (we are dealing with microchip solid state lasers where α is zero) can be rewritten as

$$\frac{dR}{dt} = ZR + \eta R(t - \theta) \cos(\phi(t - \theta) - \phi - \Omega_0 \theta), \quad (10.15)$$

$$\frac{d\phi}{dt} = \eta \frac{R(t - \theta)}{R} \sin(\phi(t - \theta) - \phi - \Omega_0 \theta), \quad (10.16)$$

$$T\frac{dZ}{dt} = P - Z - (1 + 2Z)R^2. \tag{10.17}$$

Equations (10.15)–(10.17) admit the ECMs (10.3) as the basic solutions. In terms of R, ϕ, and Z, they are given by

$$R = A, \quad \phi = (\Omega - \Omega_0)t, \quad \text{and} \quad Z = B, \tag{10.18}$$

where A, B, and $\Delta = \Omega\theta$ are constants given by (10.4), (10.5), and (10.6) with $\alpha = 0$. We investigate their linear stability properties by introducing the small perturbations u, v, and w. The linearized equations are the following equations for u, v, and w

$$\frac{du}{dt} = Bu + \eta\cos(\Delta)u(t - \theta) + \eta A\sin(\Delta)(v(t - \theta) - v) + Aw, \tag{10.19}$$

$$\frac{dv}{dt} = -\frac{\eta}{R}(u(t - \theta) - u)\sin(\Delta) + \eta\cos(\Delta)(v(t - \theta) - v) + \alpha w, \tag{10.20}$$

$$\frac{dw}{dt} = -T^{-1}\left[(1 + 2B)2Au + w(1 + 2A^2)\right]. \tag{10.21}$$

We solve these equations by looking for a solution of the form $u = a\exp(\lambda t)$, $v = b\exp(\lambda t)$, and $w = c\exp(\lambda t)$. We then obtain the following problem for the coefficients a, b, and c

$$\lambda\begin{pmatrix} a \\ b \\ c \end{pmatrix} = L\begin{pmatrix} a \\ b \\ c \end{pmatrix}, \tag{10.22}$$

where the Jacobian matrix L is defined by

$$L \equiv \begin{pmatrix} \eta\cos(\Delta)F & \eta A\sin(\Delta)F & A \\ -\dfrac{\eta\sin(\Delta)}{A}F & \eta\cos(\Delta)F & \alpha \\ -(1 + 2B)2A\varepsilon & 0 & -(1 + 2A^2)\varepsilon \end{pmatrix} \tag{10.23}$$

with

$$F \equiv \exp(-\lambda\theta) - 1 \quad \text{and} \quad \varepsilon \equiv T^{-1}. \tag{10.24}$$

A nontrivial solution is possible only if λ satisfies the condition $\det(L - \lambda I) = 0$. This condition leads to the characteristic equation for the growth rate λ

$$0 = \left[-(1 + 2A^2)\varepsilon - \lambda\right][\eta\cos(\Delta)F - \lambda]^2 + \eta^2\sin^2(\Delta)F^2$$

$$+ (1 + 2B)2A^2\varepsilon\left[\eta\cos(\Delta)F - \lambda\right]. \tag{10.25}$$

10.2.2 Low feedback rate approximation

Equation (10.25) is hard to solve even numerically. Several approximations have been investigated in the past [254]. In this section, we propose to investigate the solution of Eq. (10.25) for low values of η, which is the case for imaging through a diffuse medium since very few photons return back into the laser.

If $\eta\theta$ is small, there is only one ECM. From (10.5), (10.6), and (10.4), we find the simple approximation

$$\Delta = \Omega_0\theta + O(\eta\theta), \quad A^2 = P + O(\eta), \quad \text{and} \quad B = O(\eta). \tag{10.26}$$

If $\eta = 0$, the characteristic equation (10.25) reduces to

$$\lambda\left[\lambda^2 + \lambda(1 + 2P)\varepsilon + 2P\varepsilon\right] = 0, \tag{10.27}$$

which we recognize as the characteristic equation for the solitary laser (see Chapter 1). For small ε and $\lambda \neq 0$, Eq. (10.27) has the solution

$$\lambda = \pm i\sqrt{2P\varepsilon} - \varepsilon\frac{1 + 2P}{2} + O(\varepsilon^{3/2}). \tag{10.28}$$

The leading term is the RO frequency, defined by

$$\omega_R \equiv \sqrt{2P\varepsilon}. \tag{10.29}$$

These are the solutions for our standard rate equations reformulated in terms of the notations of the LaROFI problem.

The expression (10.28) motivates seeking a solution of (10.25) of the form

$$\lambda = \varepsilon^{1/2}\lambda_0 + \varepsilon\lambda_1 + \ldots \tag{10.30}$$

and in order to balance terms with η in Eq. (10.25), we assume η as an $O(\varepsilon)$ quantity. Specifically, we expand η as

$$\eta = \varepsilon\eta_1 + \varepsilon^{3/2}\eta_2 + \ldots \tag{10.31}$$

Introducing (10.30) and (10.31) into Eq. (10.25), taking into account (10.26), we equate to zero the coefficients of each power of $\varepsilon^{1/2}$. The first two problems are

$$O(\varepsilon^{3/2}) : 0 = -\lambda_0^3 - 2P\lambda_0, \tag{10.32}$$

$$O(\varepsilon^2) : 0 = -(3\lambda_0^2 + 2P\varepsilon)\lambda_1 + 2\lambda_0^2 \eta_1 \cos(\Omega_0 \theta) F_0$$
$$-(1 + 2P)\lambda_0^2 + 2P F_0 \eta_1 \cos(\Omega_0 \theta), \tag{10.33}$$

where

$$F_0 \equiv \exp(-\varepsilon^{1/2}\lambda_0 \theta) - 1 \tag{10.34}$$

and we have assumed $\varepsilon^{1/2}\theta = O(1)$. From Eq. (10.32) and then Eq. (10.33), we determine λ_0 and λ_1. Together, the growth rate λ is then given by

$$\lambda \simeq \pm i\omega_R + \frac{1}{2}\left[-\varepsilon(1 + 2P) - 2\sin^2(\omega_R\theta)\eta\cos(\Omega_0\theta) \right.$$
$$\left. \mp i\sin(\omega_R\theta)\eta\cos(\Omega_0\theta) \right]. \tag{10.35}$$

The LaROFI imaging technique as invented by Lacot *et al.* [273] is based on the change in the RO frequency predicted by (10.35). Specifically, they determined the modification of the relaxation oscillation frequency of the laser as the feedback rate increases. In the case of constructive interference,

$$\cos(\Omega_0\theta) = 1, \tag{10.36}$$

the ECM solution (10.26) is stable since $\text{Re}(\lambda) < 0$. The imaginary part in (10.35) provides the correction to the RO frequency ω_R due to optical feedback. This relative change of the RO frequency is thus given by

$$\frac{\omega_{OF} - \omega_R}{\omega_R} = -\frac{\eta}{2\omega_R}\sin(\omega_R\theta). \tag{10.37}$$

Furthermore, if $\omega_R\theta$ is small, we have $\sin(\omega_R\theta) \simeq \omega_R\theta$, and the expression (10.37) can be further simplified as

$$\frac{\omega_{OF} - \omega_R}{\omega_R} = -\frac{\eta\theta}{2}. \tag{10.38}$$

In terms of the original parameters (using $\eta = \gamma_c\sqrt{R_{eff}}$, where γ_c is the damping rate and R_{eff} is the effective feedback reflectivity), (10.37) and (10.38) lead to

$$F \equiv \frac{\Omega_{OF} - \Omega_R}{\Omega_R\sqrt{R_{eff}}} = -\frac{\gamma_c}{2\Omega_R}\sin(\Omega_R\tau) \tag{10.39}$$

and

$$F \equiv \frac{\Omega_{OF} - \Omega_R}{\Omega_R\sqrt{R_{eff}}} = -\frac{\gamma_c}{2}\tau = -\frac{\gamma_c}{c}d, \tag{10.40}$$

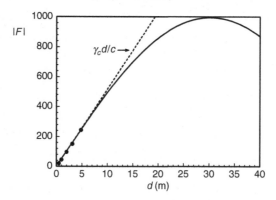

Fig. 10.12 Modified RO frequency due to optical feedback. The figure repre-
sents the relative change of the relaxation frequency $|F|$ as a function of the
laser–target distance d. Dashed line: small d approximation. Full line: exact
expression valid for arbitrary d. Dots are experimental results documented
in [271].

respectively. Ω_R and τ are defined by

$$\Omega_R \equiv \sqrt{\gamma_1 \gamma_c (P - 1)} \quad \text{and} \quad \tau \equiv \frac{2d}{c}, \tag{10.41}$$

where d is the laser–target distance and $c = 3 \times 10^8 \, \text{m s}^{-1}$ is the speed of light.
In Stoeckel's device, the population inversion damping rate is $\gamma_1 = 1/(255 \, \mu s) = 3.92 \times 10^3 \, \text{s}^{-1}$, the cavity damping rate is $\gamma_c = 1.55 \times 10^{10} \, \text{s}^{-1}$, and the pump
parameter above threshold is $P = 2$. With $R_{eff} = 10^{-4}$ and $d \sim 1$ m, we note
from (10.40) that the relative change of the RO frequency resulting from the feed-
back, $|\Omega_{OF} - \Omega_R|/\Omega_R$, is of the order of 10^{-2} which is easily measured (dots in
Figure 10.12).

To determine the two-dimensional image shown in Figure 10.10, the laser beam
is focused by a microscope objective on a French one-franc coin that is localized
2 m from the laser source. The sampling step is 100 μm in both the x and y directions (see Figure 10.13). In this experiment, the effective reflectivity was taken
very small (of the order of 10^{-7}) in order to demonstrate the high sensitivity of the
method.

However, there are theoretical and technical limits to the sensitivity enhancement if the delay becomes large. Variations of F with d as given by (10.39)
and (10.40) are shown in Figure 10.12 by the full and the broken line,
respectively. We note that the increase of $|F|$ does not remain linear but
exhibits a maximum near $d = 30$ m. This behavior results from the sine function
in (10.39).

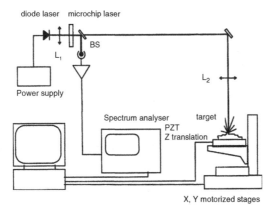

Fig. 10.13 The laser beam is focused on a target. Only photons back-scattered from points located near the center of the laser beam on the target are reinjected by mode matching into the laser. The laser dynamics is modified by the interference effects taking place between the back-scattered field and the standing wave inside the laser cavity. This interference effect depends on the reflectivity, distance, and motion of the target. The laser output power is detected by a photodetector and the laser relaxation frequency is determined by a spectrum analyzer via the intensity noise spectrum of the laser. In order to obtain an image, a micrometric translation unit combined with a PZT moves the target. Figure 5 adapted with permission from Lacot *et al.* [272]. Copyright 2001 by the American Physical Society.

10.3 Optoelectronic oscillator

High repetition rate pulse sources are usually implemented by active mode-locking of fiber or diode lasers, which requires a microwave-driving source whose phase noise determines or limits the resultant jitter.[9] Passively mode-locked lasers, on the other hand, do not need a microwave drive but they tend to have more jitter than actively mode-locked lasers. A completely different approach for obtaining sustained pulse sources is to use optoelectronic oscillators (OEOs). OEOs typically incorporate a nonlinear modulator, an optical-fiber delay line, and optical detection in a closed-loop resonating configuration. These devices can generate radiofrequency oscillations with extremely high spectral purity and low phase noise in the microwave range at up to tens of GHz. Here, we concentrate on OEOs composed by a Mach–Zehnder modulator, an optical fiber delay line, a photodiode, and a radiofrequency amplifier. See Figure 10.14. Specifically, a continuous-wave semiconductor laser provides the energy source of power P_0 (0–10 mW). It illuminates a Mach–Zehnder (MZ) modulator that produces the essential nonlinearity for the feedback loop. The transmission of the MZ modulator is a nonlinear function of

---

[9] Jitter is an unwanted variation of the signal characteristics of the laser output. Jitter may be seen in the interval between successive pulses, or the amplitude, frequency, or phase of successive cycles.

Delayed feedback dynamics

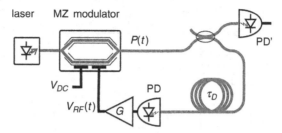

laser MZ modulator

$P(t)$

PD'

V_{DC}

PD

$V_{RF}(t)$ G

τ_D

Fig. 10.14 Schematic description of the feedback loop. It is formed by a Mach–Zehnder modulator (MZ), a fiber delay line, a photodiode (PD), and a radiofrequency amplifier. A semiconductor laser provides the energy source and an oscilloscope measures the fast time dynamics of the feedback loop. Figure 1 adapted with permission from Kouomou *et al.* [274]. Copyright 2005 by the American Physical Society.

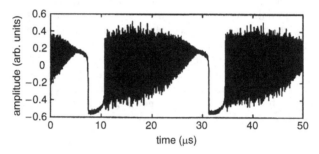

Fig. 10.15 Bursting oscillations in an OEO device. Rapid oscillations are modulated by a slowly varying envelope. Reprinted Figure 5 with permission from Kouomou *et al.* [274]. Copyright 2005 by the American Physical Society.

the applied voltage, where we independently apply a time-dependent voltage to the radiofrequency (RF) port of the device (half-wave voltage $V_{\pi RF} = 4.0$ V) and a DC voltage V_{DC} to bias it at any point on the transmission curve (half-wave voltage $V_{\pi DC} = 4.0$ V). The light power at the output of the modulator then is a sinusoidal function of V_{DC} with an amplitude that depends on the power P. The output of the modulator is injected into a long optical fiber with delay time τ_D ($\tau_D \sim 40$ ns) and a photodiode (PD) of gain g converts the light into an electrical current. Finally, the radiofrequency amplifier with gain G converts the signal from the photodiode into an electronic voltage $V_{RF}(t)$ that is fed back in the MZ modulator. This voltage, added to V_{DC}, changes the output of the modulator and the feedback loop is completed. The overall attenuation of the loop (delay line, connectors, and so on) is described by the parameter A. The electronic bandwidth of the feedback loop is assumed to result from cascaded linear first order low-pass and high-pass filters, with low and high cutoff frequencies f_L and f_H, respectively. Figure 10.15 shows experimentally observed bursting oscillations where

fast oscillations are modulated by a slowly varying envelope. The rapid oscillations operate on the ns time scale while the period of the slow envelope is on the
μs time scale. This large time-scale difference motivates asymptotic studies of the
model equations. According to our usual approach, we introduce dimensionless
variables and identify the different time scales of the problem. Introducing the
dimensionless voltage $x(t) \equiv \pi V_{RF}(t)/2V_{\pi RF}$, the dynamical response of the
system as recorded from the photodiodes PD or PD$'$ (see Figure 10.14) is well
described by the following delay-integro-differential equation ([275])

$$x + \tau \frac{dx}{dt} + \theta^{-1} \int_0^t x(s)ds = \beta \left[\cos^2 (x(t - \tau_D) + \phi) - \cos^2(\phi) \right], \quad (10.42)$$

where $\beta \equiv \pi g A G P/2V_{\pi RF}$, proportional to the source power P, measures the
feedback strength and $\phi \equiv \pi V_{DC}/2V_{\pi DC}$ is the offset phase that is proportional to
the bias voltage V_{DC} of the MZ. The time constants $\theta \equiv 1/2\pi f_L$ and $\tau \equiv 1/2\pi f_H$
are inversely proportional to the cutoff frequencies f_L and f_H of the filter. In these
experiments, the three time parameters τ, τ_D, and θ have quite different orders of
magnitude, namely

$$\tau = 25 \text{ ps}, \ \tau_D = 30 \text{ ns}, \quad \text{and} \quad \theta = 5 \text{ μs}. \quad (10.43)$$

The OEO system differs from the Ikeda delay differential equation (DDE) (4.20)
by the integral term in (10.42). Ikeda ignores the low cutoff frequency by assuming
$f_L = 0$ (equivalently, $\theta^{-1} = 0$). In the absence of delay, however, the integral term
allows for a new degree of freedom that generates new oscillatory regimes. This
can best be seen by introducing the new variable

$$z \equiv \int_0^t x(s)ds \quad (10.44)$$

and rewriting Eq. (10.42) as the following system of two first order equations

$$x + \tau \frac{dx}{dt} + \theta^{-1}z = \beta \left[\cos^2 (x(t - \tau_D) + \phi) - \cos^2(\phi) \right], \quad \text{and} \quad \frac{dz}{dt} = x. \quad (10.45)$$

Further differentiating the first equation allows the elimination of z and the
reformulation of Eq. (10.45) as a second order DDE of the form

$$\frac{dx}{dt} + \tau \frac{d^2x}{dt^2} + \theta^{-1}x = \beta \frac{d}{dt} \left[\cos^2 (x(t - \tau_D) + \phi) \right]. \quad (10.46)$$

It is now necessary to introduce a dimensionless time so that we may evaluate the
contribution of each term in Eq. (10.46). If $\beta = 1.5$–3, a stable periodic solution

is found numerically and exhibits a large period (3–10 µs), which is proportional to the slowest time constant θ. This motivates the introduction of the new time variable

$$s \equiv \theta^{-1} t \tag{10.47}$$

as our basic time. In terms of (10.47), Eq. (10.46) then becomes

$$x' + \varepsilon x'' + x = \beta \left[\cos^2 (x(s - \delta) + \phi) \right]', \tag{10.48}$$

where prime means differentiation with respect to s. This equation now exhibits two small parameters, namely

$$\varepsilon \equiv \tau \theta^{-1} = 5 \times 10^{-6} \quad \text{and} \quad \delta \equiv \tau_D \theta^{-1} = 6 \times 10^{-3}, \tag{10.49}$$

which we would like to neglect. However, the $\varepsilon x''$ term could be important if x' is changing fast like ε^{-1}, and the delay δ could lead to high frequency nearly δ-periodic solutions. We shall proceed in two stages. First we shall look for a slowly varying periodic solution and neglect the small delay δ. We then shall look for the stability of this solution with respect to the δ short time scale.

10.3.1 Slowly varying oscillations ($\delta = 0$, $\varepsilon \neq 0$)

We first concentrate on the low frequency oscillations that modulate the rapid bursting oscillations. We ignore the effect of the delay, in first approximation, and seek a time-periodic solution of Eq. (10.48) with $\delta = 0$, given by

$$x' + \varepsilon x'' + x = \beta \left[\cos^2 (x + \phi) \right]'. \tag{10.50}$$

We solve Eq. (10.50) numerically and find a limit-cycle that consists of slowly varying parabolic plateaus connected by fast transition layers. See Figure 10.16. Like the Van der Pol equation [15], we may analyze Eq. (10.50) in the phase plane for ε small (see Exercise 10.4.4). The starting point of our analysis is a reformulation of Eq. (10.50) as the following system of two first order differential equations

$$x' = y, \tag{10.51}$$

$$\varepsilon y' = -x - y \left[1 + \beta \sin (2x + 2\phi) \right]. \tag{10.52}$$

Neglecting ε, the trajectory $y = y(x)$ corresponding to the slowly varying plateaus is given by

$$y = -\frac{x}{1 + \beta \sin (2x + 2\phi)}. \tag{10.53}$$

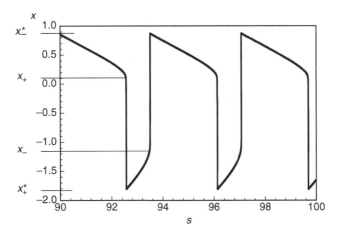

Fig. 10.16 Time-periodic numerical solution of Eq. (10.50) for $\phi = -\pi/10$, $\varepsilon = 10^{-3}$, and $\beta = 2.5$. Four points on X mark important changes in the time evolution. Approximations of these points are determined in the text.

This solution is valid until the denominator in (10.53) approaches zero. This happens when $x = x_{\pm}$, which are the first two roots of

$$1 + \beta \sin (2x + 2\phi) = 0. \tag{10.54}$$

From (10.54), we obtain

$$x_+ \equiv -\frac{1}{2} \arcsin(1/\beta) - \phi, \tag{10.55}$$

$$x_- \equiv \frac{1}{2}(-\pi + \arcsin(1/\beta)) - \phi. \tag{10.56}$$

If $\beta = 2.5$ and $\phi = -\pi/10$, we find $x_+ \simeq 0.1084$ and $x_- \simeq -1.0509$ (see Figure 10.16). They clearly mark the beginning of the fast transition layers. Assuming $y >> x$, the leading approximation of the transition layer trajectory is given by

$$\varepsilon y = -(x - x_{\pm}) + \frac{\beta}{2} \left[\cos(2x + 2\phi) - \cos(2x_{\pm} + 2\phi)\right]. \tag{10.57}$$

The transition layers end at $x = x_{\pm}^*$ when $\varepsilon y = 0$. Solving Eq. (10.57) for $\varepsilon y = 0$, we obtain $x_+^* \simeq -1.7424$ and $x_-^* \simeq 0.8000$ for $\beta = 2.5$ and $\phi = -\pi/10$ (see Figure 10.16). Finally, the leading approximation for the period is obtained by integrating Eq. (10.51) over time for the evolutions along the two plateaus.

10.3.2 Fast bursting oscillations ($\varepsilon = 0$, $\delta \neq 0$)

The rapid bursting oscillations observed numerically and found experimentally (see Figure 10.15) motivate a stability analysis of the slowly varying plateaus in Figure 10.16. We cannot ignore the delay any more, but we may neglect ε because the slowly varying plateaus do not depend on ε, in first approximation. We therefore consider the DDE (10.48) with $\varepsilon = 0$, given by

$$x' + x = \beta \left[\cos^2 (x(s - \delta) + \phi) \right]'. \tag{10.58}$$

We analyze this equation by a multi-time scale analysis where $\zeta \equiv \delta^{-1}s$ is the fast time and s is the slow time. For mathematical clarity, we summarize the main results. The leading approximation $x = x_0(\zeta, s)$ satisfies the following equation for a map

$$x_0 = \beta \cos^2 (x_0(\zeta - 1) + \phi) + F, \tag{10.59}$$

where F is the constant of integration and ζ is now a discrete time. Equation (10.59) provides the successive maxima and minima, $x_0 = x_n$, of a square-wave-like solution. The successive extrema x_n are obtained by solving

$$x_n = \beta \cos^2 (x_{n-1} + \phi) + F. \tag{10.60}$$

Note now that we have obtained (10.60) by integrating Eq. (10.58) with respect to the fast time ζ. Consequently, we need to assume that the constant of integration F is a function of the slow time s. In order to obtain an equation for F, we proceed as usual, i.e. we investigate the higher order problem and apply a solvability condition. This condition is

$$\frac{dF}{ds} = - \lim_{\zeta \to \infty} \frac{1}{\zeta} \int_0^\zeta x_0(s, \xi) d\xi. \tag{10.61}$$

The bifurcation equation is given by Eq. (10.60) where the parameter F is slowly varying according to Eq. (10.61). A Period 1 fixed point of Eq. (10.60) corresponds to $x_n = x_{n-1} = x^*$. Equations (10.60) and (10.61) then simplify as

$$x^* = \beta \cos^2 (x^* + \phi) + F, \tag{10.62}$$

$$\frac{dF}{ds} = -x^*. \tag{10.63}$$

Differentiating (10.62) with respect to s and using (10.63), we correctly obtain Eq. (10.58) with $\delta = 0$, which is the leading equation for the slowly varying plateaus. Thus, the Period 1 fixed point solution of Eqs. (10.60) and (10.61) correctly matches the slowly varying envelope of the rapid oscillations.

In order to determine if fast oscillations are possible, we need to determine if a Period 2 fixed point is possible. To this end, we propose to analyze the linear stability of $x = x^*$. We consider $x = x^*$ as our reference solution and keep F constant. F is defined by means of Eq. (10.62), i.e.

$$F = x^* - \beta \cos^2 \left(x^* + \phi \right). \tag{10.64}$$

From Eq. (10.60), we then determine the following linearized equation for $x = x^*$

$$u_n = -\beta \sin \left(2x^* + 2\phi \right) u_{n-1}, \tag{10.65}$$

where $u_n \equiv x_n - x^*$ is defined as a small perturbation. Seeking then a solution of the form $u_n = r^n$, the characteristic equation for r is

$$r = -\beta \sin \left(2x^* + 2\phi \right). \tag{10.66}$$

The solution $x = x^*$ (F constant) is stable if $|r| < 1$, i.e. when

$$|\beta \sin \left(2x^* + 2\phi \right)| < 1.$$

The critical condition $r = 1$ marks a saddle-point. The condition is $\beta \sin (2x^* + 2\phi) = -1$, which exactly matches the condition (10.54) for the onset of the fast transition layers. On the other hand, the critical condition $r = -1$ marks a Hopf bifurcation point and it is given by

$$\beta \sin \left(2x^* + 2\phi \right) = 1. \tag{10.67}$$

The solutions of Eq. (10.67) are

$$x_{H+} = \frac{1}{2} \arcsin(1/\beta) - \phi, \tag{10.68}$$

$$x_{H-} = \frac{1}{2}(-\pi - \arcsin(1/\beta)) - \phi. \tag{10.69}$$

For $\phi = -\pi/10$ and $\beta = 2.5$, we find $x_{H+} = 0.5199$ and $x_{H-} = -1.462$. Figure 10.17 shows the location of these Hopf bifurcation points as well as the bifurcation diagram of all the stable solutions of the map (10.60) obtained numerically. Note that we have ignored the slow evolution of F and treated it as fixed. If we now consider its slow variation using Eq. (10.61), the actual solution will show successive slow passages through all the bifurcations of the map.

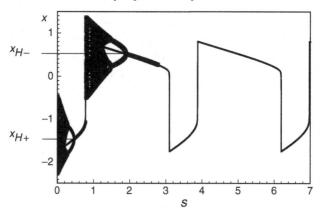

Fig. 10.17 The bifurcation diagram of the extrema of the rapid oscillations overlap the slowly varying envelope. $\phi = -\pi/10$ and $\beta = 2.5$. The pitchfork on the upper branch indicates the position of a Hopf bifurcation.

10.4 Exercises

10.4.1 Optoelectronic feedback

A feedback system where the intensity of the field and the carrier density are the only dependent variables can be realized if the intensity of the laser field is detected electronically, amplified, and then reinjected into the pumping current of the laser [276]. The laser rate equations modeling this system are given by

$$\frac{dI}{dt} = 2NI, \tag{10.70}$$

$$T\frac{dN}{dt} = P + \eta I(t - \tau) - N - (1 + 2N)I, \tag{10.71}$$

where I and N represent the intensity of the laser field and the electronic carrier density, respectively. These equations are our dimensionless SL rate equations where P is replaced by $P + \eta |Y(t - \tau)|^2$ to take into account the effect of the DC coupled optoelectronic feedback.

(1) Introducing the RO frequency $\omega_{RO} = \sqrt{2P/T}$, reformulate these equations as

$$x' = -y + \eta(1 + y(s - \theta)) - \varepsilon x\left[1 + \frac{2P}{1 + 2P}y\right], \tag{10.72}$$

$$y' = (1 + y)x, \tag{10.73}$$

where prime now means differentiation with respect to time $s = \omega_{RO}t$ and

$$\omega_{RO} = \sqrt{\frac{2P}{T}} << 1, \quad \varepsilon \equiv \omega_{RO}\frac{(1 + 2P)}{2P}, \quad \text{and} \quad \theta \equiv \omega_{RO}\tau. \tag{10.74}$$

(2) Formulate the linearized equations for $(x, y) = (0, 0)$ and then the characteristic equation for the growth rate σ. Determine the conditions for a Hopf bifurcation by introducing $\sigma = i\omega$ into the characteristic equation and by separating the real and imaginary parts. Plot the Hopf bifurcation lines in the η vs. θ parameter space. Consider the case $\varepsilon = 0$ first.

10.4.2 Delayed incoherent feedback

For a laser subject to optoelectronic feedback, the bandwidth of the electronics must be very large and flat since the chaotic dynamics of the SL can span tens of GHz. If the bandwidth of the electronics does not match the speed of the optical intensity fluctuations, the response of the laser system will be dominated by this bandwidth limitation. An alternative way to investigate this specific delayed feedback problem is to consider a SL subject to incoherent optical feedback [7, 277]. The output field of a SL is reinjected into the laser cavity after rotation of the polarization to the orthogonal state, providing a delayed feedback that affects only the carriers. The advantage is that the feedback remains purely optical. The laser device is described by the following equations for the intensity of the laser field I and the carriers N

$$\frac{dI}{dt} = 2NI, \tag{10.75}$$

$$T\frac{dN}{dt} = P - N - (1 + 2N)[I + \eta I(t - \tau)], \tag{10.76}$$

where the last term in Eq. (10.76) represents the reinjected orthogonal polarization intensity. Reformulate these equations as

$$x' = -y - \eta(1 + y(s - \theta))$$
$$-\varepsilon x\left[1 + \frac{2P}{1 + 2P}(y + \eta(y + y(s - \theta)))\right], \tag{10.77}$$

$$y' = (1 + y)x. \tag{10.78}$$

If η is $O(\varepsilon)$ small, we may neglect the $\varepsilon\eta$ term in Eq. (10.77) and we find Eq. (10.72). If $\eta = O(1)$, we may neglect the terms multiplying ε in (10.72) and (10.77) and we again obtain the same reduced problem.

10.4.3 Adler's equation with delay

The following Adler's phase equation

$$\phi' = \omega + \mu \sin(\phi - \phi(t - \tau)) \tag{10.79}$$

has been analyzed by Beta *et al.* [278] to explain the emergence of bistable branches of periodic solutions for sufficiently large delay. Introduce $\phi = \Omega t$ and analyze the implicit solution $\tau = \tau(\Omega)$. Investigate the conditions for nascent hysteresis, $d\tau/d\Omega = d^2\tau/d\Omega^2 = 0$.

Another Adler's equation has been derived by Wünsche *et al.* [279] and was motivated by synchronization experiments between two delayed coupled lasers. This equation is of the form

$$\Phi' = \frac{\Delta}{2} - \kappa \sin(\Phi + \Phi(t - \tau)). \tag{10.80}$$

The authors were interested in the case $\Delta \gg 2\kappa$. Determine an asymptotic solution for large Δ.

Hint: if $\Delta \to \infty$, the leading equation is $\Phi' = \Delta/2$ implying the solution $\Phi = s/2 + \Theta$, where Θ is a constant. Seek a solution of the form

$$\Phi = \frac{\Omega}{2}s + \Theta + \Delta^{-1}\Phi_1(s) + \Delta^{-2}\Phi_2(s) + \dots, \tag{10.81}$$

where $s = \Delta t$. Ω is expanded as

$$\Omega = 1 + \Delta^{-2}\Omega_2 + \dots \tag{10.82}$$

and takes into account possible corrections to $\Omega = 1$. We have anticipated that the first non-zero correction is $O(\Delta^{-2})$. Ω_2 is determined by requiring that the functions Φ_1 and Φ_2 are bounded 2π-periodic functions of s. Show that the averaged frequency $<\Phi'>$ has a snake-like behavior as a function of $\Delta\tau$.

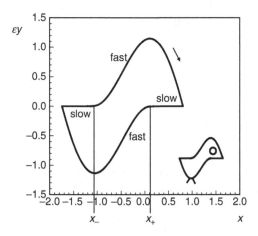

Fig. 10.18 Limit-cycle in the phase plane $(x, \varepsilon y)$. $\phi = -\pi/10$, $\varepsilon = 10^{-3}$, and $\beta = 2.5$. The two quasi-horizontal lines correspond to slow increases of the x manifold while the parabolic lines are rapid transition layers. The arrow indicates the direction of rotation.

10.4.4 Phase plane analysis

Figure 10.18 shows the bird-shaped limit-cycle solution of Eqs. (10.51) and (10.52) in the phase plane $(x, \varepsilon y)$. Determine separate approximations for the slow and fast parts of the limit-cycle orbit. The slow parts are determined by analyzing Eqs. (10.51) and (10.52) with $\varepsilon = 0$. Equations for the fast parts are determined by assuming $y \gg x$ and neglecting $-x$ in Eq. (10.52).

11

Far-infrared lasers

Far-infrared (FIR) molecular lasers have a restricted domain of application because their technology in the 100 μm to 1 mm spectral range is not yet mature.[1] This wavelength range is, however, unavoidable in radioastronomy because of the transparency windows of the Earth's atmosphere, and in semiconductor physics because of the energy domain of some lattice excitations. So far, applications of FIR lasers are limited. They have been used for checking high-voltage cable insulation [280] and, more recently, for security-screening systems [281]. On the other hand, FIR lasers are highly interesting for their instabilities and they have been studied in several laboratories.

The analogy found by Haken [99] between the Lorenz equations [282] and the laser (Maxwell–Bloch) equations for the homogeneously broadened laser triggered the search for an experimental laser system that could be well described by these equations. Haken's model of the laser is based on a semiclassical approach in which the electric polarization is explicitly considered, contrary to the standard rate equations where this variable is absent. By contrast to the laser rate equations, Haken–Lorenz equations admit sustained pulsating intensities and could be relevant for lasers that exhibit spontaneous pulsating instabilities. We have already discussed the complicated case of the ruby laser spiking. The 3.51 μm Xe laser self-pulsations were also known and investigated in detail but the mechanism responsible for this particular instability was partly masked by the difficulty in accounting for the inhomogeneous broadening,[2] which is a dominant process in this laser. In the early 1980s,

---

[1] Semiconductor lasers based on the quantum cascade effect recently appeared as an interesting alternative to the bulky molecular FIR lasers.

[2] In a homogeneously broadened laser, all the active centers (atoms, molecules, ions etc.) have the same resonant frequency. Relaxation processes are responsible for the broadening of the atomic resonances. In an inhomogeneously broadened laser, the resonance frequencies of the different atoms are spectrally spread because the latter have different velocities (Doppler effect) or because they experience different local fields (Stark effect). If the associated frequency spread is much larger than the (homogeneous) relaxation broadening, the medium is said to be inhomogeneously broadened.

researchers expended considerable effort on finding a laser system that was well described by the Haken–Lorenz equations. The search was impeded by the opposing requirements of a bad cavity operating far above the laser threshold, but it was felt that the conditions could be fulfilled using a class of laser-pumped FIR lasers [283].

Besides the possibility of realizing experiments on a system close to a Haken–Lorenz system, the FIR laser is interesting because it exhibits different instabilities depending on the operating conditions. In large-diameter lasers, the relaxation of molecules from the lower laser level is so slow that the population accumulates and the laser output power decreases at the millisecond time scale. The reasons for this "vibrational bottleneck effect" can be simply explained in terms of time scales, as we shall see in Section 11.1. More subtle dynamical outputs are observed on shorter time scales. At low pressures, the gain vs. frequency curve splits into two symmetrical components, leading to a different instability now associated with a Hopf bifurcation (see Section 11.3). The FIR laser also exhibits a very clean transition to chaotic regimes via a period-doubling cascade similar to the one observed in the Lorenz equations (Section 11.2).

11.1 Vibrational bottleneck

The processes involved in an FIR laser can be described by a model in which we consider three rotational levels, with populations N_1, N_2, and N_3 belonging to two vibrational states of a molecule (see Figure 11.1). The lower energy state is often the ground vibrational state. In order to generate a population inversion between levels 2 and 3, a strong infrared (IR) radiation resonant or quasi-resonant with levels 1 and 2 is coupled to the medium. Practically speaking, the

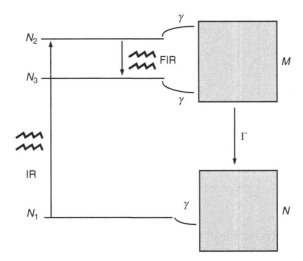

Fig. 11.1 Model of the energy levels for an FIR laser.

population inversion is created by optical pumping with an infrared laser (for example, a CO_2 laser). When the medium is set inside a properly tuned cavity, a stimulated emission can appear at the $2 \to 3$ transition frequency if the gain of the medium is sufficient. Each rotational level denoted by 1, 2, or 3 is not only coupled to IR and FIR fields but also by rotational relaxation to all other rotational levels of its own vibrational state. We assume that all rotational relaxation rates are equal to γ ($\gamma \sim 10^8$ s^{-1} $Torr^{-1}$). The two vibrational states with population densities M and N are also coupled to each other by incoherent processes represented by an average decay rate Γ independent of the rotational state. Vibrational de-excitation occurs via two processes: during wall collisions through diffusion ($\Gamma_{diff} \sim p^{-1}$, where p is the gas pressure) and by intermolecular collisions leading to the transfer of vibrational energy to translational and rotational energy ($\Gamma_{int} \sim p$). In low-pressure FIR lasers the total decay rate $\Gamma = \Gamma_{diff} + \Gamma_{int}$ may be as large as 10^3 s^{-1} $Torr^{-1}$, i.e. it has an order of magnitude much smaller than γ, which implies that the slow vibrational relaxation controls the long time evolution of the laser.

Typical signals obtained from a D_2CO laser pumped by a CO_2 laser are shown in Figure 11.2. The rise time of the emission is first very fast but then decays and

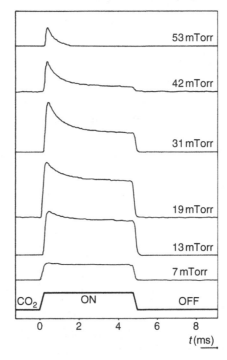

Fig. 11.2 Time dependence of FIR emission following sudden switch-on of the pump power for different pressures in a D_2CO laser. The evolution at the millisec-ond time scale is due to the slow relaxation between the reservoir populations N and M. The rise time of the observed signal is limited here by the observation technique (from Figure 8.3 of Glorieux and Dangoisse [284]).

approaches a steady state. This decay becomes more prominent as the pressure decreases (the so called "bottleneck effect") and is almost perfectly exponential (the decay is proportional to $\exp(-\Gamma t)$, where $\Gamma = \Gamma(p)$ is the vibrational relaxation rate).

11.2 Lorenz chaos in the FIR laser

Deterministic aperiodic solutions of single-mode laser equations were found in 1964 by Grasyuk and Oraevsky [285] and by Buley and Cummings [286]. However, the link of irregular laser pulsations with the more general field of nonlinear dynamics came much later when it was recognized that the laser equations are isomorphic to the Lorenz equations. In 1975, Haken [99] showed that the semiclassical equations for a single-mode, homogeneously broadened, and resonantly tuned ring laser are equivalent to three ordinary differential equations originally introduced by Lorenz as a simple model of fluid convection [282]. But the Lorenz equations would not be of interest if irregular sustained pulsating regimes had not been seen numerically. These regimes are highly sensitive to initial conditions and are globally called "chaotic". They immediately generated a series of questions of mathematical and physical relevance. For the laser community, the question was raised whether a real laser could exhibit Haken–Lorenz chaos. Haken equations are derived from the Maxwell–Bloch equations in many textbooks (see [6, 21]). But it is worth recalling the conditions for their derivation. First, the laser needs to operate between two homogeneously broadened energy levels. Because of the importance of the Doppler effect, this condition requires a long wavelength laser. Second, Haken assumed a ring cavity for mathematical convenience but most realworld lasers use a Fabry–Pérot (finite) cavity. Even if we find a laser that can reasonably be modeled by Haken equations, there are additional conditions on the laser parameters in order to observe chaotic outputs. First, a high pump power, typically 10–20 times the threshold pump power, is necessary. This requires a laser with relatively low threshold. Second, the field lifetime must be shorter than the inversion lifetime (the so-called "bad cavity" condition). To meet this condition without introducing too large losses, the relaxation rate of the population inversion should not be too large.

It was only in 1985 that Haken–Lorenz chaotic self-pulsing was observed in NH_3 FIR single-mode lasers [287, 288]. Many of the features (thresholds, perioddoubling sequences) of the chaotic pulsations for high pressures are in agreement with predictions from the Haken–Lorenz model. At lower pressures, three-level coherence effects seem to become relevant and cannot be described by the Haken–Lorenz equations.

The formal justification for modeling optically pumped FIR lasers (in the high pressure regime) by the simple Haken–Lorenz model has been widely disputed (see [4, 6]) because the coherent optical pumping appears to prevent reduction of the three-level FIR laser model to the simpler form of a two-level laser model. More complex models have been developed but most laser physicists agree that the Haken–Lorenz equations are a good starting point when interpreting experimental data from an FIR laser.

In their simplest form, the Haken–Lorenz equations are three equations for the normalized electric field x, the normalized amplitude of the polarization y, and the normalized inversion z, given by

$$\frac{dx}{dt} = \sigma(y - x),$$

$$\frac{dy}{dt} = rx - y - xz,$$

$$\frac{dz}{dt} = xy - bz. \tag{11.1}$$

Three parameters steer the behavior of these equations: σ is the cavity decay rate divided by the polarization decay rate ($\sigma = \kappa/\gamma_\perp$), b is the population inversion decay rate divided by the polarization decay rate ($b = \gamma_\parallel/\gamma_\perp$), and r is the pumping rate, where $r = 1$ gives the lasing threshold.

The standard rate equations (1.7) and (1.8), which have been the basis of our analysis of laser dynamics up to now, have been derived using a purely phenomenological approach. On the other hand, the Haken–Lorenz equations (11.1) have been deduced from first principles (Maxwell equations and quantum mechanics). The standard rate equations can also be obtained from (11.1) by eliminating adiabatically the polarization y (see Section 11.4.2).

Equations (11.1) have received extensive mathematical attention [289]. Here, we describe only the results that are relevant for interpretation of the experiments. In addition to the zero solution $(x, y, z) = (0, 0, 0)$, Eqs. (11.1) admit non-zero steady state solutions given by

$$x = y = \pm\sqrt{b(r - 1)} \quad \text{and} \quad z = r - 1. \tag{11.2}$$

The trivial solution corresponds to the laser OFF and (11.2) to the laser ON. The existence of two ON solutions is related to the invariance of the Lorenz equations under the transformation $(x, y, z) \rightarrow (-x, -y, z)$. Physically this corresponds to the invariance of the Maxwell–Bloch equations with respect to the reversal of the orientation of the electric field.

The linear stability analysis of these solutions as found in textbooks reveals that the OFF solution is stable if $r < 1$ and unstable if $r > 1$. The laser ON solutions exist only for $r > 1$ and are always stable if $\sigma - 1 - b < 0$. On the other hand, if

$$\sigma - 1 - b > 0 \qquad (11.3)$$

the ON solutions are stable in the interval

$$1 < r < r_H \equiv \sigma \frac{(\sigma + b + 3)}{(\sigma - 1 - b)}. \qquad (11.4)$$

The critical point $r = r_H$ is a Hopf bifurcation point. The frequency of the oscillations at the Hopf bifurcation is obtained from the characteristic equation and is given by

$$\omega_H \equiv \left(\frac{2\sigma b (\sigma + 1)}{\sigma - b - 1} \right)^{1/2}. \qquad (11.5)$$

Consequently, the ON steady state solution may become unstable if $r > r_H$ provided the condition (11.3) is satisfied. In terms of the original laser parameters, (11.3) implies the inequality

$$\kappa > \gamma_\perp + \gamma_\parallel, \qquad (11.6)$$

i.e. the field relaxation rate must exceed the sum of the polarization and population damping rates. This is the so-called "bad cavity condition". The function $r_H = r_H(b)$ has a minimum at $\sigma = \sigma_m = b + 1 + [2(b + 1)(b + 2)]^{1/2}$. Substituting this expression into r_H, we find that the lowest possible numerical value occurs for $b = 0$ and is $r_H = 9$. This implies that the pump parameter must be about 10 times larger than the threshold value ($r = 1$).

The experiments were performed by Weiss and coworkers on an 81µm $^{14}NH_3$ cw (FIR) laser pumped optically by an N_2O laser. Figure 11.3 shows an example of the chaotic emission where the FIR laser detuning is close to zero and in a high pressure range where homogeneous broadening dominates. In [290], the experimental pump rate is $r = 15$ and the values of b and σ were estimated at $b = 0.25$ and $\sigma = 2$. For those values of the parameters, Eqs. (11.1) admit a chaotic output similar to the one observed experimentally (see Figure 11.4). The presence of oscillations with increasing amplitudes in the temporal evolution suggests a saddle-focus behavior near the steady states which is typical of Lorenz dynamics. As mentioned earlier, it remains difficult to conclude that the observed oscillations indeed correspond to pure Lorenz chaos. This situation is complicated because of the nonproportionality between theoretical and experimental physical parameters. In other words, changing one experimental parameter usually alters several parameters of the model. For instance, changing the pressure in the active

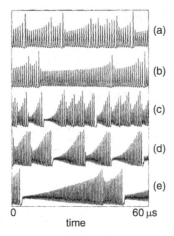

Fig. 11.3 Spiral-type pulsing of the laser intensity. The pressure varied in cases (a) to (e) from 8 to 10 Pa and the pump intensity was about 14 times above threshold. Reprinted Figure 2 with permission from Hübner *et al.* [290]. Copyright 1989 by the American Physical Society.

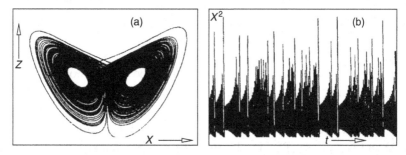

Fig. 11.4 Numerical solution of the Haken–Lorenz equations. $r = 15$, $b = 0.25$, and $\sigma = 2$. (a) Phase-plane trajectories in the z vs. x plane (about 850 loops). (b) Time trace for the laser intensity x^2 vs. time t (about 175 pulses). The average period is $T = 28.6$. Reprinted Figure 7a and 7b with permission from Hübner *et al.* [290]. Copyright 1989 by the American Physical Society.

medium mainly controls the relaxation rates γ and $\gamma_{\parallel}$, but it also modifies r and the ratio of homogeneous to inhomogeneous widths, bringing the system out of the range of validity of the model. Similarly, the pump power mainly changes r but to a minor extent alters the inhomogeneous width.[3] Consequently a part of the challenge was to find kill checks that would permit an answer to the question of

---

[3] In the case of the FIR laser, the contribution of inhomogeneous broadening is highly nontrivial. These effects have been studied in the steady state [291] and their contribution to laser dynamics was confirmed by heavy numerical simulations.

the existence of Lorenz chaos in a real laser. The Haken–Lorenz model may be considered satisfactory if it quantitatively reproduces the following features:

- the ratio of the first (laser OFF → ON) to second (cw stable → pulsing) thresholds
- the sequences of bifurcations obtained as the pump parameter is increased (for zero and non-zero detuning)
- the double-sided (symmetric) character of the Lorenz attractor.

Although the observed instability threshold closely corresponded to that predicted by the model, there was some controversy concerning the measurements because they could not distinguish between motion about two unstable steady states as shown in Figure 11.3 (a). This ambiguity arose because the intensity of the laser field emitted by the laser was recorded, rather than the field amplitude. To quell the controversy, Weiss and coworkers [290, 292] set up a laser-heterodyne detector that could measure the field amplitude. They observed that the field experienced abrupt π phase changes as shown in Figure 11.5 (left), which corresponds to the trajectory switching from one spiral to the other. Figure 11.5 (right) shows results for a lower pressure. We again observe the spiral Lorenz type pulsations

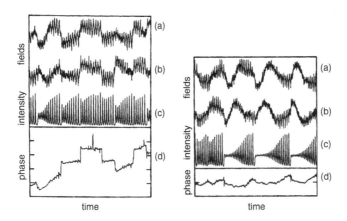

Fig. 11.5 Left: high pressure ($p = 9$ Pa) chaotic pulsing of the NH$_3$ FIR laser in the case of resonant tuning. (a) and (b) represent the in-phase and in-quadrature heterodyne signals measuring the laser field; (c) shows the laser intensity pulses; (d) gives the phase changes of the laser field and has been reconstructed from the field traces. One division of the vertical axis corresponds to a phase change of π rad. Pulsating period is 1 µs. Right: lower pressure ($p = 5$ Pa) chaotic pulsing. No more π phase jump is observed. Traces marked as in the left hand figures, with the same time and phase scales. Note that there are no π jumps at the end of each spiral. Reprinted Figures 1 and 2 with permission from Weiss *et al.* [292]. Copyright 1988 by the American Physical Society.

Far-infrared lasers

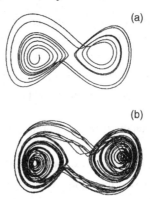

Fig. 11.6 Comparison of (a) the Lorenz attractor and (b) the attractor reconstructed from measured laser pulses. The laser field strength E is plotted as a
function of dE/dt (from Weiss and Vilaseca [4], p. 98).

in Figure 11.5 (c). However, the reconstructed phase for lower pressure measurements (Figure 11.5 (d)) shows no switching by π at the end of each spiral. In this
case the attractor is "one-sided" (asymmetric in the field amplitude) in accordance
with the predictions from the complex Lorenz equations, which are appropriate
if the detuning is not zero. Reconstruction of the attractor by the time-delayed
technique confirms that the experimental attractor for the higher-pressure measurements is symmetrical with respect to the origin and that it reproduces many
features of the Lorenz attractor [290]. An example of such a reconstruction based
on the field evolution is given in Figure 11.6. It clearly shows that for a cen-
trally tuned laser, the attractor is doubled-sided and symmetrical, showing the
same heteroclinic behavior as the Lorenz attractor. Similar experiments performed
with mid-infrared lasers confirm that other lasers may indeed exhibit Lorenz-type
chaos. A good review on the experiments and model predictions for the NH_3 FIR
is presented in [293]. The question of whether the NH_3 FIR laser is correctly
described by the Haken–Lorenz equations should not be pushed too far. As stated
by Khanin [6], many approximations which are known to be crude are required for
reducing this laser to a Haken–Lorenz system. Among these are:

- The level degeneracy: lasing occurs between levels with angular momentum J, and consequently of degeneracy $(2J + 1)$. They are not two nondegenerate levels as stated in the
  model but a transition from $(2J + 1)$ to $(2J - 1)$ levels[4] with transition moments (i.e.
  matrix elements of the dipole moment) depending on the magnetic quantum numbers
  and on the polarization of the electric fields.

---

[4] Here $J = 2, 4,$ or 7.

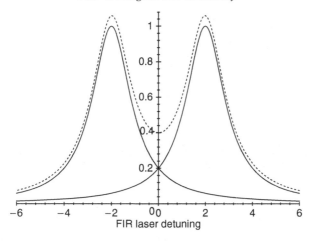

Fig. 11.7 Dual-peaked gain curve resulting from off-resonant pumping of the FIR laser at low pressures. Full lines: FIR gain associated with one velocity group and one propagation direction for the pump. Dotted line: total FIR gain (from Figure 8.21a of Glorieux and Dangoisse [284]).

- The laser field is not a plane wave and the cavity losses (up to 20%) are such that the uniform field limit is not valid.
- Most of the molecular relaxation parameters are unknown and the relaxation effects may not be reducible to the two damping phenomena accounted for by γ and Γ (see Section 11.1). Moreover there is no measurement of the different values of γ for the transitions involved in the FIR emission. The only relaxation measurements available are line broadening coefficients, which provide a value of γ for the pump transition (see, e.g., [294]).

11.3 Dual gain line instability

Quite different behaviors are observed in the laser when its frequency is detuned from the atomic transition frequency. New routes to chaos can be identified that had not been observed previously. These results show that the dynamics of optical systems are rich, and complement the studies of hydrodynamic systems.

11.3.1 Experiments

The FIR laser may exhibit a new kind of spontaneous instability at low pressures if the pump laser is sufficiently detuned from resonance. In this regime the gain curve appears as the combination of two Lorentzian profiles (see Figure 11.7). This may be interpreted as the result of inhomogeneous broadening. Detuning the pump from resonance means that molecules with a given non-zero velocity component along the laser axis are pumped, so that their Doppler effect compensates for the pump

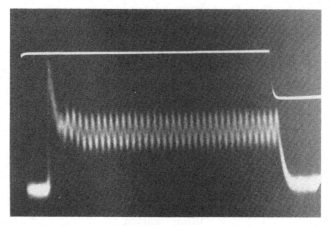

Fig. 11.8 Transient build-up of laser emission following a switch-on of the pump. The laser exhibits undamped oscillation at a frequency of about 200 kHz. Upper trace: pump power. Lower trace: FIR response. The total time interval is 50 μs (from Figure 8.18 of Glorieux and Dangoisse [284]).

frequency offset. However, as these molecules have a non-zero velocity along the cavity axis, the corresponding FIR gain curve is also Doppler shifted.[5] In a Fabry–Pérot cavity, the laser field propagates back and forth along the cavity axis so that the global gain peak for molecules of a given velocity shifts in both directions, each corresponding to a propagation direction for the FIR laser beam. In most lasers, this effect is symmetrized by the fact that the pump also propagates back and forth since it is only weakly absorbed in the low-pressure gas. As a result, two velocity groups are pumped, each of them having opposite Doppler-shifted gain curves. As a consequence the global gain curve presents a double-peaked symmetric structure as shown in Figure 11.7. An example of the intensity oscillations is given in Figure 11.8 where the laser emission is monitored just after a switch-on of the pump radiation. The laser starts with some delay (5 μs in Figure 11.8) and an initial spike as with any turn-on experiment (see Section 1.3.1), but in the present case it does not reach a stable steady state. Instead, it undergoes undamped oscillations, here at a frequency of about 100 to 200 kHz. This is a manifestation of the destabilization of the steady state through a mechanism that will be explained in the next section.

11.3.2 Model

The physical model considers a single-mode ring laser with two homogeneously broadened groups of two-level atoms with different resonance frequencies as

[5] Note that the Doppler effect on the FIR line is greatly reduced with respect to its value on the pump line since the Doppler shift is proportional to the optical (i.e. far-infrared vs. infrared) frequency. Therefore the inhomogeneous broadening of the FIR gain is negligible, except at low pressures, typically below 2 Pa.

shown on Figure 11.1 (for a discussion of the approximations leading to this model, see [295]). The evolution equations for the amplitude of the electric field X, the polarizations P_1 and P_2, and the population inversions D_1 and D_2 are given by

$$P_1' = -(1 + i\delta)P_1 + XD_1, \tag{11.7}$$

$$D_1' = -\gamma(D_1 - 1) - \frac{\gamma}{2}(XP_1^* + X^*P_1), \tag{11.8}$$

$$P_2' = -(1 - i\delta)P_2 + XD_2, \tag{11.9}$$

$$D_2' = -\gamma(D_2 - 1) - \frac{\gamma}{2}(XP_2^* + X^*P_2), \tag{11.10}$$

$$X' = -(\kappa - i\Delta)X + \frac{\kappa}{2}A(P_1 + P_2). \tag{11.11}$$

δ and Δ are the detunings of the complex polarization and the cavity, respectively, and κ and γ are the cavity and population relaxation rates, respectively. All frequency and relaxation rates have been normalized to the polarization relaxation rate. A is the pump parameter, normalized so that the pump parameter at threshold is $A = 1$. These equations were also derived by Idiatulin and Uspenskii [296], who examined how the presence of two groups of atoms could reduce instability thresholds. They are a special case of the general theory of lasers with inhomogeneously broadened atoms which considers a continuous distribution of atomic resonance frequencies.

Steady-state intensity solutions

In this section we determine the steady state intensity solutions for $\Delta = 0$ (perfectly tuned laser cavity) for which instabilities were experimentally observed. We first have the trivial OFF solution given by

$$X = P_1 = P_2 = D_1 - 1 = D_2 - 1 = 0. \tag{11.12}$$

Second, we note from Eq. (11.11) with $X' = 0$ that

$$X = \frac{A}{2}(P_1 + P_2). \tag{11.13}$$

Introducing (11.13) into Eqs. (11.7) and (11.9) with $P_1' = P_2' = 0$ then leads to a homogeneous system of two equations for P_1 and P_2. From the condition of a nontrivial solution we find that $D_1 = D_2 = D$ and

$$D = \frac{1 + \delta^2}{A} \quad (\delta \neq 0). \tag{11.14}$$

Adding Eqs. (11.8) and (11.10) with $D_1' = D_2' = 0$ and using Eq. (11.13) provides the intensity as

$$|X|^2 = A - 1 - \delta^2 \geq 0. \tag{11.15}$$

The inequality requires that $A \geq A_{th}$ where

$$A_{th} = 1 + \delta^2. \tag{11.16}$$

The solution (11.15) represents the ON state, now including the effect of the detuning δ.

Third, the particular structure of Eqs. (11.7) and (11.10) with $D_1' = D_2' = 0$ and $\Delta = 0$ allows us to determine another steady state intensity solution. Specifically, we seek a solution of the form $X = x \exp(i\mu t)$ and $P_j = p_j \exp(i\mu t)$ $(j = 1, 2)$, where x and p_j are (complex) constants. From Eqs. (11.7) and (11.10), we obtain the following equations for μ, x, p_j, and D_j

$$i\mu p_1 = -(1 + i\delta)p_1 + x D_1, \tag{11.17}$$

$$0 = D_1 - 1 + \frac{1}{2}(x p_1^* + x^* p_1), \tag{11.18}$$

$$i\mu p_2 = -(1 - i\delta)p_2 + x D_2, \tag{11.19}$$

$$0 = D_2 - 1 + \frac{1}{2}(x p_2^* + x^* p_2), \tag{11.20}$$

$$i\mu x = -\kappa x + \frac{\kappa}{2}A(p_1 + p_2). \tag{11.21}$$

From Eqs. (11.18) and (11.20), we determine D_1 and D_2 as functions of x and p_j $(j = 1, 2)$. Eliminating D_1 and D_2 in Eqs. (11.17) and (11.19), we obtain two equations for p_1, p_1^*, p_2, and p_2^*. Together with the complex-conjugate equations, we determine $p_1 = p_1(x)$ and $p_2 = p_2(x)$. Finally, we use Eq. (11.21) and obtain two conditions for xx^* and μ from the real and imaginary parts. These conditions lead to the solution

$$\mu^2 = \left(\frac{\kappa\delta}{\kappa + 1}\right)^2 - \frac{\kappa A}{2(\kappa + 1)} \geq 0, \tag{11.22}$$

$$|x|^2 = \frac{\kappa A}{2(\kappa + 1)} - \left(\frac{\delta}{\kappa + 1}\right)^2 - 1 \geq 0. \tag{11.23}$$

For this solution the laser emits radiation with an optical frequency shifted from the $\mu = 0$ solution by an offset $\pm\mu$. Its domain of existence is determined by the inequalities in (11.22) and (11.23). Equation (11.22) requires that $A \leq A_{\max}$, where

$$A_{\max} = 2\kappa\delta^2 / (\kappa + 1) \qquad (11.24)$$

and (11.23) requires that $A \geq A_{\min}$, where

$$A_{\min} = 2\left[\delta^2 + (\kappa + 1)^2\right] / [\kappa (\kappa + 1)]. \qquad (11.25)$$

The birth of this solution occurs if $A_{\max}(\delta) = A_{\min}(\delta)$. Using (11.24) and (11.25), we find that it appears if $\delta \geq \delta_c$, where

$$\delta_c \equiv \sqrt{(\kappa + 1) / (\kappa - 1)}. \qquad (11.26)$$

The critical value $A_c = A_{\max}(\delta_c) = A_{\min}(\delta_c) = 2\kappa/(\kappa - 1)$ then exactly matches A_{th} defined by (11.16). The three solutions are shown in Figure 11.9.

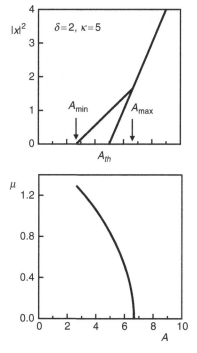

Fig. 11.9 Top: intensity of the steady state solutions vs. pumping parameter A. Bottom: the frequency μ of the solution bounded by $A_{\min}$ and $A_{\max}$.

Linear stability

For the zero intensity solution (11.12), the linearized equations for P_1, P_2, and X are of the form

$$P_1' = -(1 + i\delta)P_1 + X, \tag{11.27}$$

$$P_2' = -(1 - i\delta)P_2 + X, \tag{11.28}$$

$$X' = -\kappa X + \frac{\kappa}{2}A(P_1 + P_2). \tag{11.29}$$

The characteristic equation for the growth rate λ is then given by

$$\lambda^3 + \lambda^2(2 + \kappa) + \lambda\left(\delta^2 + 1 + 2\kappa - \kappa A\right) + \kappa\left(\delta^2 + 1 - A\right) = 0 \tag{11.30}$$

and predicts two bifurcations. A steady bifurcation point appears at $A = A_{th}$, where $\lambda = 0$, and a Hopf bifurcation is possible if $\delta > \delta_c$. It is located at $A = A_{min}$ where $\lambda = i\Omega \equiv i\sqrt{(\delta/\delta_c)^2 - 1}$. Note that because $P_2 = P_1^*$ and $X = X^*$, each root of the characteristic equation has an algebraic multiplicity of two.

For the non-zero intensity steady state (11.14) and (11.15), the linear stability analysis is harder. Partial information may, however, be obtained if we assume X real, $P_1 = P_2^* = P_1 = P_r + iP_i$, and $D_1 = D_2 = D$. Equations (11.7)–(11.11) then become

$$P_r' = -P_r + \delta P_i + XD, \tag{11.31}$$

$$P_i' = -P_i - \delta P_r, \tag{11.32}$$

$$D' = -\gamma(D - 1) - \gamma XP_r, \tag{11.33}$$

$$X' = -\kappa X + \kappa AP_r. \tag{11.34}$$

In terms of the new variables P_r, P_i, D, and X, the non-zero intensity steady state is given by

$$P_r = \frac{1}{A}\sqrt{A - 1 - \delta^2}, \quad P_i = -\delta P_r, \quad D = \frac{1 + \delta^2}{A}, \quad \text{and} \quad X = \sqrt{A - 1 - \delta^2}. \tag{11.35}$$

From the linearized equations, we then determine a fourth order characteristic equation for the growth rate λ. If $\gamma = 1$, one root is $\lambda = -1$ and the characteristic equation can be rewritten as

$$(\lambda + 1)\left[\lambda^3 + \lambda^2(2 + \kappa) + \lambda\left(A - \kappa\delta^2 + \kappa\right) + 2\kappa\left(A - 1 - \delta^2\right)\right] = 0. \tag{11.36}$$

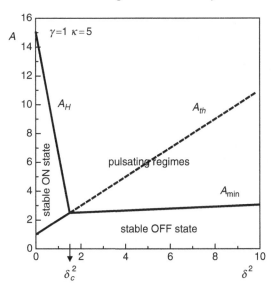

Fig. 11.10 Stability diagram. $A = A_{th}$ denotes a bifurcation point from the OFF to the ON steady state. $A = A_{min}$ and $A = A_H$ represent Hopf bifurcation points from the OFF state and from the ON state, respectively. The critical point $\delta_c^2 = 1.5$ ($\delta_c \simeq 1.23$). If $\kappa \to \infty$, A_{min} approaches the horizontal line $A = 2$ and A_H approaches the vertical line $\delta^2 = 1$.

From the cubic equation in λ, we determine a Hopf bifurcation point located at

$$A_H = \frac{\kappa(4 + \kappa(1 - \delta^2))}{\kappa - 2} \tag{11.37}$$

and admitting the frequency

$$\Omega_H^2 = \frac{2\kappa(\kappa - 1)}{\kappa - 2}(\delta^2 - \delta_c^2) > 0. \tag{11.38}$$

The stability diagram in the (δ^2, A) parameter space is shown in Figure 11.10.

11.3.3 Comparison with the experiments

The parameters for the FIR laser in which the instabilities have been observed are given in Table 11.1 and correspond to the conditions of the phase diagram shown in Figure 11.10. Only the lower left corner of Figure 11.10 was accessible in the experiment.

κ and δ are normalized to $\gamma_\perp$ and A is in units of the pump power at threshold. The agreement between the experimental observations and the model predictions

Table 11.1 *Parameters for the HCOOH FIR laser operating at 4 mTorr used for the demonstration of the dual-gain line instability. κ and δ are normalized to $\gamma_\perp$ and A is in units of the pump power at threshold.*

Physical quantities	In physical units	In reduced units
molecular relaxation	$5 \times 10^5 \, s^{-1}$ (80 kHz)	$\gamma_\perp = \gamma_\parallel = 1$
cavity loss	$3–5 \times 10^6 \, s^{-1}$	$\kappa = 6–10$
pump power	300 mW–1.1 W	$A = 1–3$
cavity detuning	0–400 kHz	$\delta = 0–5$

is evaluated through the range of parameters in which pulsating instabilities are observed. The theoretical analysis suggests two conditions on the parameters in order to observe pulsating intensities. The "bad cavity" condition $\kappa > 1$ is satisfied in the experiments since the evaluation of the laser losses leads to $\kappa \sim 6–10$, depending on the pressure. The detuning condition $\delta > \delta_c$ ($\delta_c \simeq 1.23$ for $\kappa = 5$ and $\delta_c = 1$ for $\kappa \to \infty$) is satisfied by the experimental value of $\delta \sim 1.26$. A detailed comparison is, however, complicated by the fact that the single-frequency CO_2 laser used for pumping the FIR laser admits an output power (A) that depends on the detuning (δ).[6]

Throughout the region of stable pulsing, the period T of the pulsations decreases with increasing A values (as in experiments, where $T = 8 \to 6 \, \mu s$ for $P_{CO_2} = 300 \to 1100 mW$). The instability frequency (125–160 kHz) is approximately equal to the value calculated at resonance (120 kHz) from (11.38) and the values of Table 11.1. For A values in excess of 14.5 (not accessible in these experiments), the regular pulsing breaks down into chaotic pulsing. There have been presumed observations of chaotic behavior at larger incident power but there have not been further investigations of the dual-gain instabilities discussed in the present section because other experiments gave much more characteristic chaotic signals, as explained in Section 11.2. Numerical simulations show that the intensity pulsations begin as 100% sinusoidal amplitude modulation. The initial pulsation frequency is exactly twice the value of μ for the $X_{ss}(\mu)$ solution predicted analytically. This is consistent with two equal amplitudes at optical frequencies located at $+\mu$ and $-\mu$ from the reference frequency.

Most features of the experimentally observed pulsations are well described within the framework of the model described here. Because κ can be considered a large parameter, further analytical work on Eqs. (11.7)–(11.11) is possible.

---

[6] It was assumed that $A(\delta) = A(0) \exp(-\delta^2/1.44) \exp(-(\delta-1)^2/1.44)$, where 1.44 corresponds to a Gaussian half-width at half-maximum of 0.99. The latter is necessary to obtain quantitative agreement but it is slightly different from the experimental value of 0.63.

11.3.4 FIR laser dynamics in the "bad cavity limit"

Pulsating intensities were observed for $\kappa \sim 6\text{--}10$ and suggest an analysis of Eqs. (11.7)–(11.11) in the limit κ large. If $\kappa \to \infty$, X can be eliminated from Eq. (11.11) by a quasi-steady state approximation. With $\Delta = 0$, we find

$$X = \frac{A}{2}(P_1 + P_2) \tag{11.39}$$

and inserting this expression into the remaining equations with $\gamma = 1$, we obtain

$$P_1' = -(1 + i\delta)P_1 + \frac{A}{2}(P_1 + P_2)D_1, \tag{11.40}$$

$$D_1' = -(D_1 - 1) - \frac{A}{4}\left[(P_1 + P_2)P_1^* + (P_1^* + P_2^*)P_1\right], \tag{11.41}$$

$$P_2' = -(1 - i\delta)P_2 + \frac{A}{2}(P_1 + P_2)D_2, \tag{11.42}$$

$$D_2' = -(D_2 - 1) - \frac{A}{4}\left[(P_1 + P_2)P_2^* + (P_1^* + P_2^*)P_2\right]. \tag{11.43}$$

Introducing the amplitude–phase decomposition $P_j = R_j \exp(i\phi_j)$ $(j = 1, 2)$ into Eqs. (11.40)–(11.43) leads to five equations for R_1, R_2, D_1, D_2, and $\Phi = \phi_2 - \phi_1$. The nontrivial steady state $D_1 = D_2 = D$ given by (11.14) suggests a consideration of the pure symmetric case where $R_1 = R_2 = R$ and $D_1 = D_2 = D$. This assumption reduces the five original equations to three equations for R, D, and Φ given by

$$R' = R\left[-1 + \frac{AD}{2}(1 + \cos(\Phi))\right], \tag{11.44}$$

$$D' = -(D - 1) - \frac{A}{2}R^2(1 + \cos(\Phi)), \tag{11.45}$$

$$\Phi' = 2\delta - AD\sin(\Phi), \tag{11.46}$$

where the phase equation (11.46) has meaning only if $R \neq 0$. We now hope that some of the bifurcations of the original FIR equations are well captured by Eqs. (11.44)–(11.46). Eq. (11.46) is an Adler-type equation with a restoring term proportional to the population inversion D. In addition to the zero intensity solution

$$R = D - 1 = 0 \tag{11.47}$$

there exists a non-zero intensity steady state given by

$$R = \sqrt{\frac{1+\delta^2}{A^2}(A - A_{th})}, \quad D = \frac{A_{th}}{A}, \quad \text{and} \quad \Phi = 2\arctan(\delta) \qquad (11.48)$$

if $A \geq A_{th}$, where A_{th} is defined by (11.16). We have already determined these steady states but the linear stability analysis will be simplified because we are considering the three equations (11.44)–(11.46) rather than the original five equations. We first examine the stability of the zero intensity solution (11.47). Assuming R small, Eq. (11.45) tells us that $D \to 1$ as $t \to \infty$. The long time solution is then described by the remaining equations for R and Φ with $D = 1$ given by

$$R' = R\left[-1 + \frac{A}{2}(1 + \cos(\Phi))\right], \qquad (11.49)$$

$$\Phi' = 2\delta - A\sin(\Phi). \qquad (11.50)$$

Equation (11.50) is an Adler equation for the phase Φ. If $A \geq |2\delta|$, it admits the steady state solution $\Phi = \arcsin(2\delta/A)$. Inserting this expression into the right hand side of Eq. (11.49), we note that $R = 0$ is stable if $A < A_{th}$, where A_{th} is defined by (11.16). On the other hand, if $A < |2\delta|$, the solution of Eq. (11.50) is unbounded in time. Integrating Eq. (11.49) for R using (11.50) leads to the result[7]

$$R = C\exp\left[\left(-1 + \frac{A}{2}\right)t\right]\frac{1}{\sqrt{2\delta - A\sin(\Phi)}}, \qquad (11.51)$$

where C is a constant of integration that depends on the initial conditions. The exponential in (11.51) clearly indicates a change of stability as

$$A > A_{min} = 2. \qquad (11.52)$$

---

[7] Eq. (11.49) is separable:

$$R^{-1}dR = \left(-1 + \frac{A}{2}\right)dt + \frac{A}{2}\cos(\Phi)dt$$

$$= \left(-1 + \frac{A}{2}\right)dt + \frac{A}{2}\frac{\cos(\Phi)d\Phi}{2\delta - A\sin(\Phi)}$$

$$= \left(-1 + \frac{A}{2}\right)dt - \frac{1}{2}duu^{-1},$$

where $u = 2\delta - A\sin(\Phi)$. Integrating both sides then gives

$$\ln(R) = \left(-1 + \frac{A}{2}\right)t + \ln(u^{-1/2}) + C.$$

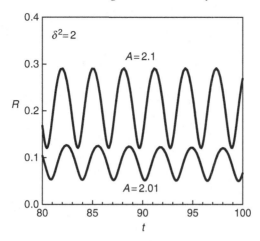

Fig. 11.11 Numerical solutions of Eqs. (11.44)–(11.46) in the vicinity of the Hopf bifurcation from $R = 0$ ($A_H = 2$).

At $A = 2$, the solution is periodic and oscillates with Adler's frequency $\omega = \sqrt{4\delta^2 - A^2}$. This is not a conventional Hopf bifurcation point because the bifurcation at $A = 2$ is from $R = 0$ but $R(t) > 0$ as soon as $A > 2$. The long time solution of Eqs. (11.44)–(11.46) near the Hopf bifurcation $A = 2$ ($\delta > 1$) is shown in Figure 11.11 for two different values of A.

The linear stability analysis of the non-zero steady state (11.48) leads to a third order polynomial for the characteristic equation. It is given by

$$\lambda^3 + a_1\lambda^2 + a_2\lambda + a_3 = 0, \tag{11.53}$$

where

$$a_1 = 2 - \delta^2, \quad a_2 = 2(A - A_{th}) + 1 - \delta^2, \quad \text{and} \quad a_3 = 2(A - A_{th}). \tag{11.54}$$

The Routh–Hurwitz stability conditions are $a_1 > 0$, $a_3 > 0$, and $a_1 a_2 - a_3 > 0$. The first two conditions require $\delta^2 < 2$ and $A > A_{th}$. The last condition simplifies as

$$a_1 a_2 - a_3 = (1 - \delta^2)\left[\left(2(A - A_{th}) + 2 - \delta^2\right)\right] > 0. \tag{11.55}$$

Together with the two previous conditions, it requires that $\delta^2 < 1$. In summary, the non-zero steady state is stable if

$$A > A_{th} \quad \text{and} \quad \delta^2 < 1. \tag{11.56}$$

The critical point $\delta = 1$ corresponds to a Hopf bifurcation with frequency equal to $\sqrt{a_2} = \sqrt{2(A - 2)}$ if $\delta = 1$ and $A > 2$. It has meaning only if we treat δ as the

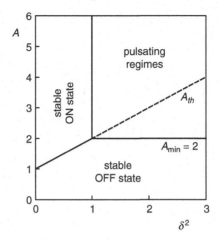

Fig. 11.12 Stability of the steady states in the large κ limit. The horizontal line $A = 2$ and vertical lines $\delta^2 = 1$ are Hopf bifurcation boundaries for the zero intensity steady state and the non-zero intensity steady states, respectively. The straight line $A = A_c = 1 + \delta^2$ corresponds to a bifurcation between the zero and non-zero intensity steady states.

bifurcation parameter (fixed A). If $\delta < 1$, we have a bifurcation at $A = A_{th}$ from the OFF to the ON state and there are no Hopf bifurcations. Figure 11.12 displays the stability diagram for both the zero and non-zero steady states. It represents the "bad cavity limit" ($\kappa \to \infty$) of general diagrams such as Figure 11.18 and provides a reasonable approximation in the accessible ($A < 4$) range of the pump parameter.

11.4 Exercises

11.4.1 Real Haken–Lorenz equations

In the case of central tuning of the laser cavity ($\delta = \Delta = 0$), Eqs. (11.57) reduce to Eqs. (11.1).

(1) Derive the characteristic equation for the stability of the steady state solutions of these equations.
(2) Find the condition linking κ, γ, and $\gamma_{\parallel}$ for the existence of instabilities of the nontrivial solution.
(3) Derive the angular frequency ω_H of the instabilities at the bifurcation point.

11.4.2 Complex Haken–Lorenz equations

The complex Haken–Lorenz equations for a single-mode laser apply when we take detunings into account. They are given by [21]

$$X' = \kappa \left[-(1 - i\Delta) X + AP \right],$$

$$P' = -(1 + i\delta) P + XD,$$

$$D' = \gamma \left[1 - D - \frac{1}{2} \left(XP^* + X^*P \right) \right] \qquad (11.57)$$

with the same notations as in Section 11.3 except for a different normalization of $\Delta \equiv (\omega_c - \Omega)/\kappa$. P and X are complex variables and D is real.

(1) Using the fact that D is real, show that $\Delta + \delta = 0$ at steady state. This is called the dispersion relation and sets the laser frequency. Investigate the linear stability of the two steady state solutions of Eqs. (11.57). Show that the trivial solution is always stable below threshold.

(2) For the nontrivial solution, the stability problem is five-dimensional. Show that one root of the characteristic equation is always 0. Interpret this in terms of a physical invariance. Derive from this an instability criterion and interpret it physically by comparison with, for example, the results of the rate equations (use the fact that the constant term of the characteristic equation is equal to the product of its roots and that it must cancel at the instability threshold).
Solution: the solution is fully documented in [21].

(3) Check that the adiabatic elimination of the polarization in Eqs. (11.57) leads to the standard rate equations. Analyze why the frequency variables disappear in the reduced equations.
Solution: the adiabatic elimination of P leads to $P = XD/(1 + i\delta)$. After inserting this expression into Eqs. (11.57), we obtain

$$X' = \kappa X \left[-(1 - i\Delta) + \frac{AD}{1 + i\delta} \right],$$

$$D' = \gamma \left[1 - D - \frac{1}{2} XX^* D \left(\frac{1}{1 + i\delta} + \frac{1}{1 - i\delta} \right) \right].$$

Introducing the intensity $I = |X|^2$, we eventually obtain

$$I' = 2\kappa I \left[-1 + \frac{AD}{1 + \delta^2} \right],$$

$$D' = \gamma \left[1 - D - \frac{DI}{1 + \delta^2} \right],$$

which are our standard rate equations with detuning δ.

12

Optical parametric oscillator

Optical Parametric Oscillators (OPOs) are based on multiwave interaction in a nonlinear medium. They have been realized in a variety of configurations, giving rise to an extended range of new dynamical problems. Like lasers, OPOs admit a steady state bifurcation at threshold and, in addition, they may exhibit bistability or Hopf bifurcations. Moreover, thermal effects may be dominant in cw oscillators leading to interesting slow–fast responses where the temperature is a new dynamical variable. Second-harmonic generation (SHG) is in a sense the inverse process of degenerate parametric amplification. Devices based on SHG are described by similar evolution equations but show different phenomena.

12.1 Parametric processes

An OPO is a light source similar to a laser, but based on optical gain from parametric amplification in a nonlinear crystal rather than from stimulated emission. Like a laser, such a device exhibits a threshold for the pump power, below which there is negligible output power. A main attraction of OPOs is that the signal and idler wavelengths, which are determined by a phase-matching condition, can be varied in wide ranges. We may thus access wavelengths (e.g. in the mid-infrared, far-infrared, or terahertz spectral region) which are difficult or impossible to obtain from any laser and we may also realize wide wavelength tunability. The downside is that any OPO requires a pump source with high intensity and relatively high spatial coherence. Therefore, we always need a laser as the pump source, generally a diode-pumped solid state laser.

The potential application areas for OPOs are diverse. Spectroscopy and many other scientific applications can profit from the ability of OPOs to cover very wide spectral regions, and to deliver outputs with narrow linewidth and high power. A common military application is the generatation of broadband high power light in the 3–5 µm region for the blinding of heat-seeking missiles when they attack airplanes. But despite their interesting capabilities, OPOs have so far not found widespread use in commercial products. One of the reasons is that an OPO system which includes pump laser, OPO, and possibly a temperature-stabilized crystal oven, is more complex to build than a pure laser system. Another reason is that a detailed understanding of the physics of parametric amplification is not very widespread in the laser industry [297].

Optical parametric processes are multiwave interactions which occur in nonlinear media. Typically, several optical waves excite a medium whose nonlinear response produces new radiation at a frequency which is a simple combination (e.g. sum or difference) of the incoming ones. Evolution equations describing these interactions require fewer approximations than those for lasers. Light–matter interaction in OPOs is ruled by the nonlinear susceptibility χ. The dielectric polarization P in the most commonly used nonlinear materials may be expanded as the following power series of the electric field(s) E

$$P = \varepsilon_0 \chi(E)E = \varepsilon_0 \left(\chi^{(1)}E + \chi^{(2)}E \oplus E + \chi^{(3)}E \oplus E \oplus E + \ldots \right), \quad (12.1)$$

where $\chi^{(n)}$ stands for the components of the nonlinear susceptibility tensors. Writing the field–matter interaction as Eq. (12.1) assumes instantaneous response so that the material variables are adiabatically eliminated. Equation (12.1) is valid as long as the material variables relax faster than the electric fields. In nonlinear optics, this approximation applies at time scales larger than 10^{-15} s.[1] In nonabsorbing materials, the first order term is a refractive index contribution and may be included by introducing $\varepsilon \equiv \varepsilon_0 \chi^{(1)}$. The higher order terms are responsible for various nonlinear parametric processes. For instance, the $\chi^{(2)}$ contributions account for second harmonic generation (SHG), optical parametric amplification (OPA), and sum and difference frequency generation (SFG and DFG), while the third harmonic generation (THG) and Raman and Kerr effects are linked to the $\chi^{(3)}$ contributions. Because $\chi^{(2)}$ effects are the most efficient and commonly used processes, this chapter concentrates on devices using only a second order nonlinearity. Practically speaking, $\chi^{(2)} \neq 0$ only in noncentrosymmetric materials, so that the parametric amplification may be realized in solid state systems such as crystals with suitable symmetry, or poled glass.

In the following section, we briefly introduce the OPO and SHG evolution equations. We then concentrate on specific dynamical phenomena.

---

[1] Subfemtosecond pulse propagation may require more sophisticated treatment.

12.1.1 Optical parametric amplification

In optical parametric amplification, an electromagnetic field experiences gain through a three-wave process. More specifically a nonlinear medium subjected to a strong field at frequency ω_p (the pump) emits two waves, the signal (s) and the idler (i) at frequencies ω_s and ω_i, respectively, such that $\omega_p = \omega_s + \omega_i$. It is sometimes said that one (pump) photon is converted into two photons with frequencies satisfying the energy conservation law.

In OPOs, the active medium is placed inside an optical cavity which may be resonant for one, two, or the three fields involved in the parametric amplification, and corresponds to singly (SROPO), doubly (DROPO), or triply (TROPO) resonant OPOs, respectively. As a consequence one, two, or three (complex) equations are required for describing the OPO, assuming that only one electromagnetic mode of the cavity is involved for each field (monomode OPOs). Multimode operation will not be treated here.

The equations for the triply resonant monomode OPO are given by

$$E_p' = -(\gamma_p + i\delta_p)E_p - \chi E_s E_i + E,$$

$$E_s' = -(\gamma_s + i\delta_s)E_s + \chi E_p E_i^*,$$

$$E_i' = -(\gamma_i + i\delta_i)E_i + \chi E_p E_s^*, \tag{12.2}$$

where E_j stands for the electric field amplitudes ($j = p, s,$ or i), γ_j and δ_j are the corresponding loss and cavity detuning coefficients, and χ is the relevant component of the dielectric tensor. E is the input pump field. All quantities are in physical units.

Equations for DROPOs and SROPOs may be derived from Eqs. (12.2) by adiabatic elimination of the nonresonant fields. Mathematically, we assume that the cavity is nonresonant (γ large) or strongly detuned (Δ large) for the nonresonant fields (see Exercise 12.10.1).

By far the most widely studied OPO problem corresponds to the situation where the cavity losses are similar for the signal and the idler, either because they have exactly the same frequency as in the degenerate OPO (called DOPO) or because their frequencies are close to each other ($\omega_s \simeq \omega_i$). One of these two fields, e.g. i, may be eliminated and equations for the triply resonant but nearly degenerate OPO become (Exercise 12.10.2)

$$A_0' = -(\gamma + i\Delta_0)A_0 - A_1^2 + E, \tag{12.3}$$

$$A_1' = -(1 + i\Delta_1)A_1 + A_1^* A_0. \tag{12.4}$$

In these equations, $\gamma \equiv \gamma_p/\gamma_s$ is the ratio of the cavity losses for the pump and signal radiation. $\Delta_0 \equiv (\omega_0 - 2\omega)/\gamma_s$ and $\Delta_1 \equiv (\omega_1 - \omega)/\gamma_s$ are the detunings in units of the cavity field damping rate γ_p. ω_0 and ω_1 are the cavity resonances closest to the pump and signal frequencies, respectively, and ω is the signal frequency. Indices 0 and 1 refer to the pump and the signal, respectively. The source terms associated with parametric amplification in the right hand sides of Eqs. (12.3) and (12.4) involve products of two fields as expected from the nature of the nonlinearity responsible for this so-called "second order process".

Comparing these equations with the rate equations of a class B laser, we find that four independent parameters control the response of a DOPO instead of two for the standard laser rate equations. Moreover, recall that A_0 and A_1 are complex variables and Eqs. (12.3) and (12.4) are then equivalent to four real equations.

12.1.2 Second harmonic generation

Crudely speaking, second harmonic generation (SHG) may be considered the reverse of degenerate optical parametric amplification. In SHG, a crystal irradiated by a laser at frequency ω emits a wave at frequency 2ω. In terms of photons, we can consider that two identical photons are "added" to generate a new photon with twice the energy of the original photons ($\omega + \omega \rightarrow 2\omega$), i.e. the opposite of degenerate optical parametric amplification. Photons interacting with a nonlinear material are effectively "combined" to form new photons with twice the energy, and therefore twice the frequency and half the wavelength of the initial photons. Historically, SHG was discovered before optical parametric amplification (OPA), and its experimental demonstration by Peter Franken and coworkers at the University of Michigan [298] was made possible by the invention of the laser, which created the required high intensity monochromatic light. Specifically, they focused a ruby laser with a wavelength of 694 nm onto a quartz sample. They sent the output light through a spectrometer, recording the spectrum on photographic paper, which indicated the production of light at 347 nm. In recent years, SHG has been extended to biological applications such as imaging molecules that are intrinsically second-harmonic-active in live cells or whose position at an interface breaks inversion symmetry.

In cw SHG, the nonlinear crystal is most often placed inside a cavity which is resonant for both the pump and the frequency-doubled wave so that the general model equations for intracavity SHG are given by [21]

$$\frac{dE_1}{dt} = -(\gamma_1 + i\delta_1)E_1 + iaE_2E_1^* + E_{ext},$$

$$\frac{dE_2}{dt} = -(\gamma_2 + i\delta_2)E_2 + ibE_1^2, \tag{12.5}$$

which, after rescaling (Exercise 12.10.3), become

$$A'_1 = -(\gamma + i\Delta_1)A_1 + A_2A_1^* + E,$$

$$A'_2 = -(1 + i\Delta_2)A_2 - A_1^2. \tag{12.6}$$

Indices 1 and 2 refer to the fundamental and harmonic radiations, respectively. The number of independent parameters and dynamical variables is the same as for the DOPO.

12.2 Semiclassical model for the DOPO

The degenerate OPO (DOPO) equations are compact and simple. They have been used to study a variety of dynamical effects such as walk-off and thermal runaway. The fact that an OPO delivers twin photons also makes it a perfect tool for quantum optics experiments [299]. Surprisingly, there have been few experiments on OPO classical dynamics. This may be due to the fact that OPOs are most often operated under pulsed pump conditions, so that dynamical effects have no time to fully develop during the pulse duration. Practically speaking, pulsed optical parametric generation is relatively easy to achieve in single-pass crystals (without a cavity) while most cw OPO cavities need highly reflecting mirrors. Indeed, the OPO signal gain is directly proportional to the pump field amplitude, which is typically 10^3–10^4 times smaller in the cw regime. Moreover, cw operation of OPOs requires special attention since it is technically difficult to simultaneously control pump and signal cavity detunings.

Like lasers, cw OPOs admit a pump threshold where the parametric gain compensates the cavity losses. The nature of the bifurcation at threshold may be anticipated qualitatively. Depending on the signs of the detunings and of the imaginary parts of the coupling term, the parametric gain may be pulling towards or pushing away from the resonance frequency of the cavity, meaning either an increase or a decrease of the gain process, so we expect to switch from subcritical or supercritical bifurcation as the detuning changes sign. In the next subsections, we analyze the steady states and their stability for the special case of the DOPO.

12.2.1 Steady state solutions and bistability

The steady state solutions of Eqs. (12.3) and (12.4) satisfy the conditions

$$-(\gamma + i\Delta_0)A_0 - A_1^2 + E = 0, \tag{12.7}$$

$$-(1 + i\Delta_1)A_1 + A_1^*A_0 = 0. \tag{12.8}$$

Consider Eq. (12.8). One solution is $A_1 = 0$ and we determine $|A_0|$ from (12.7). The OFF solution where the signal field is absent is given by

$$A_1 = 0, \quad |A_0|^2 = E^2 / \left(\gamma^2 + \Delta_0^2 \right). \tag{12.9}$$

Equation (12.8) and its complex conjugate form a homogeneous system of two equations for A_1 and A_1^*. The condition for a nontrivial solution gives

$$|A_0|^2 = 1 + \Delta_1^2. \tag{12.10}$$

Moreover, we have a relation between A_1 and A_1^* given by

$$A_1^* = (1 + i\Delta_1) A_1 A_0^{-1}. \tag{12.11}$$

Using (12.7), we determine EE^* and simplify the resulting expression using (12.11). This leads to the signal intensity for the ON solution (in implicit form)

$$E^2 = |A_1|^4 + 2 |A_1|^2 (\gamma - \Delta_0 \Delta_1) + \left(\gamma^2 + \Delta_0^2 \right) \left(1 + \Delta_1^2 \right). \tag{12.12}$$

Setting $|A_1|$ equal to zero in (12.12) gives the threshold for the ON solution as

$$E_{th}^2 = \left(\gamma^2 + \Delta_0^2 \right) \left(1 + \Delta_1^2 \right). \tag{12.13}$$

We now briefly comment on these results. In (12.13), the term $(1 + \Delta_1^2)$ is similar to what was obtained for a detuned class B laser and accounts for the efficiency loss due to the off-resonance of the cavity for the signal. Moreover, we expect a higher threshold if the cavity is detuned for the pump radiation as indicated by the second term $(\gamma^2 + \Delta_0^2)$. We also note from (12.12) that the output power increases like

$$|A_1|^2 \simeq E \tag{12.14}$$

as $E \to \infty$, implying a square-root law as we could expect for a conversion process $2\omega \to \omega + \omega$. Above threshold, the intracavity pump is clamped to its threshold value given by (12.10) and is independent of the input power. Its increase with signal detuning simply reflects the fact that a detuned cavity requires a larger pump intensity to operate.

Depending on the values of the parameters, there are either one or two ON solutions. By analyzing the quadratic equation (12.12), we find that the case of two ON solutions is possible provided that (Exercise 12.10.4)

$$\Delta_0 \Delta_1 > \gamma. \tag{12.15}$$

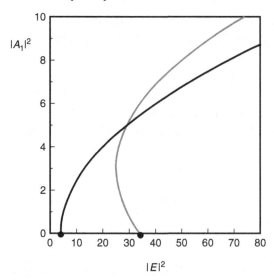

Fig. 12.1 Steady state solutions for the degenerate OPO. Single ON solution for $\Delta_0 = 0.5$, $\Delta_1 = \gamma = 1$ (black line); two ON solutions for $\Delta_0 = 4$, $\Delta_1 = \gamma = 1$ (gray line).

The linear stability analysis of the OFF solution shows that it is stable below threshold and unstable above. If (12.15) is satisfied and the upper ON solution is stable, we have a domain of coexistence of stable OFF and stable ON states (bistability). See Figure 12.1. The steady state solution (12.12) accurately fits the experimental observations in the vicinity of the threshold including the existence of bistability for detunings satisfying (12.15), as we shall see later.

12.2.2 Hopf bifurcation

We now concentrate on the case $\Delta_0 \Delta_1 < \gamma$ (no bistability). Increasing the pump rate, the ON solution may become unstable through a Hopf bifurcation to sustained oscillations [300, 301, 302, 303]. A necessary condition for this bifurcation is [303]

$$\Delta_0 \Delta_1 < -\frac{1}{2} \left(\gamma^2 + 2\gamma + \Delta_0^2 \right), \tag{12.16}$$

which ensures that there is a positive signal intensity given by

$$|A_{1H}|^2 = -\frac{\gamma \left(\gamma^2 + \Delta_0^2 \right) \left[\Delta_0^2 + (2 + \gamma)^2 \right]}{2 \left(1 + \gamma \right)^2 \left(\Delta_0^2 + 2\Delta_0 \Delta_1 + \gamma^2 + 2\gamma \right)}. \tag{12.17}$$

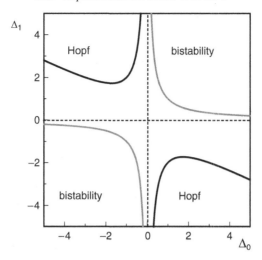

Fig. 12.2 The two domains for a Hopf bifurcation are shown in terms of the signal field detuning Δ_1 and the pump field detuning Δ_0 ($\gamma = 1$). The gray lines delimit the domains where bistability of steady states is possible.

The corresponding input field $E = E_H$ is then deduced from (12.12). The domains where the inequality (12.16) is verified are shown in Figure 12.2. The boundaries admit a minimal value at $|\Delta_1| = \sqrt{\gamma(\gamma + 2)}$. This indicates that a Hopf bifurcation is possible only for sufficiently high signal field detuning Δ_1.

OPOs are known to experience mode hops where the system jumps to the mode of lowest cavity detuning. This phenomenon significantly limits the range of accessible detunings and could prevent the experimental observations of the Hopf bifurcation [304]. The Hopf bifurcation frequency Ω is

$$\Omega^2 = 2\,|A_{1H}|^2 + \frac{\gamma^2 + \Delta_0^2}{1 + \gamma}. \qquad (12.18)$$

It is interesting to evaluate this frequency in the limiting cases $\gamma << 1$ and $\gamma >> 1$. For simplicity we assume perfectly tuned cavities ($\Delta_0 = \Delta_1 = 0$). In both limits, we found that the Hopf bifurcation frequency is proportional to $\gamma \equiv \gamma_p/\gamma_s$. Because time is scaled by the damping rate of the signal field γ_s, the oscillations appearing at the Hopf bifurcation are expected to physically occur at the damping rate of the pump field γ_p.

12.3 Experiments on TROPO-DOPO

Optical parametric amplification is a highly nonlinear process with little efficiency at low light intensities. Therefore cw oscillation needed for the observation of the steady state solutions can only be achieved in low loss cavities. Such cavities were designed only when low loss mirrors at the different wavelengths (and high gain media) became available. Practically speaking, one often takes advantage of the enhancement of the pump field inside a resonant cavity to reach a pump threshold consistent with the power levels delivered by cw lasers.[2] For these reasons, all the experiments discussed in this section have been made on TROPOs or quasi-degenerate DOPOs because they admit the lowest pump threshold.

In this section, we review some experimental results that illustrate the analytical results obtained in the previous section, namely the power law dependence of the output power, the bistability phenomenon, and the relaxation oscillations. The Hopf bifurcation was theoretically predicted in 1978 [300, 301, 302] but has not been observed experimentally. Oscillatory responses have been reported but they result from either thermal effects [305, 306] or from the interaction of transverse modes. The thermal effects are analyzed in Section 12.4.

12.3.1 Power 1/2 law for the output power

Equation (12.12) predicts the existence of a threshold for cw oscillation and a power 1/2 dependence for the evolution of the OPO signal (i.e. output) power vs. pump (i.e. input) power (see Eq. (12.14)). The evolution of the input/output characteristics in the static regime was investigated by Lee *et al.* [307] and indeed fits a square law dependence as shown in Figure 12.3.

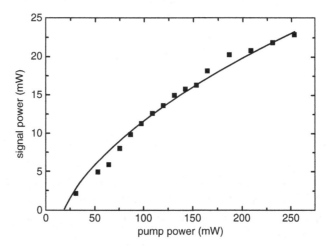

Fig. 12.3 Evolution of the signal power vs. the pump power. The solid line is a fit to a power 1/2 law. It also yields a pump threshold of 16 mW. With kind permission from Springer Science+Business Media (Figure 2 of Lee *et al.* [307]).

---

[2] Typical pump powers at threshold are of the order of 1 mW, 10–100 mW, and 1 W for TROPO, DROPO, and SROPO, respectively.

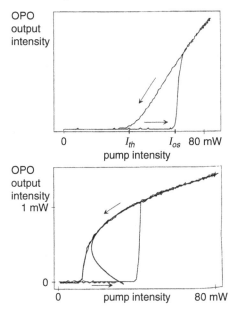

Fig. 12.4 Signal output power of a cw OPO for increasing and decreasing values of the pump power. Top: no bistability of the steady states. Bottom: bistability of the steady states. The full line superimposed on the bottom figure is a theoretical fit of the branch of steady states for $\Delta_0 = 2$ and $\Delta_1 = 2.6$. The experimental bifurcation diagrams are deformed because the control parameter is slowly changing in time and experiences delays before jumping (from Figures 4 and 6 of Richy *et al.* [308]).

12.3.2 Bistability for large detunings

Bistability in the near threshold regimes of a continuous OPO has been observed by Richy *et al.* [308] in a triply resonant, nearly degenerate OPO. By rapidly sweeping the pump power back and forth, they recorded variations of the OPO output intensity as displayed in Figure 12.4. The experimental bifurcation diagrams are deformed because the pump parameter is continuously changed in time. As a consequence, the expected transitions to a new state are delayed as seen in Figure 12.4. Passages through bifurcation points are discussed in detail in Chapter 7.

12.3.3 Relaxation oscillations

In an OPO, noise is induced by technical fluctuations and/or by purely quantum fluctuations. Noise actually drives the system by a broadband excitation and the response spectrum reflects its dynamical properties. As a consequence, the noise spectrum of the output intensity gives direct access to the relaxation oscillation (RO) frequencies and damping rates, provided that the spectrum of these driving fluctuations is wide enough and any technical (e.g. mechanical) resonance is avoided. If these conditions are met, the OPO's intensity exhibits a wide noise

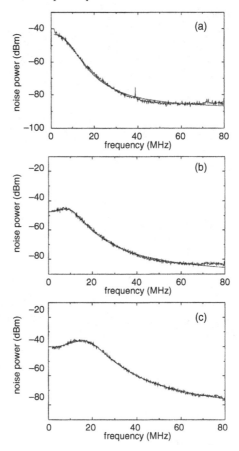

Fig. 12.5 RF intensity noise spectra of the signal measured for the OPO operated at different pump powers: (a) 3.3, (b) 4.7, (c) 15.8 times the threshold power. The full lines represent the predicted shapes of the noise spectrum. With kind permission from Springer Science+Business Media (Figures 4a–4c of Lee *et al.* [307]).

spectrum with a broad peak as shown in Figure 12.5. This peak coalesces to zero as the threshold is approached. The frequency of this peak thus corresponds to the RO frequency, which can be compared to the analytical expression obtained by the linearized equations. The RF noise spectrum of a continuous OPO was measured by Lee *et al.* [307] for several values of the pump power in the near threshold region. They observed tendency laws as shown in Figure 12.6. Such laws may be deduced from a linear stability analysis of the ON state, as we shall now see.

In order to determine expressions for the RO frequency and the RO damping rate, Lee *et al.* [307] considered the case of a triply resonant OPO at exact resonance for the three fields. Setting all $\delta_j = 0$ and introducing the decomposition $E_j = A_j \exp(i\phi_j)$ and $E = F \exp(i\Phi)$ into Eqs. (12.2), the authors further

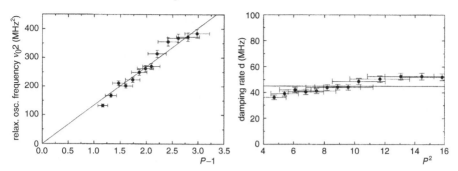

Fig. 12.6 RO frequency (left) and RO damping rate (right) as functions of the pumping rate. With kind permission from Springer Science+Business Media (Figures 5 and 6 of Lee *et al.* [307]).

assumed $\phi_p - (\phi_s + \phi_i) = 0$, which allows maximum parametric interaction. The intracavity pump field is fixed to the phase of the pump laser ($\phi_p = \Phi$). The evolution equations are then

$$A'_p = -\gamma_p A_p - \chi A_s A_i + E,$$

$$A'_s = -\gamma_s A_s + \chi A_p A_i,$$

$$A'_i = -\gamma_i A_i + \chi A_p A_s. \tag{12.19}$$

The ON steady state is given by

$$A_p = \frac{\sqrt{\gamma_s \gamma_i}}{\chi}, \quad A_s = \frac{1}{\chi}\sqrt{\gamma_i \gamma_p (P-1)}, \quad \text{and} \quad A_i = \frac{1}{\chi}\sqrt{\gamma_s \gamma_p (P-1)}, \tag{12.20}$$

where $P = E/E_{th}$ and $E_{th} = \gamma_p A_p$. To analyze its stability, we determine the characteristic equation[3] given by

$$\lambda^3 + \lambda^2(\gamma_p + \gamma_s + \gamma_i) + \lambda \gamma_p(\gamma_s + \gamma_i)P + 4\gamma_i \gamma_s \gamma_p (P-1) = 0. \tag{12.21}$$

For large values of the pump parameter above threshold P, we seek a large root of the form $\lambda = P^{1/2}\lambda_0 + \lambda_1 + \ldots$ Inserting this expression in (12.21) and equating the coefficients of each power of $P^{1/2}$ to zero gives

$$\lambda_0 = i\sqrt{\gamma_p(\gamma_s + \gamma_i)} \quad \text{and} \quad \lambda_1 = -\frac{1}{2}(\gamma_p + \gamma_s + \gamma_i) + \frac{4\gamma_i \gamma_s}{\gamma_s + \gamma_i}. \tag{12.22}$$

We conclude that, in this limit, the RO frequency is given by

$$\omega = \sqrt{\gamma_p(\gamma_s + \gamma_i)P} \tag{12.23}$$

[3] The characteristic equation derived by Lee *et al.* [307] is obtained using approximations. The exact equation has been later derived by Wei *et al.* [309].

and is proportional to the square root of the pump power. On the other hand, the damping rate of the relaxation oscillations is given by

$$\Gamma = \frac{1}{2}(\gamma_p + \gamma_s + \gamma_i) - \frac{4\gamma_i\gamma_s}{\gamma_s + \gamma_i} \tag{12.24}$$

and does not depend on P. These two scaling laws coincide with those which have been tested experimentally by Lee *et al.* [307] (see Figure 12.6).

12.4 TROPO-DOPO and temperature effects

In this section, we reexamine our results taking into account the influence of the temperature variation associated with radiation absorption inside the nonlinear crystal. We show how thermal effects produce slow ON–OFF alternations in a monomode cw OPO.

12.4.1 *Experimental results*

Slow oscillations at frequencies typically in the 20 kHz range are observed experimentally in monomode cw OPOs. They consist of periodic switching between ON and OFF states (see Figure 12.7) at a rate which is about 10^3 times slower than that of the relaxation oscillations (RO) (10–20 MHz). Therefore these oscillations are too slow to be the result of a Hopf bifurcation from the ON state. On the other hand, thermal changes of a parameter could be responsible for an ON–OFF slow oscillatory modulation. In a laser with macroscopic dimensions, thermal oscillations typically appear in the 10^{-4}–1 s range, i.e. much slower than the ROs. Such

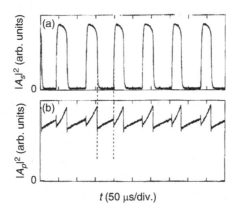

Fig. 12.7 Thermally induced cycles in the OPO. The signal (upper trace) displays ON–OFF square-wave oscillations and the pump (lower trace) switches simultaneously between two well-identified regimes. Reprinted Figure 2 with permission from Suret *et al.* [305]. Copyright 2000 by the American Physical Society.

effects are well known in lasers; for instance thermal lensing is known to be detrimental for high-power Nd^{3+}:YAG lasers. However their action in OPOs is far from trivial. Experiments performed by Suret *et al.* [305] demonstrated that the temperature rise due to residual absorption of radiation is indeed responsible for a drift of the cavity length, possibly leading to instabilities. Thermal variations of the cavity length are expected to either stabilize or destabilize the OPO output, depending on the cavity detuning, since they may shift the cavity towards resonance or away from resonance. If the thermal shift is large enough, a cyclic behavior is possible, as we shall now demonstrate.

12.4.2 Model for thermally induced cavity drift

Suret *et al.* [305] described the influence of temperature variations by introducing temperature dependent detunings. Instead of Eqs. (12.3) and (12.4), the dimensionless DOPO equations are given by

$$A'_0 = E - (\gamma + i\sigma_0(\theta))A_0 - A_1^2, \tag{12.25}$$

$$A'_1 = -(1 + i\sigma_1(\theta))A_1 + A_0 A_1^*, \tag{12.26}$$

where $\sigma_0(\theta)$ and $\sigma_1(\theta)$ are the detunings and θ is the temperature. Provided the variations of θ remain small, the detunings are assumed to be linear functions of θ of the form

$$\sigma_0(\theta) = \Delta_0 - 2\theta/\gamma \quad \text{and} \quad \sigma_1(\theta) = \Delta_1 - \theta. \tag{12.27}$$

The additional factor 2 in σ_0 comes from the fact that the pump wavelength is exactly half that of the signal wavelength in a DOPO. An equation for θ is obtained from the heat equation assuming that the main heating contribution comes from the two intracavity fields. It has the form

$$\theta' = \varepsilon(-\theta + a|A_0|^2 + b|A_1|^2), \tag{12.28}$$

where a and b are coefficients proportional to the optical absorptions at the pump and signal wavelengths, respectively. $\varepsilon \ll 1$ measures the effective thermal constant of the OPO cavity. Equation (12.28) simply says that the temperature slowly relaxes towards its equilibrium value $\theta_{eq} = a|A_0|^2 + b|A_1|^2$ fixed by the absorbed pump and signal powers inside the OPO crystal. Table 12.1 provides typical values of the parameters.[4] The dimensionless time is measured in units of the cavity decay time for the signal, $\tau = 7 \times 10^{-2}$ μs. The parameters a and ε have been chosen so that the numerical simulations correctly reproduce the time scales observed experimentally.

---

[4] The parameters σ_0, a, b, and E are related to the parameters defined in [305] as $\sigma_0 = \gamma\sigma_p$, $a = \alpha$, $b = \beta/\sqrt{\gamma}$, and $E \to \gamma E$.

Table 12.1 *Typical parameter values for a cw DOPO.*

γ	E	a	b	ε	Δ_0	Δ_1
10	40	1.5	0.06	10^{-3}	-20	2.6

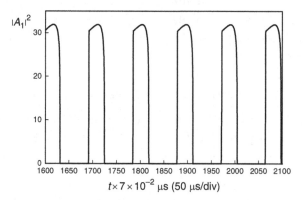

Fig. 12.8 Numerical solution of Eqs. (12.25)–(12.28) with $\gamma = 10$, $E = 40$, $a = 1.5$, $b = 0.06$, $\varepsilon = 10^{-3}$, $\Delta_0 = -20$, and $\Delta_1 = 2.6$.

Periodic switching between OFF and ON states has been obtained numerically with the parameters of Table 12.1 (see Figure 12.8). Experimental and numerical periods are close to 90 μs.

12.4.3 Thermal cycles in the single-mode OPO

Because θ is slowly varying ($\varepsilon << 1$), we first examine the bifurcation diagram of the fast subsystem (12.25) and (12.26) keeping θ as a parameter. The steady state solutions are given by (12.9) and (12.10)–(12.12) with Δ_0 and Δ_1 replaced by σ_0 and σ_1, respectively. The steady state intensities with the parameters listed in Table 12.1 are shown in Figure 12.9 with their stability properties. A subcritical steady bifurcation from the OFF to the ON state appears at $\theta_{bif} \simeq 3.55$ and allows the overlap of stable ON and OFF states for the interval $\theta_{bif} < \theta < \theta_{LP}$, where $\theta_{LP} \simeq 3.85$ corresponds to the steady state limit point.

We are now ready to consider the time dependent temperature. We note that $\theta' > 0$ if the right hand side of Eq. (12.28) is evaluated along the upper ON state while $\theta' < 0$ if it is evaluated along the OFF state. Taking into account the stability properties of the ON and OFF states, we anticipate a cyclic behavior between the slowly varying OFF and ON states. This is indeed observed in Figure 12.9 but the fast jumps between the two slowly varying states do not occur at the bifurcation and limit points. The analysis of simple slow passage through bifurcation

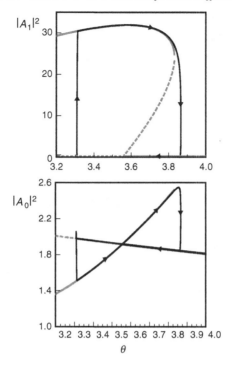

Fig. 12.9 ON and OFF steady state solutions of the DOPO equations vs. θ (gray lines) and cyclic behavior between the slowly varying OFF and ON states (black line). Same values of the parameters as in Figure 12.8.

points (see Chapter 7) indicates that the delay near a limit point is typically proportional to $\varepsilon^{2/3}$ but that it may be large and independent of ε if the slow passage is through a bifurcation point and if noise is sufficiently small. This is indeed the case in our DOPO problem and we wish to have an analytical understanding of this phenomenon.

The OPO equations are equivalent to five real ordinary differential equations but we may take advantage of the relatively large value of γ in typical OPOs (see Table 12.1). To this end, we must put the equations in a form such that γ appears as a global multiplying factor in an evolution equation. To this end, we introduce $A_p = A_0$, $A_s = \gamma^{-1/2} A_1$, $\mathcal{E} = \gamma^{-1} E$, $\sigma_p = \gamma^{-1} \sigma_0$ ($\Delta_p = \gamma^{-1} \Delta_0$), and $\sigma_s = \sigma_1$ into Eqs. (12.25)–(12.28) and obtain

$$A_p' = \gamma \left[-(1 + i\sigma_p) A_p - A_s^2 + \mathcal{E} \right], \tag{12.29}$$

$$A_s' = -(1 + i\sigma_s) A_s + A_s^* A_p, \tag{12.30}$$

$$\theta' = \varepsilon(-\theta + \alpha|A_p|^2 + \beta|A_s|^2), \tag{12.31}$$

where $\sigma_p(\theta) = \Delta_p - 2\theta/\gamma$, $\sigma_s(\theta) = \Delta_s - \theta$, $\alpha = a$, and $\beta = \gamma b$. The equations are now in a form suitable for simplification. Because γ is large, we may eliminate A_p by a quasi-steady state approximation. From (12.29), we find

$$A_p = \frac{\mathcal{E} - A_s^2}{(1 + i\sigma_p)}. \tag{12.32}$$

Inserting (12.32) into (12.30)–(12.31) leads to the following three equations for $A_s = R\exp(i\phi)$ and θ

$$R' = R\left[-1 + \frac{-R^2 + \mathcal{E}\sqrt{1 + \sigma_p^2}\cos(2\phi + \phi_0)}{1 + \sigma_p^2}\right], \tag{12.33}$$

$$\phi' = -\sigma_1 + \frac{\sigma_p R^2 - \mathcal{E}\sqrt{1 + \sigma_p^2}\sin(2\phi + \phi_0)}{1 + \sigma_p^2}, \tag{12.34}$$

$$\theta' = \varepsilon\left[-\theta + \alpha\frac{R^4 + \mathcal{E}^2 - 2\mathcal{E}R^2\cos(2\phi)}{1 + \sigma_p^2} + \beta R^2\right], \tag{12.35}$$

where $\phi_0 = \arctan(\sigma_p)$. We now concentrate on the evolution of the system in the vicinity of the slowly varying OFF state where R is almost zero. Neglecting all nonlinear terms in Eqs. (12.33)–(12.35), we obtain

$$R' = R\left[-1 + \frac{\mathcal{E}\cos(2\phi + \phi_0)}{\sqrt{1 + \sigma_p^2}}\right], \tag{12.36}$$

$$\phi' = -\sigma_1 - \frac{\mathcal{E}\sin(2\phi + \phi_0)}{\sqrt{1 + \sigma_p^2}}, \tag{12.37}$$

$$\theta' = \varepsilon\left[-\theta + \frac{\alpha\mathcal{E}^2}{1 + \sigma_p^2}\right]. \tag{12.38}$$

The last equation indicates that the OPO slowly warms up and that the heating contribution is given by the dissipation of the pump power inside the cavity. The slowly varying temperature is obtained by integrating Eq. (12.38) from the limit point θ_0 where the slow change with $R \ll 1$ starts (i.e. cooling due to the absence of radiation, hence no power dissipation). This value is set by the limit point of the ON solution. In the conditions of Figure 12.9, $\theta_0 = 3.86$. Because this cooling evolution is slow, the signal phase ϕ quickly relaxes to its quasi-steady state value, namely

$$\phi = \frac{1}{2}\arcsin\left[-\sigma_1\sqrt{1+\sigma_p^2}/\mathcal{E}\right] - \frac{\phi_0}{2} + n\pi, \tag{12.39}$$

where n is an integer. With (12.39), Eq. (12.36) for the evolution of the signal amplitude simplifies as

$$R' = R\left[-1 + \sqrt{\frac{\mathcal{E}^2}{1+\sigma_p^2} - \sigma_s^2}\right]. \tag{12.40}$$

Eq. (12.40) tells us that the signal decays due to cavity losses (first term in the right hand side) and that it is amplified with a gain (second term in the right hand side) proportional to the pump field inside the cavity. The right hand side of Eq. (12.40) is slowly varying in time because $\sigma_s(\theta)$ and $\sigma_p(\theta)$ are functions of the slowly varying temperature. The right hand side is first negative at $\theta = \theta_0$ and changes sign at $\theta = \theta_{bif}$. But the bifurcation point is not the point where R grows exponentially. Indeed the solution for R is given by $R = R(0)\exp(\varepsilon^{-1}F(\varepsilon t))$, where the growth rate F is

$$F(\varepsilon t) = \int_0^{\varepsilon t}\left[-1 + \sqrt{\frac{\mathcal{E}^2}{1+\sigma_p^2(s)} - \sigma_s^2(s)}\right]ds \tag{12.41}$$

with $\sigma_s = \sigma_s(\theta(\varepsilon t))$ and $\sigma_p = \sigma_p(\theta(\varepsilon t))$. This integral has been computed numerically and is shown in Figure 12.10. The fact that the evolution of R is ruled by

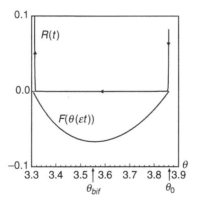

Fig. 12.10 The slowly varying growth rate $F(\theta(\varepsilon t))$ given by (12.41) is represented as a function of θ when the signal intensity $R(t)$ is exponentially close to zero. It is not the bifurcation point θ_{bif} that marks the change of stability of $R = 0$, but $F = 0$. Same parameters as in Figure 12.8.

an integral such as (12.41) means that we need to take into account the cumulative effect, i.e. the former history of the slowly varying detunings, in order to find the real change of stability of $R = 0$. Practically speaking, the delay of the bifurcation transition is not easily observed in experiments because the decrease of R associated with negative F values is limited by the background blackbody radiation and residual stray fields which set lower bounds on the value of R.

12.5 Intracavity singly resonant parametric oscillator

As mentioned earlier, the TROPO threshold is significantly lower (by one or two orders of magnitude) than for the DROPO and SROPO. A simple way to achieve a similar cavity enhancement for the pump is to put the OPO inside the cavity of the pump laser. Separate cavities (see Figure 12.11) are often used for the pump and the signal waves in order to avoid the mode-hopping problem encountered while tuning the OPO. This way, it is possible to change the OPO cavity length while keeping the pump laser cavity constant. Intracavity OPOs have been considered in the monomode and multimode regimes.

A simple model of the continuous-wave (cw) intracavity singly resonant parametric oscillators (ICOPOs) is obtained by writing the rate equations for the signal power P_s, the laser power P_p, and the gain (or population inversion) N. These equations are given by [310]

$$\tau_s P_s' = P_s(P_p - 1), \tag{12.42}$$

$$\tau_p P_p' = P_p(N - 1 - F P_s), \tag{12.43}$$

$$\tau N' = \sigma - N(1 + x P_p), \tag{12.44}$$

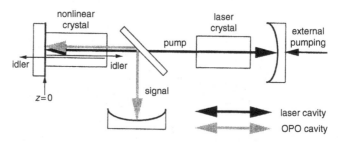

Fig. 12.11 Intracavity OPO. OPO operation inside the cavity of the pump laser allows us to take advantage of the larger intensity inside this cavity. The geometry of the device also allows separate tuning of the cavities at the pump and the OPO wavelengths (from Figure 1 of Turnbull *et al.* [310]).

where F is the ratio between the laser and signal cavity finesses,[5] and the τ_j are lifetimes. σ is the pump parameter normalized so that $\sigma = 1$ corresponds to the laser threshold, and x is a saturation parameter. Equations (12.42)–(12.44) admit the following steady state solutions

$$(1): \ P_s = 0, \ P_p = 0, \ N = \sigma, \tag{12.45}$$

$$(2): \ P_s = 0, \ P_p = \frac{\sigma - 1}{x} \geq 0, \ N = 1, \tag{12.46}$$

$$(3): \ P_s = \frac{\sigma - 1 - x}{F(1 + x)} \geq 0, \ P_p = 1, \ N = \frac{\sigma}{1 + x}. \tag{12.47}$$

Solutions (12.45), (12.46), and (12.47) describe the laser OFF state, the laser ON state, and the OPO ON state, respectively. The bifurcation diagram of the steady states (see Figure 12.12) exhibits two successive thresholds, namely the laser threshold $\sigma_{las} = 1$ and the OPO threshold $\sigma_{OPO} = 1 + x$. We now denote one of the steady states by (P_s, P_p, N) and formulate the linearized equations.

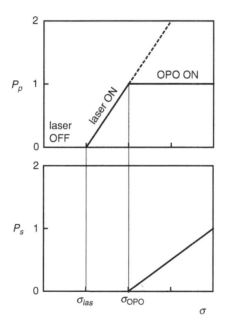

Fig. 12.12 Bifurcation diagram of the steady states ($F = x = 1$).

---

[5] The finesse of an optical cavity (or resonator) is defined as its free spectral range divided by the bandwidth (full width at half-maximum) of its resonance (http://www.rp-photonics.com/finesse.html).

The growth rates λ_j are then obtained by determining the eigenvalues of the following Jacobian matrix

$$\begin{pmatrix} (P_p - 1)\tau_s^{-1} & P_s\tau_s^{-1} & 0 \\ -FP_p\tau_p^{-1} & (N - 1 - FP_s)\tau_p^{-1} & P_p\tau_p^{-1} \\ 0 & -Nx\tau^{-1} & -(1 + xP_p)\tau^{-1} \end{pmatrix}. \qquad (12.48)$$

For the laser OFF state (12.45), we find that $\lambda_1 = -\tau_s^{-1}$, $\lambda_2 = -\tau^{-1}$, and $\lambda_3 = (\sigma - 1)\tau_p^{-1}$. The third eigenvalue indicates a change of stability at the laser threshold $\sigma = \sigma_{las}$.

For the laser ON state (12.46), we find that $\lambda_1 = (\sigma - 1 - x)(\tau_s x)^{-1}$ and that the two remaining eigenvalues satisfy the quadratic equation

$$\lambda^2 + \lambda\sigma\tau^{-1} + (\sigma - 1)\tau_p^{-1}\tau^{-1} = 0. \qquad (12.49)$$

λ_1 indicates a change of stability at the OPO threshold $\sigma = \sigma_{OPO}$. The fact that the coefficients in Eq. (12.49) are both positive if $\sigma > 1$ implies that the real parts of the two remaining eigenvalues are negative. Equation (12.49) is identical to the characteristic equation of the standard laser equations in Chapter 1. As $\tau \to \infty$, the eigenvalues approach the limit

$$\lambda_{2,3} = \pm i\sqrt{(\sigma - 1)\tau_p^{-1}\tau^{-1}} - \frac{\sigma\tau^{-1}}{2} + O(\tau^{-3/2}). \qquad (12.50)$$

The leading term exhibits the frequency of the relaxation oscillations $\omega_{RO} \equiv \sqrt{(\sigma - 1)\tau_p^{-1}\tau^{-1}}$ and the correction term gives their decay rate $\Gamma_{RO} \equiv \sigma\tau^{-1}/2$.

For the OPO ON state (12.47), we determine the characteristic equation and obtain

$$\lambda^3 + C_1\lambda^2 + C_2\lambda + C_3 = 0, \qquad (12.51)$$

where the coefficients are defined by

$$C_1 = (1 + x)\tau^{-1},$$

$$C_2 = \tau_p^{-1}(1 + x)^{-1}((\sigma - 1 - x)\tau_s^{-1} + \sigma x\tau^{-1}),$$

$$C_3 = \tau_s^{-1}\tau_p^{-1}\tau^{-1}(\sigma - 1 - x).$$

The stability conditions are given by the Routh–Hurwitz conditions, namely $C_1 > 0$, $C_3 > 0$, and $C_1 C_2 - C_3 > 0$. The first condition is always satisfied. The second condition is satisfied if $\sigma > \sigma_{OPO}$. Finally the third condition simplifies as $C_1 C_2 - C_3 = \tau^{-2}\tau_p^{-1}\sigma x > 0$, which is always satisfied.

Table 12.2 *Intracavity OPO parameters.*

laser	τ_s (µs)	τ_p (µs)	τ (µs)	x
Ti:Sapphire	0.3	0.2	3.2	3
Nd:YVO$_4$	1	1	98	7

Typical values of the parameters are listed in Table 12.2.[6]

The relatively large value of τ compared to τ_s and τ_p suggests that we look for an approximation of the growth rates. If $\tau^{-1} = 0$, the roots are $\lambda_1 = 0$ and $\lambda_{2,3} = \pm i\omega$, where

$$\omega \equiv \sqrt{\tau_p^{-1}\tau_s^{-1}(1+x)^{-1}(\sigma - 1 - x)}. \qquad (12.52)$$

For the real root λ_1 we then assume that $\lambda_1 = O(\tau^{-1})$ as $\tau^{-1} \to 0$ and find from (12.51)

$$\lambda_1 = -\tau^{-1}(1+x) + O(\tau^{-2}). \qquad (12.53)$$

For the complex roots $\lambda_{1,2}$, we assume that the real part is small as $\tau^{-1} \to 0$ and using (8.47) we find

$$\lambda = \pm i\omega - \frac{\tau^{-2}\tau_p^{-1}\sigma x}{2\omega^2} + O(i\tau^{-2}, \tau^{-3}). \qquad (12.54)$$

In summary, as $\tau \to \infty$, the relaxation oscillations (RO) exhibit a frequency proportional to $\sqrt{\tau_p^{-1}\tau_s^{-1}}$ but their damping rate is very slow and is proportional to $\tau^{-2}\tau_s$. These properties significantly differ from those of the ROs for class B lasers where the RO frequency and damping rate scale like $\tau^{-1/2}$ and τ^{-1}, respectively. Using the values of the parameters for the Nd:YVO$_4$ laser (see Table 12.2), this means that the RO period is of the order of 1 µs and that the damping time is of the order of 10 ms, which is quite surprising. When the OPO is ON, the transient response of the OPO exhibits new time scales due to the nonlinear interaction between all three variables. This couldn't have been anticipated by simple examination of the time scales τ_p, τ_s, and τ appearing in Eqs. (12.42)–(12.44).

---

[6] From [310]. For a typical OPO inside a Ti:Sapphire laser, the laser and OPO thresholds are 1 W and 4 W of argon-ion laser pump power, respectively. This implies that $\sigma_{OPO} = 1 + x = 4$ and hence $x = 3$. For the Nd:YVO$_4$ laser, the laser and OPO thresholds are 0.5 W and 4 W of diode pump power, respectively. The dimensionless OPO threshold value is then $\sigma_{OPO} = 1 + x = 8$ leading to $x = 7$. The value of F is not necessary since F can be eliminated from the rate equations by redefining P_s and τ_s.

12.6 Intracavity SHG

Although there has been no experimental evidence of pure SHG instabilities up to now, we present here the model for a cavity containing an SHG crystal as an introduction to more complicated systems which combine SHG with other nonlinearities. Intracavity SHG was among the first non-laser systems for which dynamical instabilities were predicted [300]. Schiller *et al.* [311], Marte [313], and Lodahl *et al.* [316] investigated devices suitable for the observation of these instabilities but other processes overcame them, as we shall see in the following sections.

12.6.1 Intracavity SHG model

Intracavity SHG was theoretically studied in a doubly resonant system, i.e. the purely real equivalent of the doubly resonant DOPO. Practical systems (see, for instance, Figure 12.13) are usually designed so the cavity is highly resonant for the pump field and weakly resonant for the doubled field, so that the pump radiation is trapped inside the cavity to achieve a high power and therefore a high efficiency. The output coupling at the pump frequency may be small because this radiation is not wanted for further use. Simultaneously the generated doubled frequency field may be efficiently coupled out. Of special interest is the ideal converter limit ($\gamma_1 \ll \gamma_2$ in Eqs. (12.5)), in which the cavity is perfectly resonant and only the useful radiation, i.e. the second harmonic radiation, can escape the cavity.

The ideal converter limit is obtained by setting $\gamma = \gamma_1/\gamma_2 = 0$ in Eqs. (12.6) with $\Delta_1 = \Delta_2 = 0$, leading to

$$A'_1 = A_2 A_1^* + E,$$

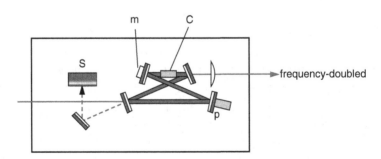

Fig. 12.13 Intracavity SHG (TA-SHG 110, TOPTICA Photonics). A single-frequency signal from the left enters a resonant doubling cavity (thick lines) and produces a frequency-doubled signal at the right. m: intensity monitor photo-diode; c: crystal; s: photodiode for cavity length stabilization; p: piezo (a control loop via the piezo actuator regulates the length of the enhancement cavity).

$$A_2' = -A_2 - A_1^2. \tag{12.55}$$

The steady state solution of Eqs. (12.55) is

$$A_1 = E^{1/3} \quad \text{and} \quad A_2 = -E^{2/3}, \tag{12.56}$$

which implies the quadratic relation $|A_2| = |A_1|^2$. We next examine its linear stability. Introducing $A_j = R_j \exp(i\phi_j)$ into Eq. (12.55) and separating real and imaginary parts, we obtain

$$R_1' = R_1 R_2 \cos(2\phi_1 - \phi_2) + E \cos(\phi_1),$$
$$R_2' = -R_2 - R_1^2 \cos(2\phi_1 - \phi_2),$$
$$\phi_1' = -R_2 \sin(2\phi_1 - \phi_2) - \frac{E}{R_1} \sin(\phi_1),$$
$$\phi_2' = -\frac{R_1^2}{R_2} \sin(2\phi_1 - \phi_2). \tag{12.57}$$

The steady state solution (12.56) is now given by

$$R_1 = E^{1/3}, \quad R_2 = E^{2/3}, \quad \phi_1 = 0, \quad \text{and} \quad \phi_2 = \pi. \tag{12.58}$$

The linearized equations lead to the following Jacobian matrix

$$\begin{pmatrix} -R_2 & -R_1 & 0 & 0 \\ 2R_1 & -1 & 0 & 0 \\ 0 & 0 & 2R_2 - \dfrac{E}{R_1} & -R_2 \\ 0 & 0 & 2\dfrac{R_1^2}{R_2} & -\dfrac{R_1^2}{R_2} \end{pmatrix},$$

which is a block diagonal matrix. This then allows us to formulate the characteristic equation as the following product of two square brackets

$$\left[\lambda^2 + (1 + E^{2/3})\lambda + 3E^{2/3} \right] \left[\lambda^2 - \lambda(E^{2/3} - 1) + E^{2/3} \right] = 0. \tag{12.59}$$

Equating the first bracket to zero leads to roots with a negative real part because all coefficients are positive. Equating the second bracket to zero provides roots with a negative real part only if

$$E < 1. \tag{12.60}$$

The corresponding critical point $E = 1$ is a Hopf bifurcation with frequency $\omega = 1$.

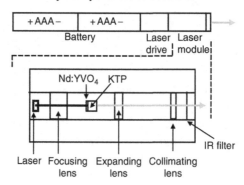

Fig. 12.14 Schematic diagram of a green laser pointer using a composite multiple crystal assembly (Nd:YVO$_4$/KTP). The wavelength of the laser beam is originally at 808 nm and then is 532 nm (green) after passing the crystal assembly.

12.6.2 Intracavity SHG

SHG is also most efficient when the doubling crystal is placed inside the cavity of the laser. This way one benefits from the cavity enhancement for both the fundamental and the doubled fields, but the nonlinear material may be damaged by higher power densities. This set-up architecture is that used in commercial lasers such as the simple green laser pointers (e.g. see Figure 12.14) and the Verdi laser (Coherent Inc.), both delivering power at 532 nm. These lasers are designed so as to avoid self-pulsing. However, spontaneous oscillations were observed in the multimode regime of YAG lasers with internal SHG [317]. It is left as an exercise to show that in the monomode regime, SHG inside the cavity of a class B laser such as the Nd:YVO$_4$ laser does not lead to instabilities (Exercise 12.10.5).

12.7 Antiphase dynamics in intracavity SHG

Antiphase dynamics is a property displayed by systems in which N (>1) oscillators synchronize with a strong phase correlation. In laser physics, this type of dynamics gave a new stimulus to the study of multimode lasers when the phenomenon was first observed in an N-mode Nd^{3+}:YAG laser with intracavity doubling crystal [318, 319]. Almost simultaneously, antiphase regimes were reported for solid state Fabry–Pérot lasers [320] and Nd-doped fiber lasers [321]. In its simplest form, antiphase dynamics implies a highly ordered state in which each modal intensity is time-periodic with the same waveform but shifted by $1/N$ of a period from its neighbor. That is, the modal intensities are of the form $I_k = I_0(t + Tk/N)$, $k = 1, \ldots N$, where I_0 is a waveform of period T. Antiphase states appear with high multiplicity because there is no preferential mode if all the modes are equally coupled. Specific antiphase solutions have been studied mathematically in the context of Josephson-junction arrays ("splay states" [322] or "ponies on a merry-go-round" [323]) and in the context of coupled laser arrays ("splay phase states" [324, 325]). For multimode lasers, the large variety of responses and frequencies has been reviewed in [21].

12.7.1 Antiphase dynamics in YAG/KTP lasers

The equations describing the evolution of the multimode laser with an intracavity doubling crystal are [317]

$$\eta \frac{dI_k}{dt} = \left[G_k - \alpha - g\varepsilon I_k - 2\varepsilon \sum_{j(\neq k)} \mu_j I_j \right] I_k, \tag{12.61}$$

$$\frac{dG_k}{dt} = \gamma - \left[1 + I_k + \beta \sum_{j(\neq k)} I_j \right] G_k, \tag{12.62}$$

where time $t = t'/\tau_f$ is normalized by fluorescence time τ_f (= 240 μs), and $\eta = \tau_c/\tau_f$ (= 8.3×10^{-7}), where τ_c (= 0.2 ns) is the cavity round-trip time. I_k and G_k are, respectively, the intensity and gain associated with the kth longitudinal mode. α (= 10^{-2}) is the cavity loss parameter, γ (= 0.05) is the gain parameter, β (= 0.7) is the cross-saturation parameter, ε (= 5×10^{-6}) is a parameter that depends on the nature of the second-harmonic generating crystal, and g (= 0.1) is a geometrical factor whose value depends on the orientation of the YAG crystal relative to the KTP doubling crystal as well as the phase delays due to birefringence (the values of the parameters are taken from [319]). Here $\mu_j = g$ for modes having the same polarization as the kth mode, while $\mu_j = 1 - g$ for modes having the opposite polarization. For simplicity, we assume that the gain γ, the cross-saturation parameter β, and the cavity loss parameter α are the same for all modes. Cross-saturation of the active medium (represented by the $\beta I_j G_k$ term) and sum-frequency generation in the intracavity nonlinear crystal (represented by the $\varepsilon \mu_j I_j I_k$ term) introduce global coupling among laser modes.

We first introduce the experimental results [318]. The laser operates above threshold and three modes oscillate simultaneously. The experimental set-up is such that two of three modes are x-polarized and one is y-polarized. To an excellent approximation, only one mode is ON at any given instant. The dynamical state coexists with another symmetry-related state. Physically, the two states are distinguished by which x-polarized mode immediately follows the y-polarized mode. Numerical simulations indicate that we are dealing with a T-periodic regime where

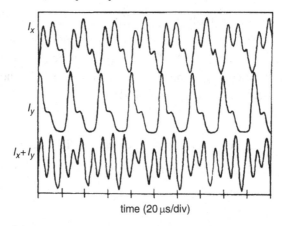

time (20 μs/div)

Fig. 12.15 Antiphase x- and y-polarized longitudinal laser mode intensities and the total intensity. Reprinted Figure 2 with permission from Wiesenfeld *et al.* [318]. Copyright 1990 by the American Physical Society.

each modal intensity exhibits the same waveform, $I_0(t)$, successively phase-shifted by a factor $T/3$. The higher complexity of the waveform of I_x compared to I_y, seen in Figure 12.15, comes from the fact that I_x is the sum of two distinct modal intensities ($I_x = I_0(t + T/3) + I_0(t + 2T/3)$) while I_y is single-mode ($I_y = I_0(t + T)$). Another remarkable feature of the antiphase collective behavior that we note from Figure 12.15 is the quite regular behavior of the total intensity $I_x + I_y$, oscillating with a larger frequency than the individual modes.

12.7.2 Analysis of the two-mode case

We shall limit our analysis of Eqs. (12.61) and (12.62) to the case of two modes ($N = 2$) exhibiting the same polarization. As we shall demonstrate, two distinct frequencies emerge from a linear stability analysis and they are respectively associated with an in-phase and an out-of-phase eigenvector of the linear stability analysis of the non-zero solution. We start from an analysis of the in-phase regimes and derive the associated relaxation frequency and damping, and then proceed to the analysis of the more general case where the two modal intensities may be different.

In-phase regimes

An in-phase regime corresponds to the solution $I_1 = I_2 = I(t)$ and $G_1 = G_2 = G(t)$. I and G satisfy the following equations

$$\eta \frac{dI}{dt} = \left[G - \alpha - 3g\varepsilon I \right] I, \tag{12.63}$$

$$\frac{dG}{dt} = \gamma - [1 + I(1 + \beta)] G. \tag{12.64}$$

The zero intensity solution $(I, G) = (0, \gamma)$ is stable if $\gamma - \alpha < 0$. On the other hand, if

$$\gamma - \alpha > 0 \tag{12.65}$$

a non-zero intensity steady state $I = I_s$ is possible and is the positive root of

$$(1 + \beta)3g\varepsilon I_s^2 + \left[\alpha(1 + \beta) + 3g\varepsilon\right] I_s + \alpha - \gamma = 0. \tag{12.66}$$

Having I_s, we determine G_s from (12.63) as

$$G_s = \alpha + 3g\varepsilon I_s. \tag{12.67}$$

From the linearized equations, we then formulate the characteristic equation for the growth rate λ. It is given by

$$\lambda^2 + \left[3g\varepsilon\eta^{-1}I_s + 1 + I_s(1 + \beta)\right]\lambda + 3g\varepsilon\eta^{-1}I(1 + I(1 + \beta))$$
$$+ (1 + \beta)G_s I_s \eta^{-1} = 0. \tag{12.68}$$

Because all coefficients are positive, the non-zero steady state is always stable. The values of $\varepsilon = 5 \times 10^{-6}$ and $\eta = 8.3 \times 10^{-6}$ are comparable in magnitude which motivates the scaling

$$\varepsilon = O(\eta). \tag{12.69}$$

With (12.69) the term multiplied by $\varepsilon\eta^{-1}$ in the coefficient of λ will contribute with the other terms to the damping of the relaxation oscillations. From (12.66), (12.67), and (12.68), we determine the following approximations for I_s, G_s, and λ

$$I_s = \frac{\gamma - \alpha}{\alpha(1 + \beta)} + O(\varepsilon), \ G_s = \alpha + O(\varepsilon), \ \lambda = \pm i\sqrt{\frac{\gamma - \alpha}{\eta}} + O(1), \quad (12.70)$$

where the $O(1)$ correction of λ is real and negative. We conclude that a small perturbation from the steady state exhibits slowly decaying relaxation oscillations with frequency

$$\omega_1 \simeq \sqrt{\frac{\gamma - \alpha}{\eta}}. \tag{12.71}$$

The frequency (12.71) is equivalent to the laser RO frequency. The cross-saturation parameter β and the crystal parameter ε do not appear in (12.71).

Out-of-phase regimes

We now ask whether other solutions are possible, in particular those with fluctuations that would be different for the two intensity components. The non-zero intensity solution $I_1 = I_2 = I_s$ and $G_1 = G_2 = G_s$, where I_s and G_s are determined from Eqs. (12.66) and (12.67), is still our basic solution. From Eqs. (12.61) and (12.62), we formulate the linearized equations for the small perturbation u_j and v_j of this steady state. Introducing then $u_j = c_j \exp(\lambda t)$ and $v_j = d_j \exp(\lambda t)$, we obtain a homogeneous system of four equations given by

$$
\begin{pmatrix}
a - \lambda & I_s \eta^{-1} & 2a & 0 \\
-G_s & b - \lambda & -\beta G_s & 0 \\
2a & 0 & a - \lambda & I_s \eta^{-1} \\
-\beta G_s & 0 & -G_s & b - \lambda
\end{pmatrix}
\begin{pmatrix}
c_1 \\ d_1 \\ c_2 \\ d_2
\end{pmatrix}
= 0,
\qquad (12.72)
$$

where $a \equiv -g \varepsilon I_s \eta^{-1}$ and $b \equiv -(1 + I_s(1 + \beta))$. We already know one solution which corresponds to the in-phase solution, i.e. $c_1 = c_2 = c$ and $d_1 = d_2 = d$, since this solution must obey the general characteristic equation. Introducing this result into (12.72) leads to a simpler problem for c and d given by

$$
\begin{pmatrix}
-3g\varepsilon I_s \eta^{-1} - \lambda & I_s \eta^{-1} \\
-(1 + \beta) G_s & -(1 + I_s(1 + \beta)) - \lambda
\end{pmatrix}
\begin{pmatrix} c \\ d \end{pmatrix}
= 0.
\qquad (12.73)
$$

A nontrivial solution is possible only if the determinant of the 2×2 homogeneous matrix is zero. This leads to Eq. (12.68). We next take advantage of the symmetry of the matrix in (12.72) by seeking a solution of the form $c_1 = -c_2$ and $d_1 = -d_2$. We again discover that the problem for (c_1, d_1) or for (c_2, d_2) is identical and is of the form

$$
\begin{pmatrix}
g\varepsilon I_s \eta^{-1} - \lambda & I_s \eta^{-1} \\
-(1 - \beta) G_s & -(1 + I_s(1 + \beta)) - \lambda
\end{pmatrix}
\begin{pmatrix} c_1 \\ d_1 \end{pmatrix}
= 0.
\qquad (12.74)
$$

The condition for a nontrivial solution now leads to the other characteristic equation for λ

$$
\lambda^2 + \lambda \left[-g \frac{\varepsilon}{\eta} I_s + 1 + I_s(1 + \beta) \right] + \frac{(1 - \beta)}{\eta} I_s G_s = 0,
\qquad (12.75)
$$

which we analyze as usual.

From (12.75), we note a change of stability through a Hopf bifurcation if

$$I_s > I_H \equiv \frac{1}{g\frac{\varepsilon}{\eta} - 1 - \beta} \qquad (12.76)$$

provided that

$$g\frac{\varepsilon}{\eta} - 1 - \beta > 0. \qquad (12.77)$$

The location of the Hopf bifurcation point is at $\gamma = \gamma_H \equiv \alpha + \alpha(1 + \beta)I_H$. The frequency of the oscillations at the Hopf bifurcation point is

$$\omega_2 \simeq \sqrt{\frac{1 - \beta}{1 + \beta} \frac{\gamma_H - \alpha}{\eta}} \qquad (12.78)$$

and depends on the cross-saturation parameter β. Moreover, the linearized theory indicates that the periodic solution at the Hopf bifurcation point exhibits two intensities phase-shifted by half the period (π/ω_2). Figure 12.16 shows the

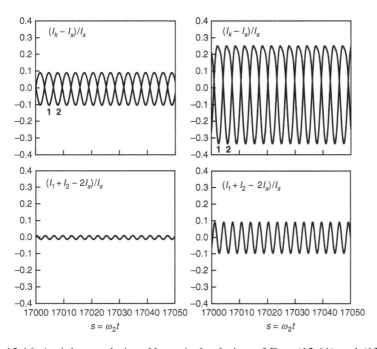

Fig. 12.16 Antiphase solution. Numerical solution of Eqs. (12.61) and (12.62) with $\eta = 8.3 \times 10^{-7}$, $\varepsilon = 5 \times 10^{-6}$, $\alpha = 10^{-2}$, $\beta = 0.7$, $g = 0.4$, and $\gamma = 0.036$ (left) or $\gamma = 0.06$ (right). The period T is close to $T = 2\pi$. Top: the two intensities ($k = 1$ and 2) are T-periodic and are phase-shifted by $T/2$. Bottom: the total intensity is $2T$-periodic and admits a much smaller amplitude than either of the individual intensities.

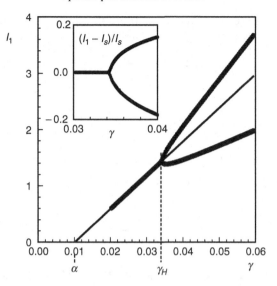

Fig. 12.17 Bifurcation diagram of the steady states and the antiphase oscillations of Eqs. (12.61) and (12.62). The figure shows the extrema of I_1 as a function of γ. The values of the fixed parameters are the same as in Figure 12.16. The steady and Hopf bifurcations are located at $\gamma = \alpha = 0.01$ and $\gamma = \gamma_H \simeq 0.034$, respectively. The inset shows the typical parabolic change of the amplitude near the Hopf bifurcation point.

antiphase regime close to and far from the Hopf bifurcation point $(\gamma > \gamma_H)$. Close to the Hopf bifurcation, the oscillations are nearly harmonic and the total intensity is nearly constant. Far from the Hopf bifurcation point, the oscillations are no longer harmonic (the minimum is larger in magnitude than the maximum) but the two intensities remain synchronized with a phase shift of half the period.

Figure 12.17 represents the bifurcation diagram of the antiphase periodic solutions. Close to the bifurcation point, Hopf bifurcation theory tells us that the intensities are of the form

$$I_1 - I_s = \sqrt{\gamma - \gamma_H}\left[B\exp(i\omega_2 t) + c.c.\right] + O(\gamma - \gamma_H), \qquad (12.79)$$

$$I_2 - I_s = -\sqrt{\gamma - \gamma_H}\left[B\exp(i\omega_2 t) + c.c.\right] + O(\gamma - \gamma_H). \qquad (12.80)$$

The leading expression in the $O(\gamma - \gamma_H)$ correction admits three terms multiplying $B\overline{B}$, $B^2\exp(2i\omega_2 t)$, and its complex conjugate. They are identical for the two intensities. Consequently, the oscillations of the individual intensities admit a leading amplitude proportional to $\sqrt{\gamma - \gamma_H}$ and a frequency ω_2 while the total intensity exhibits a smaller amplitude, proportional to $\gamma - \gamma_H$, and an oscillation frequency equal to $2\omega_2$. These different properties are illustrated by the numerical solutions displayed in Figure 12.16.

In summary, our analysis has showed that the general response of a two-mode Nd^{3+}:YAG laser with intracavity doubling crystal exhibits oscillations with two distinct frequencies. Sustained in-phase solutions are not possible but sustained antiphase solutions are possible and result from a Hopf bifurcation phenomenon.

12.8 Frequencies

The determination of the oscillation frequencies is considerably simplified for an arbitrary number of modes if we take into account the small value of η. As illustrated in other chapters, it is mathematically worthwhile to introduce the relaxation oscillation basic time and rescale the dependent variables so that η multiplying the left hand side of Eq. (12.61) can be removed. This is realized by introducing the time s and the deviations F_k and J_k defined by

$$G_k = \alpha + \eta^{1/2} F_k, \ I_k = I(1 + J_k), \quad \text{and} \quad t = \eta^{1/2} s. \tag{12.81}$$

Inserting (12.81) into Eqs. (12.61) and (12.62), where

$$I \equiv \frac{\gamma \alpha^{-1} - 1}{1 + \beta(N - 1)} \tag{12.82}$$

is the leading expression of the steady state intensity for ε small, we obtain

$$\frac{dJ_k}{ds} = \left[F_k + O\left(\varepsilon \eta^{-1/2}\right)\right](1 + J_k), \tag{12.83}$$

$$\frac{dF_k}{ds} = -\alpha I \left(J_k + \beta \sum_{j(\neq k)} J_j \right) + O\left(\eta^{1/2}\right). \tag{12.84}$$

Assuming the scaling (12.69), Eqs. (12.83) and (12.84) reduce to

$$\frac{dJ_k}{ds} = F_k(1 + J_k), \tag{12.85}$$

$$\frac{dF_k}{ds} = -\alpha I \left(J_k + \beta \sum_{j(\neq k)} J_j \right). \tag{12.86}$$

Some of the properties of these equations are analyzed in [326]. But we are only interested in the behavior of a small perturbation from the steady state. The linearized equations for $J_k = F_k = 0$ are

$$\frac{dJ_k}{ds} = F_k, \tag{12.87}$$

$$\frac{dF_k}{ds} = -\alpha I \left(J_k + \beta \sum_{j(\neq k)} J_j \right) \tag{12.88}$$

or, equivalently, after eliminating F_k, the following coupled second order differential equations for J_k ($k = 1, \ldots, N$)

$$\frac{d^2 J_k}{ds^2} = -\alpha I \left(J_k + \beta \sum_{j(\neq k)} J_j \right). \tag{12.89}$$

We next anticipate that these equations only admit periodic solutions and introduce $J_k = c_k \exp(i\omega s)$ into Eq. (12.89). The resulting problem then forms an $N \times N$ homogeneous system for the coefficients c_k. The condition for a nontrivial solution is given by equating the determinant of all the coefficients to zero

$$\begin{vmatrix} \omega^2 - \alpha I & -\alpha I \beta & -\alpha I \beta & \cdots & -\alpha I \beta \\ -\alpha I \beta & \omega^2 - \alpha I & -\alpha I \beta & \cdots & -\alpha I \beta \\ -\alpha I \beta & -\alpha I \beta & \omega^2 - \alpha I & \cdots & -\alpha I \beta \\ \cdots & \cdots & \cdots & \cdots & \cdots \\ -\alpha I \beta & -\alpha I \beta & -\alpha I \beta & \cdots & \omega^2 - \alpha I \end{vmatrix} = 0. \tag{12.90}$$

Adding the rows $k = 2, \ldots N$ to the first row leaves the determinant unchanged. The elements of the first row are then all identical, which allows us to factorize the common term. The determinant (12.90) reduces to

$$0 = \left[\omega^2 - \alpha I (1 + (N-1)\beta) \right]$$

$$\times \begin{vmatrix} 1 & 1 & 1 & \cdots & 1 \\ -\alpha I \beta & \omega^2 - \alpha I & -\alpha I \beta & \cdots & -\alpha I \beta \\ -\alpha I \beta & -\alpha I \beta & \omega^2 - \alpha I & \cdots & -\alpha I \beta \\ \cdots & \cdots & \cdots & \cdots & \cdots \\ -\alpha I \beta & -\alpha I \beta & -\alpha I \beta & \cdots & \omega^2 - \alpha I \end{vmatrix}. \tag{12.91}$$

We next subtract the first column from the second in the $(N - 1) \times (N - 1)$ determinant. The determinant remains unchanged after this operation and (12.91) becomes

$$0 = \left[\omega^2 - \alpha I (1 + (N - 1)\beta) \right]$$

$$\times \begin{vmatrix} 1 & 0 & 1 & \cdots & 1 \\ -\alpha I \beta & \omega^2 - \alpha I + \alpha I \beta & -\alpha I \beta & \cdots & -\alpha I \beta \\ -\alpha I \beta & 0 & \omega^2 - \alpha I & \cdots & -\alpha I \beta \\ \cdots & \cdots & \cdots & \cdots & \cdots \\ -\alpha I \beta & 0 & -\alpha I \beta & \cdots & \omega^2 - \alpha I \end{vmatrix}.$$

Repeating sequentially the same operation with column $k = 3, \ldots N$ will progressively reduce the problem to

$$0 = \left[\omega^2 - \alpha I (1 + (N - 1)\beta) \right]$$

$$\times \begin{vmatrix} 1 & 0 & 0 & \cdots & 0 \\ -\alpha I \beta & \omega^2 - \alpha I + \alpha I \beta & 0 & \cdots & 0 \\ -\alpha I \beta & 0 & \omega^2 - \alpha I + \alpha I \beta & \cdots & 0 \\ \cdots & \cdots & \cdots & \cdots & \cdots \\ -\alpha I \beta & 0 & 0 & \cdots & \omega^2 - \alpha I + \alpha I \beta \end{vmatrix}.$$

The determinant is now easily evaluated and leads to a characteristic equation for ω^2 given by

$$\left[\omega^2 - \alpha I (1 + (N - 1)\beta) \right] \left[\omega^2 - \alpha I (1 - \beta) \right]^{N-1} = 0. \qquad (12.92)$$

Equation (12.92) admits two possible solutions for ω^2. The first solution is

$$\omega^2 = \alpha I (1 + (N - 1)\beta) \qquad (12.93)$$

and, from the homogeneous system for the c_k, we find the eigenvector $(1, 1, \ldots 1)^t$ corresponding to an in-phase regime. We already know that it does not lead to a bifurcation but only to slowly decaying oscillations. The second solution of Eq. (12.92) is $(N - 1)$-degenerate and is given by

$$\omega^2 = \alpha I (1 - \beta). \qquad (12.94)$$

From the homogeneous system for the c_k, we find that the $N - 1$ eigenvectors satisfy the single condition

$$\sum_{1}^{N} c_k = 0. \qquad (12.95)$$

Since the c_k multiply the periodic function $\exp(i\omega s)$, it is worthwhile to determine c_k in amplitude–phase complex form. Using a table of sums of trigonometric functions [327], Eq. (12.95) is satisfied if

$$c_k = \exp(ik2\pi/N). \tag{12.96}$$

For example, if $N = 3$, we have $(c_1, c_2, c_3) = (\exp(i2\pi/3), \exp(i4\pi/3), 1)$ and $(c_1, c_2, c_3) = (\exp(i4\pi/3), \exp(i2\pi/3), 1)$ as the two linearly independent eigenvectors. In order to determine whether a Hopf bifurcation branch is possible, we would need to examine the higher order problem and apply solvability conditions.

12.9 Antiphase dynamics in a fiber laser

As mentioned at the beginning of this chapter, self-organized collective behaviors such as antiphase oscillations are observed in several multimode laser systems. The Nd^{3+}-doped optical fiber laser pumped by a laser diode and exhibiting two polarization modes is perhaps the simplest system exhibiting the antiphase phenomenon. As the pump parameter is increased from zero, the laser exhibits a first threshold, above which laser radiation is emitted in a linear polarization state. Above a second threshold, radiation is also emitted in the orthogonal polarization. The relative position of these two thresholds can be adjusted by rotating the pump polarization or changing the stress applied to the fiber. Figure 12.18 shows

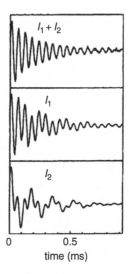

Fig. 12.18 Experimental recording of the response of an optical fiber laser to a pulse excitation. Reprinted Figure 4a with permission from Otsuka *et al.* [320]. Copyright 1992 by the American Physical Society.

power spectra (arb. units)

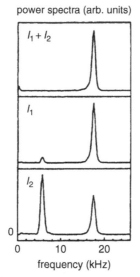

Fig. 12.19 Power spectra of the oscillations shown in Figure 12.18. Reprinted Figure 4b with permission from Otsuka *et al.* [320]. Copyright 1992 by the American Physical Society.

the response of the individual and total intensities following a small square pulse of the pump. The decaying oscillations of the polarization intensities are irregular and out of phase but the oscillatory decay of the total intensity is smooth and sinusoidal. The power spectra shown in Figure 12.19 reveal that the oscillations of the individual intensities depend on two distinct frequencies but that the total intensity only decays with respect to the largest frequency. The rate equations analyzed in [321] (without the spontaneous emission term) are given by

$$I'_k = (D_k + \beta D_j - 1)I_k, \tag{12.97}$$

$$D'_k = \gamma \left[p_k - (1 + I_k + \beta I_j)D_k \right], \tag{12.98}$$

where $k = 1$ or 2 and $j = 3 - k$. I_k and D_k denote the intensities and population inversions of the two laser modes, respectively. Prime means differentiation with respect to the dimensionless time $t = t'/\tau_c$, where τ_c is the photon lifetime in the cavity. p_1 and $p_2 = \alpha p_1$ are the pump parameters associated with mode 1 and mode 2, respectively. The asymmetry parameter $\alpha = 0.86$ is fixed by the pump polarization and remains unchanged during the experiment. $\gamma \equiv \tau_f/\tau_c = 6.7 \times 10^{-4}$, where τ_f is defined as the population inversion relaxation time. $\beta = 0.43$ is the cross-saturation coefficient that describes the coupling of the laser field k with the population inversion $j = 3 - k$. The rate equations (12.97) and (12.98) have been used to explain the antiphase polarization dynamics in fiber lasers [321, 328] and Nd^{3+}:YAG lasers [329].

12.9.1 Steady state solutions

The steady state solutions are given by (1) the zero intensity steady state $I_1 = I_2 = 0$, the two pure mode solutions,

(2) $I_2 = 0$

$$D_1 = \frac{p_1}{1 + I_1}, \quad D_2 = \frac{\alpha p_1}{1 + \beta I_1},$$

$$p_1 = \frac{(1 + I_1)(1 + \beta I_1)}{[1 + \beta I_1 + \beta \alpha(1 + I_1)]}, \tag{12.99}$$

(3) $I_1 = 0$

$$D_1 = \frac{p_1}{1 + \beta I_2}, \quad D_2 = \frac{\alpha p_1}{1 + I_2},$$

$$p_1 = \frac{(1 + \beta I_2)(1 + I_2)}{[\alpha(1 + \beta I_2) + \beta(1 + I_2)]}, \tag{12.100}$$

and a possible bimode solution of the form

(4) $D_1 = D_2 = D = \dfrac{1}{1 + \beta}$,

$$I_1 = \frac{1}{1 - \beta^2}(p_1(1 + \beta)(1 - \alpha\beta) - 1 + \beta) \geq 0,$$

$$I_2 = \frac{1}{1 - \beta^2}(p_1(1 + \beta)(\alpha - \beta) - 1 + \beta) \geq 0. \tag{12.101}$$

Figure 12.20 shows the bifurcation diagram of these steady states. P_1 and P_2 denote primary bifurcation points to pure mode solutions. They are obtained by evaluating p_1 in (12.99) for $I_1 = 0$ and p_1 in (12.101) for $I_2 = 0$. They are given by

$$P_1 : p_1 = \frac{1}{1 + \alpha\beta}, \tag{12.102}$$

$$P_2 : p_1 = \frac{1}{\alpha + \beta}. \tag{12.103}$$

On the other hand, the point S marks a secondary bifurcation from the $I_1 \neq 0$ single-mode solution. From (12.101), the condition $I_2 = 0$ gives

$$S : p_1 = \frac{1 - \beta}{(1 + \beta)(\alpha - \beta)}. \tag{12.104}$$

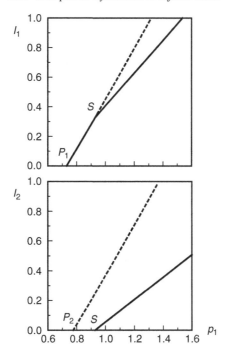

Fig. 12.20 Bifurcation diagram of the steady states given by (12.99)–(12.101). p_1 is the control parameter. P_1 and $P_2 > P_1$ denote two successive primary bifurcations to pure mode solutions. $S > P_2$ marks a secondary bifurcation point to a bimode solution. Full and broken lines represent stable and unstable solutions, respectively. The values of the parameters are $\alpha = 0.86$ and $\beta = 0.43$.

Using (12.102)–(12.104) with $\alpha = 0.86$ and $\beta = 0.43$, we determine the successive bifurcation points P_1 at $p_1 = 0.73$, P_2 at $p_1 = 0.78$, and S at $p_1 = 0.92$. Note that as $\alpha \to 1$, all three points coalesce to the same point at $p_1 = (1 + \beta)^{-1}$.

12.9.2 Stability analysis

Small perturbations from the stable steady states slowly decay with relaxation oscillations. We wish to determine the frequencies in the simplest possible way, and again take advantage of the small value of γ.[7] Specifically, we introduce the new variables x_k, y_k, and s defined by

$$I_k = I_{ks}(1 + y_k), \quad D_k = D_{ks} + \sqrt{\gamma}x_k, \quad \text{and} \quad s = \sqrt{\gamma}t, \tag{12.105}$$

where $(I_k, D_k) = (I_{ks}, D_{ks})$ represents the stable steady state. For the pure mode solution (12.99), we introduce $I_2 = y_2$. Inserting (12.105) into Eqs. (12.97) and (12.98) and taking the limit $\gamma \to 0$ leads to the following reduced problem

---

[7] By changing the time variable in Eq. (12.61) from t to t/η, we note that γ plays the same role as η for the Nd^{3+}:YAG laser.

$$y'_1 = (x_1 + \beta x_2)(1 + y_1),$$

$$x'_1 = -(I_1 y_1 + \beta y_2)d_1,$$

$$y'_2 = (-1 + D_2 + \beta D_1)y_2,$$

$$x'_2 = -(y_2 + \beta I_1 y_1)d_2 \tag{12.106}$$

for the pure mode solution (12.99), and

$$y'_1 = (x_1 + \beta x_2)(1 + y_1),$$

$$x'_1 = -(I_1 y_1 + \beta I_2 y_2)d_1,$$

$$y'_2 = (x_2 + \beta x_1)(1 + y_2),$$

$$x'_2 = -(I_2 y_2 + \beta I_1 y_1)d_2 \tag{12.107}$$

for the two-mode solution (12.101). We have omitted the subscript s from I_k and D_k, and prime now means differentiation with respect to time s. From (12.106), we first note that $y_2 \to 0$ if

$$-1 + D_2 + \beta D_1 < 0. \tag{12.108}$$

We then consider the remaining equations with $y_2 = 0$. From the linearized equations for the zero solution, we determine the characteristic equation for the growth rate. One solution is $\lambda_1 = 0$ and the other solutions satisfy $\lambda^2 = -I_1 D_1 (1 + \beta^2)$, which implies purely imaginary roots and the frequency

$$\omega_1 = \sqrt{I_1 D_1 (1 + \beta^2)}. \tag{12.109}$$

Next, we consider Eqs. (12.107). From the linearized equations for the zero solution, we determine the characteristic equation for the growth rate. We find

$$\lambda^4 + \lambda^2 (1 + \beta^2) D(I_1 + I_2) + I_1 I_2 (1 - \beta)^2 = 0, \tag{12.110}$$

where $D = (1 + \beta)^{-1}$. Equation (12.110) admits two pairs of purely imaginary solutions. Near the secondary bifurcation point S where $I_2 = 0$, they admit the simple approximations $\lambda_1^2 \simeq -(1+\beta^2)DI_1$ and $\lambda_2^2 \simeq -I_2(1-\beta)^2(1+\beta^2)^{-1}D^{-1}$. This then leads to two frequencies given by

$$\omega_1 \simeq \sqrt{(1 + \beta^2)DI_1}, \tag{12.111}$$

$$\omega_2 \simeq \sqrt{I_2(1 - \beta)^2(1 + \beta^2)^{-1}D^{-1}}. \tag{12.112}$$

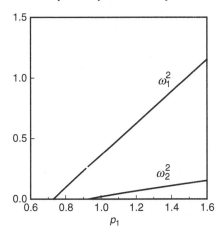

Fig. 12.21 Square of the relaxation oscillation frequency for the single-mode regime with $I_2 = 0$ (Eq. (12.109)) and for the bimode regime (Eq. (12.110)). These are almost straight lines for the interval of p_1 under consideration. ω_1^2 for the single-mode regime is shown only when it is stable ($0.73 < p_1 < 0.92$). ω_1^2 for the bimode regime emerges at $p_1 = 0.92$ and admits a slope close to the one for the single-mode regime.

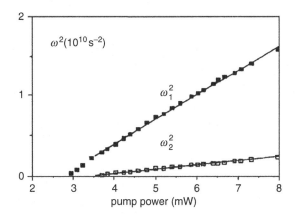

Fig. 12.22 Experimental oscillation frequencies as functions of the pump power. The thresholds of the single-mode and bimode regimes are at 3 mW and 3.5 mW, respectively. Reprinted Figure 5 with permission from Otsuka *et al.* [320]. Copyright 1992 by the American Physical Society.

Figure 12.21 represents the square of the frequency (12.109) and the frequencies (12.111) and (12.112) obtained numerically from solving the quadratic equation (12.110). The values of the parameters are the same as previously, i.e. $\alpha = 0.86$, $\beta = 0.43$, and $\gamma = 6.7 \times 10^{-4}$. The diagram exhibits two nearly straight lines that are well confirmed experimentally (see Figure 12.22).

In summary, a Hopf bifurcation is not possible for the optical fiber laser problem. Subject to a small pump pulse perturbation, the laser, however, exhibits slowly decaying oscillations with frequencies ω_1 and ω_2. From the full linearized equations, the intensities are of the form

$$I_1 - I_{1s} \simeq A \exp(i\omega_1 s) + B \exp(i\omega_2 s) + c.c., \tag{12.113}$$

$$I_2 \simeq A \exp(i\omega_1 s) - B \exp(i\omega_2 s) + c.c., \tag{12.114}$$

where the amplitudes A and B are exponentially decaying functions of $\sqrt{\eta}s$. From (12.113) and (12.114), we conclude that the oscillations of the individual intensities are quasi-periodic with frequencies ω_1 and ω_2 but that the total intensity only exhibits oscillations with frequency ω_1. These properties have indeed been observed experimentally as shown previously (see Figures 12.18 and 12.19).

12.10 Exercises

12.10.1 Thresholds for SROPO and DROPO

Derive the evolution equations for the two cases not investigated in Section 12.1.1.

(1) the DROPO with $\gamma_i \gg \gamma_p, \gamma_s$ as it occurs, for example, in OPOs with quite different frequencies for the signal and the idler.
(2) the SROPO with γ_i and $\gamma_p \gg \gamma_s$.

Determine and compare the OPO thresholds for these two cases.

12.10.2 Rescaling the OPO equations

Find the change of variables to reduce the original DOPO equations

$$\frac{dE_p}{dt} = -(\gamma_p + i\Delta_p)E_p - \chi E_s^2 + E,$$

$$\frac{dE_s}{dt} = -(\gamma_s + i\Delta_s)E_s + \chi E_p E_s^*$$

to Eqs. (12.3) and (12.4).

12.10.3 Rescaling the SHG equations

Find the change of variables to reduce the SHG equations (12.5) to Eqs. (12.6).
   Answer: $A_1 = (\sqrt{ab}/\gamma_2)E_1$, $A_2 = ia E_2/\gamma_2$, $\tau = \gamma_2 t$; the new parameters then are $E = E_{ext}\sqrt{ab}/\gamma_2^2$, $\Delta_{1,2} = \delta_{1,2}/\gamma_2$, and $\gamma = \gamma_1/\gamma_2$.

12.10.4 Linear Stability of the DOPO

Investigate the linear stability of the steady states of Eqs. (12.3) and (12.4). Show that the OFF solution is stable below the threshold and unstable above. Find the width of the bistability region. Derive the expression (12.18) for the frequency of the periodic solution appearing at the Hopf bifurcation obtained if $\Delta_0 \Delta_1 < \gamma$.

12.10.5 SHG inside the laser cavity

The single-mode laser equations for second harmonic generation inside a laser cavity are

$$\eta \frac{dI}{dt} = (G - \alpha - g\varepsilon I)I, \qquad (12.115)$$

$$\frac{dG}{dt} = \gamma - (1 + I)G, \qquad (12.116)$$

which differ from the class B laser equations by the extra term $-g\varepsilon I^2$ which describes the SHG losses. Show that these equations admit a stable non-zero intensity steady state.

References

[1] R.W. Boyd, M.G. Raymer, and L.M. Narducci, eds., *Optical Instabilities: Proceedings of the International Meeting on Instabilities and Dynamics of Lasers and Nonlinear Optical Systems*. Cambridge Studies in Modern Optics, vol. **4**, Cambridge University Press, Cambridge (1986)

[2] N.B. Abraham, L.A. Lugiato, and L.M. Narducci, eds., Instabilities in active optical media, *J. Opt. Soc. Am. B* **2**, Number 1 (1985)

[3] D.K. Bandy, A.N. Oraevsky, and J.R. Tredicce, eds., Nonlinear dynamics of lasers, *J. Opt. Soc. Am. B* **5**, Number 5 (1988)

[4] C.O. Weiss and R. Vilaseca, *Dynamics of Lasers*, VCH, Weinheim, Germany (1991)

[5] A.C. Newell and J.V. Moloney, *Nonlinear Optics*, Addison-Wesley, New York (1992)

[6] Y.I. Khanin, *Principles of Laser Dynamics*, Elsevier, Amsterdam (1995)

[7] K. Otsuka, *Nonlinear Dynamics in Optical Complex Systems*, KTK Scientific Publishers, Tokyo; Kluwer Academic Publishers, Boston (1999)

[8] S.H. Strogatz, *Nonlinear Dynamics and Chaos: with Applications in Physics, Biology, Chemistry, and Engineering*, Addison-Wesley, Reading, MA (1994)

[9] I.R. Epstein and J.A. Pojman, *An Introduction to Nonlinear Chemical Dynamics*, Oxford University Press, New York (1998)

[10] J. Keener and J. Sneyd, *Mathematical Physiology*, Springer-Verlag, New York (1998)

[11] J.D. Murray, *Mathematical Biology*, 3rd edn, Interdisciplinary Applied Mathematics, vol. **17**, Springer-Verlag, Berlin (2002)

[12] A. Beuter, L. Glass, M. Mackey, and M. Titcombe, eds., *Nonlinear Dynamics in Physiology and Medicine*, Springer-Verlag, New York (2003)

[13] D. Lenstra, ed., Fundamental nonlinear dynamics of semiconductor lasers, *Quantum and Semiclass. Opt.* **9**, Number 5 (1997)

[14] B. Krauskopf and D. Lenstra, eds., *Fundamental Issues of Nonlinear Laser Dynamics*, AIP Conference Proceedings, vol. **548**, American Institute of Physics, New York (2000)

[15] C.M. Bender and S.A. Orszag, *Advanced Mathematical Methods for Scientists and Engineers*, McGraw-Hill, New York (1978)

[16] J. Kevorkian and J.D. Cole, Perturbation methods in applied mathematics, *Appl. Math. Sci.* **34**, Springer, New York (1981)

[17] J. Kevorkian and J.D. Cole, Multiple scale and singular perturbation methods, *Appl. Math. Sci.* **114**, Springer, New York (1996)

[18] N.B. Abraham, P. Mandel, and L.M. Narducci, Dynamical instabilities, and pulsations in lasers, *Prog. in Opt.* **25**, 3–190 (1988)

[19] A. Yariv, *Quantum Electronics*, 2nd edn, Wiley, New York (1975)

[20] A.E. Siegman, *Lasers*, University Science Books, Mill Valley, CA (1986)

[21] P. Mandel, *Theoretical Problems in Cavity Nonlinear Optics.* Cambridge Studies in Modern Optics, Cambridge University Press, Cambridge (1997)

[22] D. Dangoisse, D. Hennequin, and V. Zehnlé-Dhaoui, *Les Lasers*, Dunod, Paris (1998)

[23] J.R. Tredicce, F.T. Arecchi, G.L. Lippi, and G.P. Puccioni, Instabilities in lasers with an injected signal, *J. Opt. Soc. Am. B* **2**, 173–183 (1985)

[24] C.C. Lin and L.A. Segel, *Mathematics Applied to Deterministic Problems in the Natural Sciences*, Macmillan, New York (1974)

[25] P.L. Gourley, Nanolasers, *Scientific American*, March 1998, 40–45

[26] W.E. Boyce, R.C. DiPrima, *Elementary Differential Equations*, 4th edn, John Wiley, New York (1986)

[27] G. Nicolis, *Introduction to Nonlinear Science*, Cambridge University Press, Cambridge (1995)

[28] R. Haberman, *Mathematical Models: Mechanical Vibrations, Population Dynamics, and Traffic Flow. An Introduction to Applied Mathematics*, Prentice-Hall, Englewood Cliffs, NJ (1977), reprinted SIAM Classic Series **21**, Society for Industrial and Applied Mathematics, Philadelphia, PA (1998)

[29] J.V. Uspensky, *Theory of Equations*, McGraw-Hill, New York (1948)

[30] B. Pariser and T.C. Marshall, Time development of a laser signal, *Appl. Phys. Lett.* **6**, 232–234 (1965)

[31] F.T. Arecchi and V. Degiorgio, Statistical properties of laser radiation during a transient buildup, *Phys. Rev. A* **3**, 1108–1124 (1971)

[32] R. Roy, A.W. Yu, and S. Zhu, Quantum fluctuations, pump noise, and the growth of laser radiation, *Phys. Rev. Lett.* **55**, 2794–2797 (1985)

[33] G.P. Agrawal and N.K. Dutta, *Long-Wavelength Semiconductor Lasers*, Van Nostrand Reinhold, New York (1986)

[34] K. Petermann, *Laser Diode Modulation and Noise*, Kluwer Academic Publishers, Dordrecht, The Netherlands (1988, reprinted 1991)

[35] I.J. Sola, J.C. Martin, J.M. Álvarez, and S. Jarabo, Erbium doped fibre characterisation by laser transient behaviour analysis, *Opt. Commun.* **193**, 133–140 (2001)

[36] M. Sargent III, M.O. Scully, and W.E. Lamb Jr., *Laser Physics*, Addison-Wesley, Reading, MA (1974)

[37] H. Haken, *Light*, vol. 1 and 2, North Holland, Amsterdam (1981)

[38] G. Iooss and D.D. Joseph, *Elementary Stability and Bifurcation Theory*. Undergraduate Texts in Mathematics, Springer-Verlag, New York (1980; 2nd edn 1990)

[39] H.A. Haus and S. Kawakami, On the "excess spontaneous emission factor" in gain-guided laser amplifiers, *IEEE J. Quantum Electron.* **QE-21**, 63–69 (1985)

[40] C.H. Henry, Theory of the linewidth of semiconductor lasers, *IEEE J. Quantum Electron.* **QE-18**, 259–264 (1982)

[41] H.S. Sommers, Jr., Spectral characteristics of single-mode injection lasers: the power-gain curve from weak stimulation to full output, *J. Appl. Phys.* **53**, 156–160 (1982)

[42] M. Corti and V. Degiorgio, Analogy between the laser and second order phase transitions: measurement of "coexistence curve" and "susceptibility" for a single-mode laser near threshold, *Phys. Rev. Lett.* **36**, 1173–1176 (1976)

[43] F.T. Arecchi, W. Gadomski, R. Meucci, and J.A. Roversi, Dynamics of laser build up from quantum noise, *Phys. Rev. A* **39**, 4004–4015 (1989)

[44] G.H.M. van Tartwijk and D. Lenstra, Semiconductor lasers with optical injection and feedback, *J. Opt. B Quantum Semiclassical Opt.* **7**, 87–143 (1995)

[45] J. Piprek and J.E. Bowers, Analog modulation of semiconductor lasers. In W. Chang, ed., *RF Photonic Technology in Optical Fiber Links*, Cambridge University Press, Cambridge (2002), Chapter 3

[46] *Encyclopedia of Laser Physics and Technology*, http://www.rp-photonics.com/laser_dynamics.html

[47] J.C. Garreau, P.-Y. Wang, and P. Glorieux, Bit correlation and memory effects in high speed pump modulation of a fiber laser, *IEEE J. Quantum Electron.* **30**, 1058–1074 (1994)

[48] B. Ségard, P. Glorieux, and T. Erneux, Delayed pulse dynamics in single mode class B lasers, *Appl. Phys. B* **81**, 989–992 (2005)

[49] Michael Peil, Dynamics and synchronization phenomena of semiconductor lasers with delayed optical feedback: utilizing nonlinear dynamics for novel applications. Ph.D. thesis, University of Darmstadt (2006)

[50] Kam Y. Lau and A. Yariv, Ultra-high speed semiconductor lasers, *IEEE J. Quantum Electron.* **QE-21**, 121–138 (1985)

[51] W.T. Silfvast, *Laser Fundamentals*, Cambridge University Press, Cambridge (2004)

[52] M.J. Digonnet, ed., *Selected Papers on Rare-Earth-Doped Fiber Laser Sources and Amplifiers*, SPIE Milestones Series, vol. **MS 37**, SPIE, Bellingham, WA (1992)

[53] N.G. Basov and A.M. Prokhorov, About possible methods for obtaining active molecules for a molecular oscillator, *Zh. Eksp. Teor. Fiz.* **28/2**, 249–250 (1955), translated in *Sov. Phys. JETP* **1**, 184–185 (1956)

[54] N. Bloembergen, Proposal for a new type solid state maser, *Phys. Rev.* **104**, 324–327 (1957)

[55] J.R. Tucker, Absorption saturation and gain in pulsed CH_3F lasers, *Opt. Commun.* **16**, 209–212 (1976)

[56] A. Kaminskii, *Crystalline Lasers: Physical Processes and Operating Schemes*, CRC Press, New York (1996)

[57] I. Burak, P.L. Houston, D.G. Sutton, and J.I. Steinfeld, Mechanism of passive Q-switching in CO_2 lasers, *IEEE J. Quantum Electron.* **QE-7**, 73–82 (1971)

[58] J. Dupré, F. Meyer, and C. Meyer, Influence des phénomènes de relaxation sur la forme des impulsions fournies par un laser CO_2 déclenché par un absorbant saturable, *Rev. Phys. Appl.* **10**, 285–293 (1975)

[59] H.T. Powell and G.J. Wolga, Repetitive passive Q-switching of single-frequency lasers, *IEEE J. Quantum Electron.* **QE-7**, 213–219 (1971)

[60] E. Arimondo, F. Casagrande, L.A. Lugiato, and P. Glorieux, Repetitive passive Q-switching and bistability in lasers with saturable absorbers, *Appl. Phys. B* **30**, 57–77 (1983)

[61] M. Tachikawa, K. Tanii, M. Kajita, and T. Shimizu, Undamped undulation superposed on to the passive Q-switching pulse of a CO_2 laser, *Appl. Phys. B* **39**, 83–90 (1986)

[62] K. Tanii, M. Tachikawa, M. Kajita, and T. Shimizu, Sinusoidal self-modulation in the output of a CO_2 laser with an intracavity saturable absorber, *J. Opt. Soc. Am. B* **5**, 24–28 (1988)

[63] M. Tachikawa, K. Tanii, and T. Shimizu, Comprehensive interpretation of passive Q-switching and optical bistability in a CO_2 laser with an intracavity saturable absorber, *J. Opt. Soc. Am. B* **4**, 387–395 (1987)

[64] M. Tachikawa, K. Tanii and T. Shimizu, Laser instability and chaotic pulsation in a CO_2 laser with intracavity saturable absorber, *J. Opt. Soc. Am. B* **5**, 1077–1082 (1988)

[65] M. Lefranc, D. Hennequin, and D. Dangoisse, Homoclinic chaos in a laser containing a saturable absorber, *J. Opt. Soc. Am. B* **8**, 239–249 (1991)

[66] N.J. van Druten, Y. Lien, C. Serrat, S.S.R. Oemrawsingh, M.P. van Exter, and J.P. Woerdman, Laser with thresholdless intensity fluctuations, *Phys. Rev. A* **62**, 053808 (2000)

[67] Y. Lien, S.M. de Vries, M.P. van Exter, and J.P. Woerdman, Lasers as Toda oscillators, *J. Opt. Soc. Am. B* **19**, 1461–1466 (2002)

[68] F.T. Arecchi, W. Gadomski, R. Meucci, and J.A. Roversi, Swept dynamics of a CO_2 laser near threshold: two- versus four-level model, *Opt. Commun.* **65**, 47–51 (1988)

[69] G.-L. Oppo, J.R. Tredicce, and L.M. Narducci, Dynamics of vibro-rotational CO_2 laser transitions in a two-dimensional phase space, *Opt. Commun.* **69**, 393–397 (1989)

[70] M. Ciofini, A. Politi, and R. Meucci, Effective two-dimensional model for CO_2 lasers, *Phys. Rev. A* **48**, 605–610 (1993)

[71] V. Zehnlé, D. Dangoisse, and P. Glorieux, Behavior of a CO_2 laser under loss modulation: critical analysis of different theoretical models, *Opt. Commun.* **90**, 99–105 (1992)

[72] D.J. Gauthier, D.W. Sukow, H.M. Concannon, and J.E.S. Socolar, *Phys. Rev. E* **50**, 2343 (1994)

[73] J.E.S. Socolar, D.W. Sukow, and D.J. Gauthier, *Phys. Rev. E* **50**, 3245 (1994)

[74] J.L. Boulnois, A. Van Lerberghe, P. Cottin, F.T. Arecchi, and G.P. Puccioni, Self-pulsing in a CO_2 ring laser with an injected signal, *Opt. Commun.* **58**, 124–129 (1986)

[75] P. Glorieux and A. Le Floch, Nonlinear polarization dynamics in anisotropic lasers, *Opt. Commun.* **79**, 229–234 (1990)

[76] O. Emile, J. Poirson, F. Bretenaker, and A. Le Floch, He-Xe laser magnetometry, *J. Appl. Phys.* **83**, 4994–4996 (1998)

[77] H.L. Stover and W.H. Steier, Locking of laser oscillators by light injection, *Appl. Phys. Lett.* **8**, 91–93 (1966)

[78] M. Möller, B. Forsmann, and W. Lange, Amplitude instability in coupled Nd:YVO_4 microchip lasers, *Chaos, Solitons & Fractals* **10**, 825–829 (1999)

[79] K.S. Thornburg, Jr., M. Möller, R. Roy, T.W. Carr, R.-D. Li, and T. Erneux, Chaos and coherence in coupled lasers, *Phys. Rev. E* **55**, 3865–3869 (1997)

[80] G. Bouwmans, B. Segard, D. Dangoisse, and P. Glorieux, Modeling coupled microchip lasers requires complex coupling coefficients, *J. Opt. Soc. Am. B* **17**, 781–789 (2000)

[81] M. Born and E. Wolf, *Principles of Optics: Electromagnetic Theory of Propagation, Interference and Diffraction of Light*, Pergamon Press, Oxford (1975)

[82] J.C. Cotteverte, F. Bretenaker, and A. Le Floch, Study of the dynamical behavior of the polarization of a quasi-isotropic laser in the earth magnetic field, *Opt. Commun.* **79**, 321–327 (1990)

[83] J.C. Cotteverte, F. Bretenaker, A. Le Floch, and P. Glorieux, Vectorial non-linear dynamics in lasers with one or two stable eigenstates, *Phys. Rev. A* **49**, 2868–2880 (1994)

[84] R. Adler, A study of locking phenomena in oscillators, *Proc. IRE* **34**, 351–357 (1946), reprinted in *Proc. IEEE* **61**, 1380–1385 (1973)

[85] R. Herrero, M. Figueras, F. Pi, and G. Orriols, Phase synchronization in bidirectionally coupled optothermal devices, *Phys. Rev. E* **66**, 036223 (2002)

[86] C.D. Nabors, A.D. Farinas, T. Day, S.T. Yang, E.K. Gustafson, and R.L. Byer, Injection-locking of a 13-W cw Nd:YAG ring laser, *Opt. Lett.* **14**, 1189–1191 (1989)

[87] W.W. Chow, J. Gea-Banacloche, L.M. Pedrotti, V.E. Sanders, W. Schleich, and M.O. Scully, The ring laser gyro, *Rev. Mod. Phys.* **57**, 61–104 (1985)

[88] R. Quintero-Torres, M. Navarro, M. Ackerman, and J.-C. Diels, Scatterometer using a bidirectional ring laser, *Opt. Commun.* **241**, 179–183 (2004)

[89] C. Etrich, P. Mandel, R. Centeno Neelen, R.J.C. Spreeuw, and J.P. Woerdman, Dynamics of a ring-laser gyroscope with back scattering, *Phys. Rev. A* **46**, 525–536 (1992)

[90] D. Botez and D.R. Scifres, eds., *Diode Laser Arrays*, Cambridge University Press, Cambridge (1994)

[91] L. Fabiny, P. Colet, R. Roy, and D. Lenstra, Coherence and phase dynamics of spatially coupled solid-state lasers, *Phys. Rev. A* **47**, 4287–4296 (1993)

[92] J. Terry, K.S. Thornburg, Jr., D.J. DeShazer, G.D. Van Wiggeren, S. Zhu, P. Ashwin, and R. Roy, Synchronization of chaos in an array of three lasers, *Phys. Rev. E* **59**, 4036–4043 (1999)

[93] F. Bretenaker, J.P. Taché, and A. Le Floch, Reverse Sagnac effect in ring lasers, *Europhys. Lett.* **21**, 291–297 (1993)

[94] G.E. Stedman, Ring laser tests of fundamental physics and geophysics, *Rep. Prog. Phys.* **60,** 615–688 (1997), available at http://www.phys.canterbury.ac.nz/research/laser/files/ringlaserrpp.pdf

[95] W. Lauterborn and T. Kurz, *Coherent Optics*, 2nd edn, Springer, New York (2003)

[96] R.J. Collins, D.F. Nelson, A.L. Schawlow, W. Bond, C.G.B. Garett, and W. Kaiser, Coherence, narrowing, directionality, and relaxation oscillations in the light emission from ruby, *Phys. Rev. Lett.* **5**, 303–305 (1960)

[97] D.F. Nelson and W.S. Boyle, A continuously operating ruby optical laser, *Appl. Opt.* **1**, 181–183 (1962)

[98] L.W. Casperson, Spontaneous coherent pulsations in ring-laser oscillators, *J. Opt. Soc. Am. B* **2**, 62–72 (1985)

[99] H. Haken, Analogy between higher instabilities in fluids and lasers, *Phys. Lett.* **53A**, 77–78 (1975)

[100] A. Arimoto, M. Ojima, N. Chinone, A. Oishi, T. Gotoh, and N. Ohnuki, Optimum conditions for the high frequency noise reduction method in optical videodisc players, *Appl. Opt.* **25**, 1398–1403 (1986)

[101] J.P. Goedgebuer, L. Larger, and H. Porte, Optical cryptosystem based on synchronization of hyperchaos generated by a delayed feedback tunable laser diode, *Phys. Rev. Lett.* **80**, 2249–2252 (1998)

[102] O. Gurel and O.E. Rössler, eds., *Bifurcation Theory and Applications in Scientific Disciplines*, Ann. N.Y. Acad. Sci. **316**, New York Academy of Sciences, New York (1979)

[103] D. Aubin and A. Dahan Dalmedico, Writing the history of dynamical systems and chaos: longue durée and revolution, disciplines and cultures, *Historia Mathematica* **29**, 273–339 (2002)

[104] E. Hopf, Abzweigung einer periodischen Lösung von einer stationären Lösung eines Differentialsystems, *Berichte Math.-Phys. Kl. Sächs. Akad. Wiss. Leipzig.* **94**, 3–32 (1942)

[105] J.E. Marsden and M. McCracken, The Hopf bifurcation and its applications, *Appl. Math. Sci.* **19**, Springer, New York (1976)

[106] L.N. Howard, Nonlinear oscillations. In F.C. Hoppensteadt, ed., *Nonlinear Oscillations in Biology*, Lect. Appl. Math. **17**, AMS, Providence, RI (1979), pp. 1–67

[107] *LEDs & Laser Diodes*, http://www.jaycar.com.au/images_uploaded/ledlaser.pdf, Electus Distribution Reference Data Sheet (2001)

[108] F.T. Arecchi, W. Gadomski, and R. Meucci, Generation of chaotic dynamics by feedback on a laser, *Phys. Rev. A* **34**, 1617–1620 (1986)

[109] F.T. Arecchi, R. Meucci, and W. Gadomski, Laser dynamics with competing instabilities, *Phys. Rev. Lett.* **58**, 2205–2208 (1987)

[110] R. Meucci, R. McAllister, and R. Roy, Chaotic function generator: complex dynamics and its control in loss-modulated Nd:YAG laser, *Phys. Rev. E* **66**, 026216 (2002)

[111] P.-Y. Wang, A. Lapucci, R. Meucci, and F.T. Arecchi, Onset of subcritical bifurcation in a CO_2 laser with feedback, *Opt. Commun.* **80**, 42–46 (1990)

[112] A. Varone, A. Politi, and M. Ciofini, CO_2 laser dynamics with feedback, *Phys. Rev. A* **52**, 3176–3182 (1995)

[113] K. Ikeda, Multiple-valued stationary state and its instability of the transmitted light by a ring cavity system, *Opt. Commun.* **30**, 257–261 (1979)

[114] K. Ikeda, H. Daido, and O. Akimoto, Optical turbulence: chaotic behavior of transmitted light from a ring cavity, *Phys. Rev. Lett.* **45**, 709–712 (1980)

[115] K. Ikeda and O. Akimoto, Instability leading to periodic and chaotic self-pulsations in a bistable optical cavity, *Phys. Rev. Lett.* **48**, 617–620 (1982)

[116] J.-P. Goedgebuer, M. Li, and H. Porte, Demonstration of bistability and multistability in wavelength with a hybrid acousto-optic device, *IEEE J. Quantum Electron.* **QE-23**, 153–157 (1987)

[117] J.-P. Goedgebuer, L. Larger, H. Porte, and F. Delorme, Chaos in wavelength with feedback tunable laser diode, *Phys. Rev. E* **57**, 2795–2798 (1998)

[118] L. Larger, J.-P. Goedgebuer, and J.-M. Merolla, Chaotic oscillator in wavelength: a new setup for investigating differential difference equations describing nonlinear dynamics, *IEEE J. Quantum Electron.* **34**, 594–601 (1998)

[119] L. Laurent, M.W. Lee, J.-P. Goedgebuer, W. Elflein, and T. Erneux, Chaos in coherence modulation: bifurcations of an oscillator generating optical delay fluctuations, *J. Opt. Soc. Am. B* **18**, 1063–1068 (2001)

[120] T. Erneux, L. Larger, M.W. Lee, and J.P. Goedgebuer, Ikeda Hopf bifurcation revisited, *Phys. D* **194**, 49–64 (2004)

[121] S.M. Baer and T. Erneux, Singular Hopf bifurcation to relaxation oscillations, *SIAM J. Appl. Math.* **46**, 721–739 (1986)

[122] S.M. Baer and T. Erneux, Singular Hopf bifurcation to relaxation oscillations II, *SIAM J. Appl. Math.* **52**, 1651–1664 (1992)

[123] G. Kozyreff and T. Erneux, Singular Hopf bifurcation to strongly pulsating oscillations in lasers containing a saturable absorber, *Eur. J. Appl. Math.* **14**, 407–420 (2003)

[124] T. Erneux and G. Kozyreff, Nearly vertical Hopf bifurcation for a passively Q-switched microchip laser, *J. Stat. Phys.* **101**, 543–552 (2000)

[125] J.J. Swetits and A.M. Buoncristiani, Shilnikov instabilities in laser systems, *Phys. Rev. A* **38**, 5430–5432 (1988)

[126] L.-X. Chen, C.-F. Lei, Q.-S. Hu, J.-F. Li, and N.B. Abraham, Stability and dynamical behavior of a CO_2 laser with feedback control of the cavity length, *J. Opt. Soc. Am. B* **5**, 1160–1165 (1988)

[127] J.Y. Gao, H.Z. Zhang, X.Z. Guo, G.X. Jin, and N.B. Abraham, Small-signal response of a laser to cavity-length modulation: a diagnostic for dynamical models and parameter values, *Phys. Rev. A* **40**, 6339–6348 (1989)

[128] C. Bissel, A.A. Andronov and the development of Soviet control engineering, *IEEE Control Systems Magazine* **18**, 56–62 (1998)

[129] P.S. Hagan and D.S. Cohen, Josephson point-contact resonance responses, *J. Appl. Phys.* **50**, 5408 (1979)

[130] N.G. Basov, V.N. Morosov, and A.N. Oraevsky, Theory of pulsating conditions for lasers, *IEEE J. Quantum Electron.* **QE-2**, 542–548 (1966)

[131] V.N. Morozov, Theory of the dynamics of semiconductor injection lasers, *JOSA B* **5**, 909–915 (1988)

[132] G.-L. Oppo and A. Politi, Toda potentials in laser equations, *Z. Phys. B* **59**, 111–115 (1985)

[133] F.J. Bourland and R. Haberman, The modulated phase shift for strongly nonlinear, slowly varying, and weakly damped oscillators, *SIAM J. Appl. Math.* **48**, 737–748 (1988)

[134] F.J. Bourland, R. Haberman, and W.L. Kath, Averaging methods for the phase shift of arbitrarily perturbed strongly nonlinear oscillators with an application to capture, *SIAM J. Appl. Math.* **51**, 1150–1167 (1991)

[135] J.C. Celet, D. Dangoisse, P. Glorieux, G. Lythe, and T. Erneux, Slowly passing through resonance strongly depends on noise, *Phys. Rev. Lett.* **81**, 975–978 (1998)

[136] F.T. Arecchi, R. Meucci, G. Puccioni, and J. Tredicce, Experimental evidence of subharmonic bifurcations, multistability, and turbulence in a Q-switched gas laser, *Phys. Rev. Lett.* **49**, 1217–1220 (1982)

[137] D.V. Ivanov, Ya.I. Khanin, I.I. Matorin, and A.S. Pikovsky, Chaos in a solid-state laser with periodically modulated losses, *Phys. Lett. A* **89**, 229–230 (1982)

[138] T. Erneux, S.M. Baer, and P. Mandel, Subharmonic bifurcation and bistability of periodic solutions in a periodically modulated laser, *Phys. Rev. A* **35**, 1165–1171 (1987)

[139] I.B. Schwartz, Infinite primary saddle-node bifurcation in periodically forced systems, *Phys. Lett. A* **126**, 411–418 (1988)

[140] H.G. Solari, E. Eschenazi, R. Gilmore, and J.R. Tredicce, Influence of coexisting attractors on the dynamics of a laser system, *Opt. Commun.* **64**, 49–53 (1987)

[141] J.R. Tredicce, F.T. Arecchi, G.P. Puccioni, A. Poggy, and W. Gadomski, Dynamic behavior and onset of low-dimensional chaos in a modulated homogeneously broadened single-mode laser: experiments and theory, *Phys. Rev. A* **34**, 2073–2081 (1986)

[142] A.N. Pisarchik, Y.O. Barmenkov, and A.V. Kir'yanov, Experimental characterization of the bifurcation structure in an erbium-doped fiber laser with pump modulation, *IEEE J. Quantum Electron.* **39**, 1567–1571 (2003)

[143] T. Erneux and P. Glorieux, *Exercises, Problems and Supplementary Material*, http://www.ulb.ac.be/sciences/ont/

[144] I.B. Schwartz and T. Erneux, Subharmonic hysteresis and period doubling bifurcations for a periodically driven laser, *SIAM J. Appl. Math.* **54**, 1083–1100 (1994)

[145] K. Wiesenfeld and B. McNamara, Period-doubling systems as small-signal amplifiers, *Phys. Rev. Lett.* **55**, 13–16 (1985)

[146] K. Wiesenfeld and B. McNamara, Small-signal amplification in bifurcating dynamical systems, *Phys. Rev. A* **33**, 629–642 (1986)

[147] B. Derighetti, M. Ravani, R. Stoop, P.F. Meier, E. Brun, and R. Badii, Period-doubling lasers as small-signal detectors, *Phys. Rev. Lett.* **55**, 1746–1748 (1985)

[148] C. Lepers, J. Legrand, and P. Glorieux, Experimental observation of noise deamplification in a period-doubling laser, *Phys. Lett. A* **161**, 493–498 (1992)

[149] T.C. Newell, A. Gavrielides, V. Kovanis, D. Sukow, T. Erneux, and S.A. Glasgow, Unfolding of the period-two bifurcation in a fiber laser pumped with two modulation tones, *Phys. Rev. E* **56**, 7223–7231 (1997)

[150] E. Brun, B. Derighetti, D. Meier, R. Holzner, and M. Ravani, Observation of order and chaos in a nuclear spin-flip laser, *J. Opt. Soc. Am. B* **2**, 156–167 (1985)

[151] R. Corbalán, J. Cortit, A.N. Pisarchik, V.N. Chizhevsky, and R. Vilaseca, Investigation of a CO_2 laser response to loss perturbation near period doubling, *Phys. Rev. A* **51**, 663–668 (1995)

[152] G. Ahlers, M.C. Cross, P.C. Hohenberg, and S. Safran, The amplitude equation near the convective threshold: application to time-dependent heating experiments, *J. Fluid. Mech.* **110**, 297–334 (1981)

[153] C.W. Meyer, G. Ahlers, and D.S. Cannell, Initial stages of pattern formation in Rayleigh-Bénard convection, *Phys. Rev. Lett.* **59**, 1577–1580 (1987)

[154] T. Erneux and J.-P. Laplante, Jump transition due to a time-dependent bifurcation parameter in the bistable ioadate-arsenous acid reaction, *J. Chem. Phys.* **90**, 6129–6134 (1989)

[155] P. Strizhak and M. Menzinger, Slow passage through a supercritical Hopf bifurcation: time-delayed response in the Belousov-Zhabotinsky reaction in a batch reactor, *J. Chem. Phys.* **105**, 10 905–10 910 (1996)

[156] M.T.M. Koper and B.D. Aguda, Experimental demonstration of delay and memory effects in the bifurcations of nickel electrodissolution, *Phys. Rev. E* **54**, 960–963 (1996)

[157] A.H. Nayfeh and D.T. Mook, *Nonlinear Oscillations*, John Wiley & Sons, New York (1979)

[158] Q. Ding and A.Y.T. Leung, Non-stationary processes of rotor/bearing system in bifurcations, *J. Sound Vib.* **268**, 33–48 (2003)

[159] Y. Terumichi, M. Ohtsuka, M. Yoshizawa, Y. Fukawa, and Y. Tsujioka, Non-stationary vibrations of a string with time-varying length and a mass-spring system attached at the lower end, *Nonlinear Dynamics* **12**, 39–55 (1997)

[160] R. Haberman, Slowly-varying jump and transition phenomena associated with algebraic bifurcation problems, *SIAM J. Appl. Math.* **37**, 69–106 (1979)

[161] E. Benoît, ed., *Dynamic Bifurcations* (Proceedings, Luminy, France, 1990), Lecture Notes in Mathematics, vol. **1493**, Springer-Verlag, Berlin (1991)

[162] A. Goldbeter and L.A. Segel, Control of developmental transitions in the cyclic AMP signalling system of *Dictyostelium discoideum*, *Differentiation* **17**, 127–135 (1980)

[163] B. Novák and J.J. Tyson, Modelling the controls of the eukaryotic cell cycle, *Biochem. Soc. Trans.* **31**, 1526–1529 (2003)

[164] J.R. Tredicce, G.L. Lippi, P. Mandel, B. Charasse, A. Chevalier, and B. Picqué, Critical slowing down at a bifurcation, *Am. J. Phys.* **72**, 799–809 (2004)

[165] P. Jung, G. Gray, and R. Roy, Scaling law for dynamical hysteresis, *Phys. Rev. Lett.* **65**, 1873–1876 (1990)

[166] A. Hohl, H.J.C. van der Linden, R. Roy, G. Goldsztein, F. Broner, and S.H. Strogatz, Scaling laws for dynamical hysteresis in a multidimensional laser system, *Phys. Rev. Lett.* **74**, 2220–2223 (1995)

[167] P. Glorieux and D. Dangoisse, Dynamical behavior of a laser containing a saturable absorber, *IEEE J. Quantum Electron.* **QE-21**, 1486–1490 (1985)

[168] N.E. Fettouhi, B. Ségard, and J. Zemmouri, Scaling of hysteresis in a multi-dimensional all-optical bistable system, *Eur. Phys. J. D* **6**, 425–429 (1999)

[169] P. Mandel and T. Erneux, Laser-Lorenz equations with a time-dependent parameter, *Phys. Rev. Lett.* **53**, 001818 (1984)

[170] W. Scharpf, Delayed bifurcation at threshold in argon-ion lasers, Ph.D. Thesis, Drexel University (1988)

[171] W. Scharpf, M. Squicciarini, D. Bromley, C. Green, J.R. Tredicce, and L.M. Narducci, Experimental observation of a delayed bifurcation at the threshold of an argon laser, *Opt. Commun.* **63**, 344–348 (1987)

[172] D. Bromley, E.J. D'Angelo, H. Grassi, C. Mathis, J.R. Tredicce, and S. Balle, Anticipation of the switch-off and delay of the switch-on of a laser with a swept parameter, *Opt. Commun.* **99**, 65–70 (1993)

[173] D. Dangoisse, P. Glorieux, and D. Hennequin, Chaos in a CO_2 laser with modulated parameters: experiments and numerical simulations, *Phys. Rev. A* **36**, 4775–4791 (1987)

[174] R. Kapral and P. Mandel, Bifurcation structure of the nonautonomous quadratic map, *Phys. Rev. A* **32**, 1076–1081 (1985)

[175] S.M. Baer, T. Erneux, and J. Rinzel, The slow passage through a Hopf bifurcation: delay, memory effects, and resonance, *SIAM J. Appl. Math.* **49**, 55–71 (1989)

[176] J. Sulc, H. Jelinkova, K. Nejezchleb, and V. Skoda, Nd:YAG/V:YAG microchip laser operating at 1338 nm, *Laser Phys. Lett.* **1**, 1–6 (2005)

[177] K. Shimoda, In J. Fox, ed., *Optical Masers*, Polytechnic Press, Brooklyn, NY (1963), p. 95

[178] V.N. Lisitsyn and V.P. Chebotaev, Hysteresis and "hard" excitation in a gas laser, *JETP Lett.* **7**, 1–3 (1968)

[179] Yu.V. Brzhazovskii, L.S. Vasilenko, S.G. Rautian, G.S. Popova, and V.P. Chebotaev, Theoretical and experimental investigation of radiation pulsation from a CO_2 laser with a nonlinear absorbing cell, *Sov. Phys. JETP* **34**, 265–270 (1972)

[180] E. Arimondo, P. Bootz, P. Glorieux, and E. Menchi, Pulse shape and phase diagram in the passive Q switching of CO_2 lasers, *J. Opt. Soc. Am. B* **2**, 193–201 (1985)

[181] H. Kawaguchi, *Bistabilities and Nonlinearities in Laser Diodes*, Artech House, Boston (1994)

[182] M. Yamada, A theoretical analysis of self-sustained pulsation phenomena in narrow-stripe semiconductor lasers, *IEEE J. Quantum Electron.* **29**, 1330–1336 (1993)

[183] D.R. Jones, P. Rees, I. Pierce, and H.D. Summers, Theoretical optimization of self-pulsating 650-nm-wavelength AlGaInP laser diodes, *IEEE J. Sel. Top. Quantum Electron.* **5**, 740–744 (1999)

[184] P. Peterson, A. Gavrielides, M.P. Sharma, and T. Erneux, Dynamics of passively Q-switched microchip lasers, *IEEE J. Quantum Electron.* **35**, 1247–1256 (1999)

[185] T. Erneux, P. Peterson, and A. Gavrielides, The pulse shape of a passively Q-switched microchip laser, *Eur. Phys. J. D* **10**, 423–431 (2000)

[186] T.W. Carr and T. Erneux, Dimensionless rate equations and simple conditions for self-pulsing in laser diodes, *IEEE J. Quantum Electron.* **37**, 1171–1177 (2001)

[187] S. Ruschin and S.H. Bauer, Bistability, hysteresis and critical behavior of a CO_2 laser, with SF_6 intracavity as a saturable absorber, *Chem. Phys. Lett.* **66**, 100–103 (1979)

[188] D. Dangoisse, A. Bekkali, F. Papoff, and P. Glorieux, Shilnikov dynamics in a passive Q-switching laser, *Europhys. Lett.* **6**, 335–340 (1988)

[189] A.J. De Maria, D.A. Stetser, and H. Heynau, Self mode-locking of lasers with saturable absorbers, *Appl. Phys. Lett.* **8**, 174–176 (1966)

[190] Y. Shimony, Z. Burshtein, A.B.-A. Baranga, Y. Kalisky, and M. Strauss, Repetitive Q-switching of a CW Nd:YAG laser using Cr^{4+}: YAG saturable absorbers, *IEEE J. Quantum Electron.* **32**, 305–310 (1996)

[191] G.J. Spühler, R. Paschotta, R. Fluck, B. Braun, M. Moser, G. Zhang, E. Gini, and U. Keller, Experimentally confirmed design guidelines for passively Q switched microchip lasers using semiconductor saturable absorbers, *J. Opt. Soc. Am. B* **16**, 376–388 (1999)

[192] T. Erneux, Q-switching bifurcation in a laser with a saturable absorber, *J. Opt. Soc. Am. B* **5**, 1063–1069 (1988)

[193] A. Szabo and R.A. Stein, Theory of laser giant pulsing by a saturable absorber, *J. Appl. Phys.* **36**, 1562–1566 (1965)

[194] A. Jacques and P. Glorieux, Observation of bistability in a CO_2 laser exhibiting passive Q-switching, *Opt. Commun.* **40**, 455–460 (1982)

[195] E. Arimondo and P. Glorieux, The observation of transients in a CO_2 or N_2O ir laser by modulation of an intracavity absorber, *Appl. Phys. Lett.* **33**, 49–51 (1978)

[196] F. de Tomasi, D. Hennequin, B. Zambon, and E. Arimondo, Instabilities and chaos in an infrared laser with saturable absorber: experiments and vibrorotational model, *J. Opt. Soc. Am. B* **6**, 45–57 (1989)

[197] F.-L. Hong, M. Tachikawa, R. Oda, and T. Shimizu, Chaotic passive Q-switching pulsation in a N_2O laser with a saturable absorber, *J. Opt. Soc. Am. B* **6**, 1378–1382 (1989)

[198] R. Gilmore and M. Lefranc, *The Topology of Chaos: Alice in Stretch and Squeezeland*, Wiley, New York (2002)

[199] F. Papoff, A. Fioretti, E. Arimondo, G.B. Mindlin, H.G. Solari, and R. Gilmore, Structure of chaos in the laser with a saturable absorber, *Phys. Rev. Lett.* **68**, 1128–1131 (1992)

[200] E. Schöll and H.G. Schuster, eds., *Handbook of Chaos Control*, 2nd edn, Wiley-VCH, Weinheim, Germany (2007)

[201] J. Gu, F. Zhou, W. Xie, S.C. Tam, and Y.L. Lam, Passive Q-switching of a diode-pumped Nd:YAG laser with a GaAs output coupler, *Opt. Commun.* **165**, 245–249 (1999)

[202] L.E. Erikson and A. Szabo, Spectral narrowing of dye laser output by injection of monochromatic radiation into the laser cavity, *Appl. Phys. Lett.* **18**, 433–435 (1971)

[203] P. Gallion, H. Nakajima, G. Debarge, and C. Chabran, Contribution of spontaneous emission to the linewidth of an injected-locked semiconductor laser, *Electron Lett.* **21**, 626–628 (1995)

[204] G. Stéphan, Spectral properties of an injected laser, *Phys. Rev. A* **58**, 2467–2471 (1998)

[205] S. Kobayashi and T. Kimura, Coherence of injection phase-locked AlGaAs semiconductor laser, *Electron. Lett.* **16**, 668–670 (1980)

[206] P. Spano, S. Piazzolla, and M. Tamburrini, Frequency and intensity noise in injected-locked semiconductor lasers: theory and experiments, *IEEE J. Quantum Electron.* **QE-22**, 427–435 (1986)

[207] N. Schunk and K. Petermann, Noise analysis of injection-locked semiconductor injection lasers, *IEEE J. Quantum Electron.* **QE-22**, 642–650 (1986)

[208] O. Lidoyne, P.B. Gallion, C. Chabran, and G. Debarge, Locking range, phase noise and power spectrum of an injection-locked semiconductor laser, *IEE Proc. J: Optoelectronics* **137**, 147–153 (1990)

[209] K. Iwashita and K. Nakagawa, Suppression of mode partition by laser diode light injection, *IEEE J. Quantum Electron.* **QE-18**, 1669–1674 (1982)

[210] A. Furuzawa, Amplitude squeezing of a semi-conductor laser with light injection, *Opt. Lett.* **21**, 2014–2016 (1996)

[211] J. Genest, M. Chamberland, P. Tremblay, and M. Têtu, Microwave signals generated by optical heterodyne between injection-locked semiconductor lasers, *IEEE J. Quantum Electron.* **33**, 989–998 (1997)

[212] Y.K. Seo, A. Kim, J.T. Kim, and W.Y. Choi, Optical generation of microwave signals using a directly modulated semiconductor laser under modulated light injection, *Microw. Opt. Tech. Lett.* **30**, 369–370 (2001)

[213] G.D. VanWiggeren and R. Roy, Communication with chaotic lasers, *Science* **279**, 1198–1200 (1998)

[214] Y. Liu, H.F. Chen, J.M. Liu, P. Davis, and T. Aida, Synchronization of optical-feedback-induced chaos in semiconductor lasers by optical injection, *Phys. Rev. A* **63**, 031802 (2001)

[215] A. Murakami and J. Ohtsubo, Synchronization of feedback-induced chaos in semiconductor lasers by optical injection, *Phys. Rev. A* **65**, 033826 (2002)

[216] S. Wieczorek, B. Krauskopf, T.B. Simpson, and D. Lenstra, The dynamical complexity of optically injected semiconductor lasers, *Phys. Rep.* **416**, 1–128 (2005)

[217] M. Lax, Classical noise vs. noise in self-substained oscillators, *Phys. Rev.* **160**, 290–307 (1967)

[218] H. Haug and H. Haken, Theory of noise in semiconductor laser emission, *Z. Phys. A* **204**, 262–275 (1967)

[219] T.B. Simpson, Phase-locked microwave-frequency modulations in optically-injected laser diodes, *Opt. Commun.* **170**, 93–98 (1999)

[220] D. Lenstra, G.H.M. van Tartwijk, W.A. van der Graaf, and P.C. De Jagher, Multi-wave mixing dynamics in a diode laser. In *Chaos in Optics, Proc. SPIE* **2039**, 11–22 (1993)

[221] F. Mogensen, H. Olesen, and G. Jacobsen, Locking conditions and stability properties for a semiconductor laser with external optical injection, *IEEE J. Quantum Electron.* **QE-21**, 784–793 (1985)

[222] I. Petitbon, P. Gallion, G. Debarge, and C. Chabran, Locking bandwith and relaxation oscillations of an injected-locked semiconductor laser, *IEEE J. Quantum Electron.* **QE-24**, 148–154 (1988)

[223] E.K. Lee, H.S. Pang, J.D. Park, and H. Lee, Bistability and chaos in an injection-locked semiconductor laser, *Phys. Rev. A* **47**, 736–739 (1993)

[224] V. Kovanis, A. Gavrielides, T.B. Simpson, and J.M. Liu, Instabilities and chaos in optically injected semiconductor lasers, *Appl. Phys. Lett.* **67**, 2780–2782 (1995)

[225] T.B. Simpson, J.M. Liu, K.F. Huang, and K. Tai, Nonlinear dynamics induced by external optical injection in semiconductor lasers, *J. Opt. B Quantum Semiclassical Opt.* **9**, 765–784 (1997)

[226] S. Valling, T. Fordell, and A.M. Lindberg, Maps of the dynamics of an optically injected solid-state laser, *Phys. Rev. A* **72**, 033810 (2005)

[227] S. Blin, P. Besnard, O. Vaudel, and S. LaRochelle, Optical injection in semiconductor or fiber laser: a comparison, the influence of coherence. In D. Lenstra, G. Morthier, T. Erneux, and M. Pessa, eds., *Semiconductor Lasers and Laser Dynamics*, SPIE Europe Proc. **5452**, 534–545 (2004)

[228] S. Wieczorek, T.B. Simpson, B. Krauskopf, and D. Lenstra, Global quantitative predictions of complex laser dynamics, *Phys. Rev. E* **65**, 045207(R) (2002)

[229] A. Gavrielides, V. Kovanis, and T. Erneux, Analytical stability boundaries for a semiconductor laser subject to optical injection, *Opt. Commun.* **136**, 253–256 (1997)

[230] B. Simpson, J.M. Liu, A. Gavrielides, V. Kovanis, and P.M. Alsing, Period-doubling route to chaos in a semiconductor laser subject to external optical injection, *Appl. Phys. Lett.* **64**, 3539 (1994)

[231] T.B. Simpson, J.M. Liu, A. Gavrielides, V. Kovanis, and P.M. Alsing, Period-doubling cascades and chaos in a semiconductor laser with optical injection, *Phys. Rev. A* **51**, 4181–4185 (1995)

[232] S.-C. Chan, S.-K. Hwang, and J.-M. Liu, Period-one oscillation for photonic microwave transmission using an optically injected semiconductor laser, *Opt. Express* **15**, 14 921–14 935 (2007)

[233] T. Erneux, V. Kovanis, A. Gavrielides, and P.M. Alsing, Mechanism for period-doubling bifurcation in a semiconductor laser subject to optical injection, *Phys. Rev. A* **53**, 4372–4380 (1996)

[234] S. Valling, B. Krauskopf, T. Fordell, and A. Lindberg, Experimental bifurcation diagram of a solid state laser with optical injection, *Opt. Commun.* **271**, 532–542 (2007)

[235] T. Fordell, S. Valling, and A.M. Lindberg, Modulation and the linewidth enhancement factor of a diode-pumped Nd:YVO$_4$ laser, *Opt. Lett.* **30**, 3036–3038 (2005)

[236] P.G.R. King and G.J. Steward, Metrology with an optical maser, *New Sci.* **17**, 180–182 (1963)

[237] D.E.T.F. Ashby and D.F. Jephcott, Measurement of plasma density using a gas laser as an infrared interferometer, *Appl. Phys. Lett.* **3**, 13 (1963)

[238] C. Lu, J. Wang, and K. Deng, Imaging and profiling surface microstructures with noninterferometric confocal laser feedback, *Appl. Phys. Lett.* **66**, 2022–2024 (1995)

[239] *Encyclopedia of Laser Physics and Technology*, http://www.rp-photonics.com/beam_divergence.html

[240] A. Gavrielides and D.W. Sukow, Experimental observations. In D. Kane and A. Shore, eds., *Unlocking Dynamical Diversity: Optical Feedback Effects on Semiconductor Lasers*, Wiley InterScience, Hoboken, NJ (2005), Chapter 4

[241] J. Houlihan, G. Huyet, and J. McInerney, Dynamics of a semiconductor laser with incoherent optical feedback, *Opt. Commun.* **199**, 175–179 (2001)

[242] M. Peil, I. Fischer, W. Elsässer, S. Bakic, N. Damaschke, C. Tropea, S. Stry, and J. Sacher, Rainbow refractometry with a tailored incoherent semiconductor laser source, *Appl. Phys. Lett.* **89**, 091106 (2006)

[243] R. Lang and K. Kobayashi, External optical feedback effects on semiconductor injection laser properties, *IEEE J. Quantum Electron.* **QE-16**, 347–355 (1980)

[244] T. Heil, I. Fischer, W. Elsässer, and A. Gavrielides, Dynamics of semiconductor lasers subject to delayed optical feedback: the short cavity regime, *Phys. Rev. Lett.* **87**, 243901 (2001)

[245] T. Erneux, *Applied Delay Differential Equations*, Springer, in press (2009)

[246] T. Heil, I. Fischer, and W. Elsässer, Stabilization of feedback induced instabilities in semiconductor lasers, *J. Opt. B: Quantum Semiclassical Opt.* **2**, 413–420 (2000)

[247] A. Hohl and A. Gavrielides, Bifurcation cascade in a semiconductor laser subject to optical feedback, *Phys. Rev. Lett.* **82**, 1148–1151 (1999)

[248] N. Schunk and K. Petermann, Numerical analysis of the feedback regimes for a single-mode semiconductor laser with external feedback, *IEEE J. Quantum Electron.* **QE-24**, 1242–1247 (1988)

[249] J. Mork and B. Tromborg, The mechanism of mode selections for an external cavity laser, *IEEE Photon. Tech. Lett.* **2**, 21–23 (1990)

[250] J. Mork, B. Tromborg, and P.L. Christiansen, Bistability and low frequency fluctuations in semiconductor lasers with optical feedback: a theoretical analysis, *IEEE J. Quantum Electron.* **QE-24**, 123–133 (1988)

[251] T. Sano, Antimode dynamics and chaotic itinerancy in the coherence collapse of semiconductor lasers with optical feedback, *Phys. Rev. A* **50**, 2719–2726 (1994)

[252] G.H.M. van Tartwijk, A.M. Levine, and D. Lenstra, Sisyphus effect in semiconductor lasers with optical feedback, *IEEE J. Sel. Top. Quantum Electronics* **1**, 466–472 (1995)

[253] H. Olesen, J.H. Osmundsen, and B. Tromborg, Nonlinear dynamics and spectral behavior for an external cavity laser, *IEEE J. Quantum Electron.* **QE-22**, 762–773 (1986)

[254] T. Erneux, Asymptotic methods applied to semiconductor laser model. In R. Binder, P. Blood, and M. Osinski, eds., *Physics and Simulations of Optoelectronic Devices VIII*, *Proc. SPIE* **3944**, 588–601 (2000)

[255] A.M. Levine, G.H.M. van Tartwijk, D. Lenstra, and T. Erneux, Diode lasers with optical feedback: stability of the maximum gain mode, *Phys. Rev. A* **52**, R3436–R3439 (1995)

[256] T. Heil, I. Fischer, and W. Elsässer, Coexistence of low-frequency fluctuations and stable emission on a single high-gain mode in semiconductor lasers with external optical feedback, *Phys. Rev. A* **58**, R2672–R2675 (1998)

[257] T. Heil, I. Fischer, and W. Elsässer, Influence of amplitude–phase coupling on the dynamics of semiconductor lasers subject to optical feedback, *Phys. Rev. A* **60**, 634–641 (1999)

[258] R.L. Davidchack, Y.-C. Lai, A. Gavrielides, and V. Kovanis, Dynamical origin of low frequency fluctuations in external cavity semiconductor lasers, *Phys. Lett. A* **267**, 350–356 (2000)

[259] A.A. Tager and B.B. Elenkrig, Stability regimes and high-frequency modulation of laser diodes with short external cavity, *IEEE J. Quantum Electron.* **29**, 2886–2890 (1993)

[260] A.A. Tager and K. Petermann, High-frequency oscillations and self-mode locking in short external cavity laser diodes, *IEEE J. Quantum Electron.* **30**, 1553–1561 (1994)

[261] T. Erneux, F. Rogister, A. Gavrielides, and V. Kovanis, Bifurcation to mixed external cavity mode solutions for semiconductor lasers subject to optical feedback, *Opt. Commun.* **183**, 467–477 (2000)

[262] D. Pieroux, T. Erneux, B. Haegeman, K. Engelborghs, and D. Roose, Bridges of periodic solutions and tori in semiconductor lasers subject to delay, *Phys. Rev. Lett.* **87**, 193901 (2001)

[263] M. Wolfrum and D. Turaev, Instabilities of lasers with moderately delayed optical feedback, *Opt. Commun.* **212**, 127–138 (2002)

[264] J. Sieber, Numerical bifurcation analysis for multi-section semiconductor lasers, *SIAM J. Appl. Dyn. Sys.* **1**, 248–270 (2002)

[265] S. Bauer, O. Brox, J. Kreissl, G. Sahin, and B. Sartorius, Optical microwave source, *Electron. Lett.* **38**, 334–335 (2002)

[266] O. Brox, S. Bauer, M. Radziunas, M. Wolfrum, J. Sieber, J. Kreissl, B. Sarorius, and H.-J. Wünsche, High-frequency pulsations in DFB lasers with amplified feedback, *IEEE J. Quantum Electron.* **39**, 1381–1387 (2003)

[267] S. Bauer, O. Brox, J. Kreissl, B. Sartorius, M. Radziunas, J. Sieber, H.-J. Wünsche, and F. Henneberger, *Phys. Rev. E* **69**, 016206 (2004)

[268] O. Ushakov, S. Bauer, O. Brox, H.-J. Wünsche, and F. Henneberger, Self-organization in semiconductor lasers with ultrashort optical feedback, *Phys. Rev. Lett.* **92**, 043902 (2004)

[269] T. Erneux, A. Gavrielides, and M. Sciamanna, Stable microwave oscillations due to external-cavity-mode beating in laser diodes subject to optical feedback, *Phys. Rev. A* **66**, 033809 (2002)

[270] E. Lacot, R. Day, and F. Stoeckel, Laser optical feedback tomography, *Opt. Lett.* **24**, 744–746 (1999)

[271] E. Lacot, R. Day, J. Pinel, and F. Stoeckel, Laser relaxation-oscillation frequency imaging, *Opt. Lett.* **26**, 1483–1485 (2001)

[272] E. Lacot, R. Day, and F. Stoeckel, Coherent laser detection by frequency-shifted optical feedback, *Phys. Rev. A* **64**, 043815 (2001)

[273] E. Lacot, O. Hugon, and F. Stoeckel, Hopf amplification of frequency-shifted optical feedback, *Phys. Rev. A* **67**, 053806 (2003)

[274] Y.C. Kouomou, P. Colet, L. Larger, and N. Gastaud, Chaotic breathers in delayed electro-optical systems, *Phys. Rev. Lett.* **95**, 203903 (2005)

[275] M. Peil, M. Jacquot, Y. Kouomou Chembo, L. Larger, and T. Erneux, Routes to chaos and multiple time scale dynamics in broadband bandpass nonlinear delay electro-optic oscillators, *Phys. Rev. E* **79**, 026208 (2009)

[276] P. Saboureau, J.P. Foing, and P. Schanne, Injection-locked semiconductor lasers with delayed optoelectronic feedback, *IEEE J. Quantum Electron.* **33**, 1582–1591 (1997)

[277] D. Pieroux, T. Erneux, and K. Otsuka, Minimal model of a class-B laser with delayed feedback: cascading branching of periodic solutions and period-doubling bifurcation, *Phys. Rev. A* **50**, 1822–1829 (1994)

[278] C. Beta, M. Bertram, A.S. Mikhailov, H.H. Rotermund, and G. Ertl, Controlling turbulence in a surface chemical reaction by time-delay autosynchronization, *Phys. Rev. E* **67**, 046224 (2003)

[279] H.-J. Wünsche, S. Bauer, J. Kreissl, O. Ushakov, N. Korneyev, F. Henneberger, E. Wille, H. Erzgräber, M. Peil, W. Elsässer, and I. Fischer, Synchronization of delay-coupled oscillators: a study of semiconductor lasers, *Phys. Rev. Lett.* **94**, 163901 (2005)

[280] A.J. Cantor, P.K. Cheo, M.C. Foster, and L.A. Newman, Application of sub-millimeter wave lasers to high voltage cable inspection, *IEEE J. Quantum Electron.* **QE-17**, 477–489 (1981)

[281] B. Marx, Terahertz screeners are compact, *Laser Focus World* **40**, (Nov. 2004)

[282] E. Lorenz, Deterministic non periodic flows, *J. Atmos. Sci.* **10**, 130–141 (1963)

[283] C.O. Weiss and W. Klische, On observability of Lorenz instabilities in lasers, *Opt. Commun.* **51**, 47–48 (1984)

[284] P. Glorieux and D. Dangoisse, Transients and instabilities in FIR lasers. In P.K. Cheo, ed., *Handbook of Molecular Lasers*, Marcel Dekker, New York (1987), pp. 573–644

[285] A.Z. Grasyuk and A.N. Oraevsky, Transient processes in molecular oscillators, *Radio Eng. Electron. Phys.* **9**, 424–428 (1964)

[286] E.R. Buley and F.W. Cummings, Dynamics of a system of N atoms interacting with a radiation field, *Phys. Rev.* **134**, A1454–A1460 (1964)

[287] E.H.M. Hogenboom, W. Klische, C.O. Weiss, and A. Godone, Instabilities of a homogeneously broadened laser, *Phys. Rev. Lett.* **55**, 2571–2574 (1985)

[288] C.O. Weiss and J. Brock, Evidence of Lorenz-type chaos in a laser, *Phys. Rev. Lett.* **57**, 2804–2806 (1986)

[289] C. Sparrow, The Lorenz equations: bifurcations, chaos, and strange attractors, *Appl. Math. Sci.* **41**, Springer, New York (1982)

[290] U. Hübner, N.B. Abraham, and C.O. Weiss, Dimensions and entropies of chaotic intensity pulsations in a single-mode far-infrared NH_3 laser, *Phys. Rev. A* **40**, 6354–6365 (1989)

[291] D. Seligson, M. Ducloy, J.R. Rios Leite, A. Sanches, and M.S. Feld, Quantum mechanical features of optically pumped CW FIR lasers, *IEEE J. Quantum Electron.* **QE-13**, 468–472 (1977)

[292] C.O. Weiss, N.B. Abraham, and U. Hübner, Homoclinic and heteroclinic chaos in a single-mode laser, *Phys. Rev. Lett.* **61**, 1587–1590 (1988)

[293] C.O. Weiss, R. Vilaseca, N.B. Abraham, R. Corbalán, E. Roldán, G.J. de Valcárcel, J. Pujol, U. Hübner, and D.Y. Tang, Models, predictions, and experimental measurements of far-infrared NH$_3$ laser dynamics and comparisons with the Lorenz-Haken model, *Appl. Phys. B* **61**, 223–242 (1995)

[294] G. Baldacchini, G. Buffa, and O. Tarrini, A review of experiments and theory for collisional line shape effects in the rotovibrational ammonia spectrum, *Nuovo Cimento D* **13**, 719–733 (1991)

[295] N.B. Abraham, D. Dangoisse, P. Glorieux, and P. Mandel, Observation of undamped pulsations in a low-pressure, far-infrared laser and comparison with a simple theoretical model, *J. Opt. Soc. Am. B* **2**, 23–34 (1985)

[296] V.S. Idiatulin and A.V. Uspenskii, The possibility of the existence of a pulsating mechanism related to inhomogeneous broadening of the line of the working transition, *Rad. Eng. Electron. Phys.* **18**, 422–425 (1973)

[297] *Encyclopedia of Laser Physics and Technology*, http://www.rp-photonics.com/optical_parametric_oscillators.html

[298] P.A. Franken, A.E. Hill, C.W. Peters, and G. Weinreich, Generation of optical harmonics, *Phys. Rev. Lett.* **7**, 118–119 (1961)

[299] C. Fabre, Classical and quantum aspects of $\chi^{(2)}$ cw interaction in a cavity. In A. Boardman, ed., *Advanced Photonics with Second-Order Optically Nonlinear Processes*, Proceedings of the NATO Advanced Study Institute, Sozopol, Bulgaria, Kluwer Academic Publishers (1998), p. 273

[300] K.J. McNeill, P.D. Drummond, and D.F. Walls, Self-pulsing in second-harmonic generation, *Opt. Commun.* **27**, 292–294 (1978)

[301] P.D. Drummond, K.J. McNeill, and D.F. Walls, Non-equilibrium transitions in sub-second harmonic generation, I: Semiclassical theory, *J. Mod. Opt.* **27**, 321–335 (1980)

[302] P.D. Drummond, K.J. McNeill, and D.F. Walls, Non-equilibrium transitions in sub-second harmonic generation, II: Quantum theory, *J. Mod. Opt.* **28**, 211–225 (1981)

[303] L.A. Lugiato, C. Oldano, C. Fabre, E. Giacobino, and R.J. Horowicz, Bistability, self-pulsing and chaos in optical parametric oscillators, *Nuovo Cimento D* **10**, 959–977 (1988)

[304] A. Amon and M. Lefranc, Mode hopping strongly affects observability of dynamical instabilities in optical parametric oscillators, *Eur. Phys. J. D* **44**, 1434–6060 (2007)

[305] P. Suret, D. Derozier, M. Lefranc, J. Zemmouri, and S. Bielawski, Self-pulsing instabilities in an optical parametric oscillator: experimental observation and modeling of the mechanism, *Phys. Rev. A* **61**, 021805 (2000)

[306] P. Suret, M. Lefranc, D. Derozier, J. Zemmouri, and S. Bielawski, Periodic mode hopping induced by thermo-optic effects in continuous-wave optical parametric oscillators, *Opt. Lett.* **26**, 1415–1417 (2001)

[307] D.H. Lee, M.E. Klein, and K.-J. Boller, Intensity noise of pump enhanced continuous-wave optical parametric oscillators, *Appl. Phys. B* **66**, 747–753 (1998)

[308] C. Richy, K.I. Petsas, E. Giacobino, C. Fabre, and L. Lugiato, Observation of bistability and delayed bifurcation in a triply resonant optical parametric oscillator, *J. Opt. Soc. Am. B* **12**, 456–461 (1995)

[309] X. Wei, N. Wu, X. Liu, and S. Li, Theoretical study of relaxation oscillation in triply resonant OPOs, *Appl. Phys. B* **79**, 841–844 (2004)

[310] G. Turnbull, M.H. Dunn, and M. Ebrahimzadeh, Continuous-wave, intra-cavity optical parametric oscillators: analysis of power characteristics, *Appl. Phys. B* **66**, 701–710 (1998)

[311] S. Schiller and R. Byer, Quadruply resonant optical parametric oscillation in a monolithic total-internal reflection resonator, *J. Opt. Soc. Am. B* **10**, 1696–1707 (1993)

[312] S. Schiller, G. Breitenbach, R. Paschotta, and J. Mlynek, Subharmonic-pumped continuous-wave parametric oscillator, *Appl. Phys. Lett.* **68**, 3374–3376 (1996)

[313] M.A.M. Marte, Sub-Poissonian twin beams via competing nonlinearities, *Phys. Rev. Lett.* **74**, 4815–4818 (1995)

[314] M.A.M. Marte, Competing nonlinearities, *Phys. Rev. A* **49**, R3166–R3169 (1994)

[315] M.A.M. Marte, Nonlinear dynamics and quantum noise for competing $\chi^{(2)}$ nonlinearities, *J. Opt. Soc. Am. B* **12**, 2296–2303 (1995)

[316] P. Lodahl and M. Saffman, Pattern formation in singly resonant second-harmonic generation with competing parametric oscillation, *Phys. Rev. A* **60**, 3251–3261 (1999)

[317] T. Baer, Large-amplitude fluctuations due to longitudinal mode coupling in diode-pumped intracavity-doubled Nd:YAG lasers, *J. Opt. Soc. Am. B* **3**, 1175–1180 (1986)

[318] K. Wiesenfeld, C. Bracikowski, G. James, and R. Roy, Observation of antiphase states in a multimode laser, *Phys. Rev. Lett.* **65**, 1749–1752 (1990)

[319] C. Bracikowski and R. Roy, Energy sharing in a chaotic multimode laser, *Phys. Rev. A* **43**, 6455–6457 (1991)

[320] K. Otsuka, P. Mandel, S. Bielawski, D. Derozier, and P. Glorieux, Alternate time scale in multimode lasers, *Phys. Rev. A* **46**, 1692–1695 (1992)

[321] S. Bielawski, D. Derozier, and P. Glorieux, Antiphase dynamics and polarization effects in the Nd-doped fiber laser, *Phys. Rev. A* **46**, 2811–2811 (1992)

[322] K.Y. Tsang, R.E. Mirollo, S.H. Strogatz, and K. Wiesenfeld, Dynamics of a globally coupled oscillator array, *Physica D* **48**, 102–112 (1991)

[323] D.G. Aronson, M. Golubitsky, and J. Mallet-Paret, Ponies on a merry-go-round in large arrays of Josephson junctions, *Nonlinearity* **4**, 903–910 (1991)

[324] M. Silber, L. Fabiny, and K. Wiesenfeld, Stability results for in-phase and splay-phase states of solid-state laser arrays, *J. Opt. Soc. Am. B* **10**, 1121–(1993)

[325] W.-J. Rappel, Dynamics of a globally coupled laser model, *Phys. Rev. E* **49**, 2750–2755 (1994)

[326] T. Erneux and P. Mandel, Minimal equations for antiphase dynamics in multimode lasers, *Phys. Rev. A* **52**, 4137–4144 (1995)

[327] A. Jeffrey and D. Zwillinger, eds., *Table of Integrals, Series, and Products*, 7th edn, Academic Press (2007)

[328] E. Lacot, F. Stoeckel, and M. Chenevier, Dynamics of an erbium-doped fiber laser, *Phys. Rev. A* **49**, 3997–4008 (1994)

[329] E. Cabrera, O.G. Calderon, and J.M. Guerra, Experimental evidence of antiphase population dynamics in lasers, *Phys. Rev. A* **72**, 043824 (2005)

Index

Adiabatic elimination, 61
Adler's equation
  close to locking, 81
  critical slowing down, 65
  driven
    near resonance, 117
    strongly, 73
    weakly modulated, 115
    weakly nonlinear, 112
  exact solution, 80
  locking range, 65
  optically injected laser, 218
  period, 65
  phase-locking or drift, 64
  third order, 234
  two coupled lasers, 78
  with delay, 269
Adler, Robert, 59
Antiphase dynamics
  fiber laser, 328
  intracavity SHG, 319
Asymptotic approximation
  Adler's equation near locking, 81
  close to a Hopf bifurcation point, 101
  close to the laser threshold, 22, 38
  driven Adler's equation, 74
  eigenvalues, 49, 56
  large α, 234
  optically injected laser
    weak injection limit, 217
  periodically modulated laser
    primary resonance, 124
    subharmonic resonance, 130
  quasi-steady state, 40
  turn-on time, 15, 30
Asymptotic method
  adiabatic elimination, 6, 40
  change of variables
    Adler's equation, 116
    class B laser, 119
  high-frequency modulation, 74
  matched asymptotic expansions, 82, 143

multiple time scales, 22, 38, 74, 113, 117, 167, 208, 264
slow–fast turn-on, 14, 29
weakly nonlinear, 124, 130

Bad cavity limit
  FIR dynamics, 289
Bifurcation
  Bogdanov–Takens, 238
  delayed, 156
  fold-Hopf, 233
  homoclinic, 186
  Hopf, 69, 85, 92, 98, 99, 300
  period-doubling, 132, 147, 168
  saddle-node, 92, 99, 104, 225
  saddle-node of limit cycles, 186
  steady, 7
  steady imperfect, 27
  supercritical or subcritical, 21, 86
  torus, 231
Bifurcation bridges, 252
Bistability
  generalized, 137
  of periodic solutions, 127
  optical, 156, 178
  tristability, 105
Bogdanov–Takens bifurcation, 238

Chaos, 280
Class A laser, 6, 11
  optically injected, 68
Class B laser, 4, 20, 26, 35
  conservative oscillations, 119
  driven, 119
  singular limit, 9
  slow passage, 167
Class C laser, 276
CO_2 laser, 44, 52
Coherence collapse, 250
Coupled lasers, 75
Critical slowing down

Adler's equation, 65
 laser equations, 24

Driven laser
 class B laser, 119
 dual tone modulation
  effect of the phase, 151
  small-signal detector, 149
 primary resonance
  hysteresis, 124
 pump or loss modulation, 122
 strongly modulated, 136
  map, 144
  period doubling, 143
  subharmonic periodic solutions, 140
 subharmonic modulation
  period doubling, 130

Far-infrared lasers, 272
 inhomogeneous broadening, 281
Feedback
 cavity length, 106
 delayed incoherent, 269
 delayed optical, 241
  Ikeda equation, 96
  map, 98
 electrical, 87
Fold-Hopf bifurcation, 238
Frequency pulling
 optically injected laser, 219

Haken, Hermann, 84, 272
Haken–Lorenz equations, 276
 complex, 292
Hopf bifurcation, 85
 approximation, 101
 double
  eye bifurcation diagram, 107
 intracavity SHG, 316
 laser with a saturable absorber, 186, 192
 optically injected laser, 226
  close to threshold, 237
 second harmonic generation, 300
Hopf, Eberhard, 86
Hysteresis
 periodically modulated laser, 124

Ikeda equation, 96

Lang–Kobayashi equations, 246
Laser conservative oscillations, 120
Laser gyros, 70
 dead band, 71
 dither control, 73
 high-frequency modulation, 74
Laser subject to a magnetic field, 60
 periodically modulated
  strongly nonlinear, 115
  weakly nonlinear, 112
Laser subject to feedback
 bifurcation diagrams, 252

cavity length, 106
coherence collapse, 250
delayed incoherent feedback, 269
delayed optical feedback, 241
delayed optoelectronic feedback, 261
 bursting oscillations, 263
 low pass-high pass filtering, 261
 slow–fast oscillations, 264
electrical, 87
 bifurcation diagrams, 94
 feedback, 92
 steady states, 90
external cavity modes, 246
 ellipse, 248
 maximum gain mode, 248
 mode-beating, 252
imaging, 255
low frequency fluctuations, 243
Laser threshold
 laser with a saturable absorber, 179
 single mode laser, 7
Laser with a saturable absorber, 175
 bifurcation diagrams, 195
 bursting oscillations, 199
 four-variable rate equations, 201
 Hopf bifurcation, 186
  Powell and Wolga, 193
 optical bistability, 178
 passive Q-switching, 180
  period and maxima, 188
 saturability coefficient, 177
 three-variable rate equations, 182
  linear stability, 191
 two-variable rate equations, 184
Laser with an injected signal
 Bogdanov–Takens bifurcation, 238
 class A, 68
  stability boundaries, 82
 experimental stability diagram, 220
 fold-Hopf bifurcation, 238
 four-wave mixing, 218
 frequency pulling, 219
 Hopf bifurcation, 226
 large α, 234
 linear stability, 224
 locking range, 218
 semiconductor lasers, 213
 solid state laser, 239
 steady states, 222
 torus bifurcation, 231
Linear stability
 class B laser, 8
 CO_2 laser, 47
 intracavity SHG, 316
 laser with a saturable absorber
  four variables, 203
  three variables, 191
  two variables, 186
 optical parametric oscillator, 314
Locking range
 Adler's equation, 65

Locking range (cont.)
    coupled lasers, 78
    dead band, 71
    optically injected laser, 69, 218
Lorenz equations, 275
    chaos, 277
Low frequency fluctuations, 243

Map
    fixed points, 99
    Hopf bifurcation, 101
    slowly varying in time, 265

NMR laser, 149
Normal form equation
    Adler's equation, 80
    laser rate equations, 21

Optical bistability
    laser with a saturable absorber, 178
Optical feedback
    delayed optical, 255
    delayed optoelectronic, 261
Optical parametric oscillator, 294
    DOPO, 296
    intracavity, 312
    thermal effects, 306
Oscillations
    bursting
        laser with a saturable absorber, 201
        optoelectronic feedback, 265
    harmonic to pulsating, 103

Passive Q-switching, 180
    bursting oscillations, 199
    in CO_2 lasers, 197
Period doubling
    CO_2 laser, 168
    NMR laser, 149
    strongly modulated laser, 143
    weakly modulated laser, 132
Phase locking, 59
Phase plane
    laser conservative oscillations, 121
    relaxation oscillations, 25
    saddle-point, 35
Polarization dynamics
    due to a magnetic field, 60

Quasi-periodic oscillations, 233

Rate equations
    class A, 6, 11
        optically injected, 68
    class B, 3, 5, 20
    CO_2 laser, 46
        four-level model, 52
        three-level model, 44
    critical slowing down, 24
    delayed optical feedback
        LK equations, 246

dimensionless formulation, 4
    four-level lasers, 49
    imperfect bifurcation, 26
    laser power and gain, 35
    laser threshold, 7
    laser with a saturable absorber
        four variables, 201
        three variables, 182
        two variables, 184
    linear stability, 8
    optical parametric oscillator, 296
        Hopf, 300
        intracavity, 312
        steady states, 298
        thermal effects, 307
    optically injected laser, 216
    phase plane, 25
    pitchfork bifurcation, 20
    relaxation oscillations, 9, 36
    ruby laser, 42
        reduction to class B, 44
    second harmonic generation, 297
        intracavity, 316, 319
    semiconductor lasers, 31
    singular limit, 9
    solid state laser
        reduction to class B, 52
    spontaneous emission, 26, 29
    steady state bifurcation, 9, 20
    steady states, 7
    three-level laser, 41
    two coupled lasers, 77
Relaxation oscillations
    close to the laser threshold, 36
    damping rate, 10, 13, 34
    effect of optical feedback, 255
    frequency, 10, 12, 34, 38
    high-speed lasers, 37
    optically injected laser, 227
    three level lasers, 49
Ring laser, 82
Routh–Hurwitz conditions, 92, 225
    CO_2 laser, 48
    optical parametric oscillator, 314
Ruby laser, 42, 84

Sagnac effect, 82
Second harmonic generation, 297
    intralaser cavity, 318
        antiphase dynamics, 318
Secular term, 135
Semiconductor lasers
    design of high-speed lasers, 37
    linewidth enhancement factor, 33
    rate equations, 31
Singular perturbation
    high-frequency modulation, 74
    laser turn-on, 14, 29
    passive Q-switching oscillations, 188
    relaxation oscillations, 9
    singular Hopf bifurcation, 103

singular limit, 119
slow passage, 160
two-time solution, 38
Slow passage
  delay, 160
    effect of noise, 160
  forward and backward transition, 164
  optical parametric oscillator, 308
  slowly varying solutions, 167
  through a Hopf bifurcation, 170
  through a PD bifurcation, 168
  through bifurcation points, 158
  through limit points, 156
Solvability condition
  driven Adler's equation, 113
  weakly modulated laser, 125, 131, 141
Spontaneous emission, 26
  imperfect bifurcation, 26
  reduced turn-on time, 29
Square waves, 98

Time scales
  for different lasers, 5
  lasers with a saturable absorber, 178
Transfer function, 17
  amplitude and phase, 17
Turn-on experiment, 11
  pump square pulse, 35
  turn-on time, 14
    approximation, 15
Two-time solution
  Adler's equation, 117
  driven Adler's equation
    near resonance, 113
  high-frequency modulation, 74
  optoelectronic feedback, 264
  singular Hopf bifurcation, 103
  slow passage, 160
Two-photon laser, 57, 211

Vertical cavity surface emitting laser, 214

Printed in the United States
by Baker & Taylor Publisher Services